Climate Change Management

Series Editor

Walter Leal Filho, International Climate Change Information and Research Programme, Hamburg University of Applied Sciences, Hamburg, Germany

The aim of this book series is to provide an authoritative source of information on climate change management, with an emphasis on projects, case studies and practical initiatives—all of which may help to address a problem with a global scope, but the impacts of which are mostly local. As the world actively seeks ways to cope with the effects of climate change and global warming, such as floods, droughts, rising sea levels and landscape changes, there is a vital need for reliable information and data to support the efforts pursued by local governments, NGOs and other organizations to address the problems associated with climate change.

This series welcomes monographs and contributed volumes written for an academic and professional audience, as well as peer-reviewed conference proceedings. Relevant topics include but are not limited to water conservation, disaster prevention and management, and agriculture, as well as regional studies and documentation of trends. Thanks to its interdisciplinary focus, the series aims to concretely contribute to a better understanding of the state-of-the-art of climate change adaptation, and of the tools with which it can be implemented on the ground.

More information about this series at http://www.springer.com/series/8740

Walter Leal Filho
Editor

Managing Climate Change Adaptation in the Pacific Region

Springer

Editor
Walter Leal Filho
HAW Hamburg
Hamburg, Germany

ISSN 1610-2002 ISSN 1610-2010 (electronic)
Climate Change Management
ISBN 978-3-030-40551-9 ISBN 978-3-030-40552-6 (eBook)
https://doi.org/10.1007/978-3-030-40552-6

This Springer imprint is published by the registered company Springer Nature Switzerland AG
The registered company address is: Gewerbestrasse 11, 6330 Cham, Switzerland

Contents

Climate Change Adaptation in the Agriculture and Land Use Sectors: A Review of Nationally Determined Contributions (NDCs) in Pacific Small Island Developing States (SIDS)

Krystal Crumpler and Martial Bernoux

Abstract Climate change is already altering the natural resource base upon which global food security and nutrition depend, with disproportionate impacts on rural and coastal communities in Small Island Developing States (SIDS) (IPCC in An IPCC special report on the impacts of global warming of 1.5 °C above pre-industrial levels and related global greenhouse gas emission pathways, in the context of strengthening the global response to the threat of climate change, sustainable development, and efforts to eradicate. IPCC, 2018). The agriculture and land use sectors (crops, livestock, forestry, fisheries and aquaculture) lie at the heart of the global response to climate change, with the unique capacity to protect ecosystem integrity and promote the livelihoods and resilience of the poor and vulnerable (FAO in State of food and agriculture: climate change, agriculture and food security. FAO, Rome, 2016a). The Food and Agriculture Organization (FAO) of the United Nations (UN) has developed a methodology (Crumpler et al. in Working paper no. 76, FAO, Rome, 2019) and analysis of the role of the agriculture and land use sectors in the climate change adaptation components set forth in the Nationally Determined Contributions (NDCs) of Pacific SIDS under the Paris Agreement (FAO in regional analysis of the Nationally Determined Contributions in the Pacific: gaps and opportunities in the agriculture and land use sectors. FAO, Rome, 2020). The analysis aims to provide a synthesis of the extent to which countries in the Pacific region include agriculture and land use in their adaptation components, as well as identify "gaps" and "opportunities" for enhancing adaptation ambitions by addressing the major climate-related impacts, hazards and vulnerabilities reported in ecosystems and social systems. Overall, around 90% of Pacific SIDS identify adaptation measures in ocean and coastal zone ecosystems and agroecosystems, with mangrove conservation and replanting and water storage and harvesting amongst the most frequently prioritized adaptation options. Adverse health, loss of productive infrastructure and assets, and food insecurity and malnutrition constitute the greatest climate-related risks in social systems reported. Over

K. Crumpler (✉) · M. Bernoux
Climate and Environment Division, Food and Agriculture Organization of the United Nations, Viale delle Terme di Caracalla, 00153 Rome, Italy
e-mail: krystal.crumpler@fao.org

M. Bernoux
e-mail: martial.bernoux@fao.org

W. Leal Filho (ed.), *Managing Climate Change Adaptation in the Pacific Region*, Climate Change Management, https://doi.org/10.1007/978-3-030-40552-6_1

two-thirds of countries promote health information and services as a cross-cutting adaptation priority, and half prioritize resilient infrastructure. However, high adaptation policy coverage gaps emerge around climate-related losses in ecosystem service provision, particularly biological control, soil erosion control, moderation of extreme events and the maintenance of genetic diversity and abundance. In social systems, high policy coverage gaps are found around climate-related migration and displacement. The analysis aims to inform 2020 NDC revision processes, as well as guide country support and investment options in the region.

Keywords Climate change · Adaptation · Mitigation · Agriculture · Land use · Nationally determined contributions · NDC · Pacific · SIDS

Introduction

Climate change and variability are already altering the natural resource base on which food and agriculture systems rely, with disproportionately higher risk of adverse impacts on the world's most vulnerable and food insecure (IPCC 2014). Some of the worst impacts on sustainable development are expected to be felt among agricultural and coastal livelihoods in Small Island Developing States (SIDS), where both natural and human systems face hard limits to adaptation (IPCC 2018). The main climate- and ocean-related drivers of change in small islands include variations in air and ocean temperatures, ocean chemistry, rainfall patterns, wind strength and direction and sea levels; as well as climate extremes including tropical cyclones, drought and storm swell events (IPCC 2014). All pose cascading risks to ecosystems and people through an expected increase in the price of food, income and asset loss, foregone livelihood opportunities, adverse health impacts and population displacements (IPCC 2014). Overall, climate change disproportionately impacts the poor and vulnerable, due to higher levels of exposure to risk and lower coping capacities, with an additional 100 million people expected to fall into extreme poverty by 2030 (Hallegatte et al. 2016). At the household level, women and children may bear more of the burden of climate-related shocks and stresses on health, education and paid work, with intergenerational impacts on nutrition and poverty outcomes (FAO 2018).

Climate-related impacts on human systems and ecosystem services are already being observed in the Pacific (IPCC 2019a). It is expected that global warming will result in the irreversible loss of marine and coastal ecosystems and reduce the productivity of fisheries and aquaculture through shifts in distribution and abundance of species, with grave consequences on the income, livelihoods and food security and nutrition of marine resource-dependent communities in the region (Hanich et al. 2018; IPCC 2019a). The decline in warm water coral reefs is projected to greatly compromise the services they provide to communities, including food provision, coastal protection and tourism in small islands (IPCC 2019a). At the global scale, the biomass of marine animals across the food web is projected to decrease by up 15% and the maximum catch potential of fisheries by 24% under high emission scenarios by the end of the 21st century (IPCC 2019a). Maximum catch potential was projected to decrease by up to 50% in the Pacific, with largest impacts in the western Pacific warm pool (Asch et al. 2018). When combined with increasing aridity and a decrease in freshwater availability, sea level rise will likely leave several atoll islands uninhabitable (IPCC 2018). The Pacific islands are confronted with the highest disaster risk globally in terms of per capita loss, and due to data gaps, is likely underestimated (Edmonds and Noy 2018). There is overall high confidence that a slower rate of climate-related ocean and land change associated with more ambitious mitigation action globally would provide greater adaptation opportunities (IPCC 2019b), with particular relevance to those communities and ecosystems in small islands in the Pacific where the biophysical limits to adaptation, characterized by high levels of exposure to current and future hazards, are compounded by financial, technological, institutional and other barriers.

The adoption of the Paris Agreement in 2015 constitutes a landmark achievement in the global response to climate change, when developed and developing countries alike committed to do their part in the transition to a low-emission and climate-resilient future. The Agreement seeks to limit global warming to below a 2 °C rise above pre-industrial levels and pursue efforts to stay within 1.5 °C, as well as sets a global goal on adaptation within the context of sustainable development. The Alliance of Small Island States (AOSIS), in particular, successfully advocated for the role of adaptation in the Paris Agreement as a key factor in the global response to climate change. Underpinning the Agreement are the (Intended) Nationally Determined Contributions, (I)NDCs, which for the purpose of this document are referred to as NDCs thereafter. NDCs represent the main national policy framework, under the United Nations Framework Convention on Climate Change (UNFCCC), by which Parties communicate their commitment to reducing national greenhouse gas emissions (GHG) and adapting to the impacts of climate change, based on national priorities, circumstances and capabilities (Article 4).

The success of the Paris Agreement rests upon the enhanced ambition of Parties to progressively revise and strengthen their respective mitigation and adaptation plans over time (Article 4.2). In 2023, and every five years thereafter, Parties will periodically take stock of the implementation of the Agreement to assess the collective progress towards achieving its purpose and long-term goals (Article 14). The outcome of the "Global Stocktake" will inform Parties in updating and enhancing, in a nationally determined manner, their actions and support in accordance with the relevant provisions of this Agreement, as well as in enhancing international cooperation for climate action. The tracking of NDC implementation will take place under the Enhanced Transparency Framework (Article 13), which provides a foundation for building mutual trust and confidence. The "Paris Rulebook" requires Parties to report reliable, transparent and comprehensive information on GHG emissions, climate actions and support, with built-in flexibility for developing countries under the principle of common but differentiated responsibilities and respective capabilities (Article 13).

The agriculture and land use sectors (crops, livestock, fisheries and aquaculture, and forestry) feature prominently in the NDCs, with up to 97 and 89% of developing countries prioritizing climate change adaptation and mitigation, respectively, in one or more agricultural sub-sectors (FAO 2016b). The main objective of this paper is to provide a sector-specific synthesis of the climate change adaptation priorities set forth in the NDCs of countries in the Pacific and to identify opportunities for governments to strengthen adaptation ambitions, capture mitigation co-benefits and accelerate progress on the sustainable development agenda. Furthermore, a better understanding of national adaptation priorities, barriers to implementation and support needs in the agriculture and land use sectors can inform targeted programming and investments in the region. This analysis is directed at national policy makers and practitioners in the region with a stake in ensuring that future adaptation policies and programmes are clear, measurable, transparent and ambitious. It also aims to guide international development and civil society organizations, committed to providing the country support required for scaling up climate action in the agriculture and land use sectors and co-delivering on the 2030 Agenda for Sustainable Development and Sendai Framework for Disaster Risk Reduction.

Methodology

The NDCs are the product of diverse national priorities, capacities and processes, meaning that they vary greatly in terms of style, format, scale and level of detail. A common framework was developed to facilitate the synthesis and analysis of the NDCs in the agriculture and land use sectors. The framework provides a structure for assessing the clarity, measurability, transparency and ambition of NDCs over time. Each NDC was analyzed in full within the bounds of this common framework. The framework was based on a stocktaking of 184 NDCs to quantify and qualify the types of climate change mitigation and adaptation contributions in the agriculture and land use sectors by means of a common set of categories and sub-categories. The full methodological notes are contained in Crumpler et al. (2019). In order to fill the information gap whereby some countries opted to make reference to existing national climate change plans and vulnerability analyses, a review of National Communications (NCs) and National Greenhouse Gas Inventories (NGHGIs) was carried out to supplement the information contained in the NDCs. This paper is based on the information reported in the most recently submitted NDCs, NCs and Technical Needs Assessments (TNA) of non-Annex I Parties in the Pacific to the UNFCCC as of 1 August 2019. Table 1 contains a list of the documents analyzed, which are all publicly available on the UNFCCC website (www.unfccc.int).

For this analysis, the Pacific refers to 14 independent countries, in three geographic areas in Oceania: Melanesia (Fiji, Papua New Guinea, Solomon Islands and Vanuatu), Micronesia (Kiribati, Marshall Islands, Micronesia (Federated States of), Nauru, Palau) and Polynesia (Cook Islands, Niue, Samoa, Tonga and Tuvalu) (UNSD n.d.). All 14 countries are SIDS and four (Solomon Islands, Vanuatu, Kiribati and Tuvalu) are Least Developed Countries (LDC). The assignment of countries or areas

Table 1 Source of national data for analysis

Country name	NDC	NC	TNA
Fiji	2016	2014	
Papua New Guinea	2016	2014	
Solomon Islands	2016	2017	
Vanuatu	2016	2014	
Kiribati	2016	2013	
Marshall Islands	2016	2015	
Micronesia (federated states of)	2016	2015	
Nauru	2016	2014	
Palau	2016	2002	
Cook Islands	2016	2011	
Niue	2016	2014	2003
Samoa	2016	2010	
Tonga	2016	2012	
Tuvalu	2016	2015	

to specific groupings is for statistical convenience and does not imply any assumption regarding political or other affiliation of countries or territories by the authors.

A systematic analysis of the NDCs presents a number of methodological challenges, owing to their aggregate volume and heterogeneity in terms of content, scope and detail. Due to lack of a standard template for NDC formulation, and capacity constraints, not all information was necessarily made available, nor equal in level of detail. For instance, many countries decided to make reference to their existing national climate change adaptation policies rather than explicitly integrate them into their NDCs. For this reason, the information contained in the NDC is supplemented by information from other sources, including the NCs and TNAs. Nonetheless, the information is not always comparable in absolute terms, constituting a limitation to the methodology presented. It should also be noted that the adaptation policy coverage gap analysis serves as a broad review of the coverage of adaptation priority sectors and measures mentioned by each country in the documents analyzed and does not constitute an assessment of their strength, which could be further analyzed in terms of type (e.g. action, policy, project, programme or framework), scale, comprehensiveness and timeframe. The adaptation policy gap analysis, therefore, serves as an initial stocktaking of policy coverage and does not necessarily indicate policy effectiveness.

Results and Discussion

Climate-Related Hazards, Slow Onset Events and Impacts in Ecosystems

In order to contextualize the adaptation priorities of countries in the region as set forth in the NDCs, we reviewed the types of climate-related impacts, vulnerabilities and risks found in national reports, or NCs, using definitions adapted from the IPCC (2014) and/or EM-DAT (n.d.). Countries often include a description of observed and/or expected climate-related hazards, including hydro-meteorological, climatological and biological processes or phenomenon that may cause loss of life, injury, or other health impacts, as well as damage and loss to property, infrastructure, livelihoods, service provision, ecosystems, and environmental resources, along with longer-term chemical, biological, and physical changes, leading to slow onset events. Countries also report climate-related impacts, vulnerability and risk in social systems that are observed or expected in the future, as well as non-climatic environmental, social, economic, cultural, political and institutional variables, or stressors, that can affect individual adaptive capacity to respond, as well as the level of exposure to climate change, creating new or exacerbating existing vulnerabilities to climate change.

All countries in the Pacific report the occurrence of storms amongst observed and/or projected climate-related hazards, followed by drought (93% of countries), floods (64%) and invasion by pests and non-native species in agriculture (29%), amongst others. Salt-water intrusion and water stress are reported most amongst observed and/or projected climate-related slow onset events in terrestrial and freshwater ecosystems (64%, respectively), while sea level rise and coastal erosion are reported most frequently amongst those in marine and coastal ecosystems (100 and 86%, respectively). Figure 1 illustrates the share of countries in the region that report observed and/or projected climate-related slow onset events, by type of risk.

Agro-ecosystems are considered the most vulnerable of ecosystems by all countries, followed by oceans and coastal zones (71%). In particular, marine fisheries and crops are considered the most vulnerable of agricultural sub-sectors to climate change (79 and 64%), followed by livestock (36%) and forestry (29%). Figure 2 illustrates the share of countries in the region that report observed and/or expected climate-related impacts in agro-ecosystems, by sub-sector/land use category.

Overall, genetic resources, primarily in agro-ecosystems, are reported most frequently amongst natural resource impacts (93% of countries), followed by land and soil resources in coastal zone ecosystems (86%) and water resources across all ecosystems (79%). While loss of primary production and productivity are most frequently reported amongst ecosystem service impacts (93% of countries), primarily in the marine fisheries and crops sub-sectors, followed by changes in water availability and quality across all ecosystems and coastal erosion (79%, respectively), as well as biodiversity loss, primarily in ocean and coastal zone ecosystems (57%), amongst

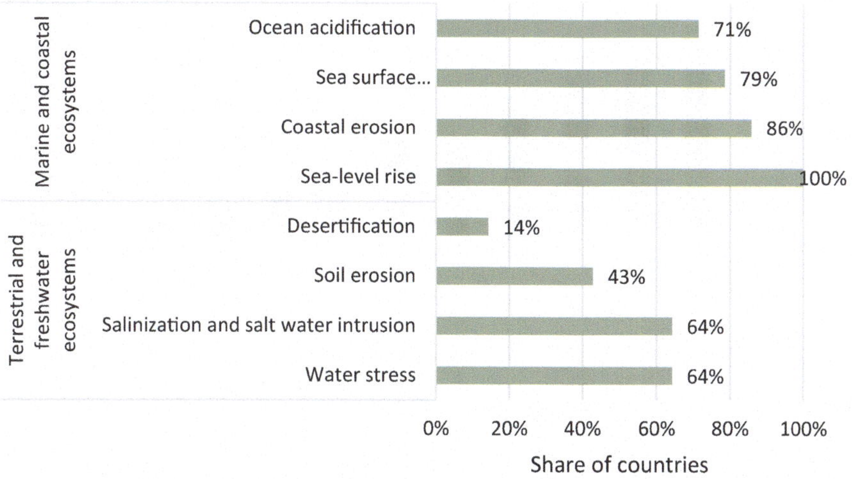

Fig. 1 Observed and/or projected climate-related slow onset events reported in the Pacific, by type (share of countries)

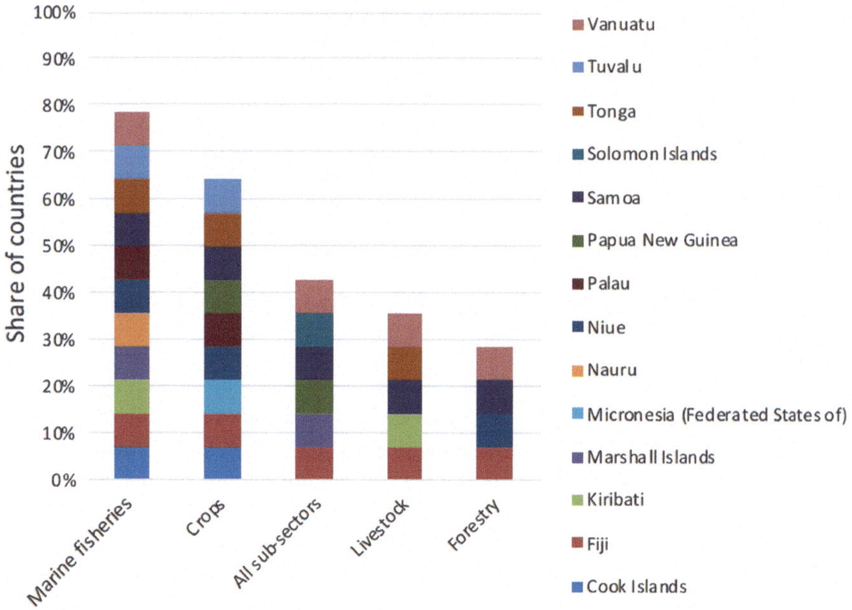

Fig. 2 Observed and/or projected climate-driven impacts in agro-ecosystems reported in the Pacific, by sub-sector/land use category (share of countries)

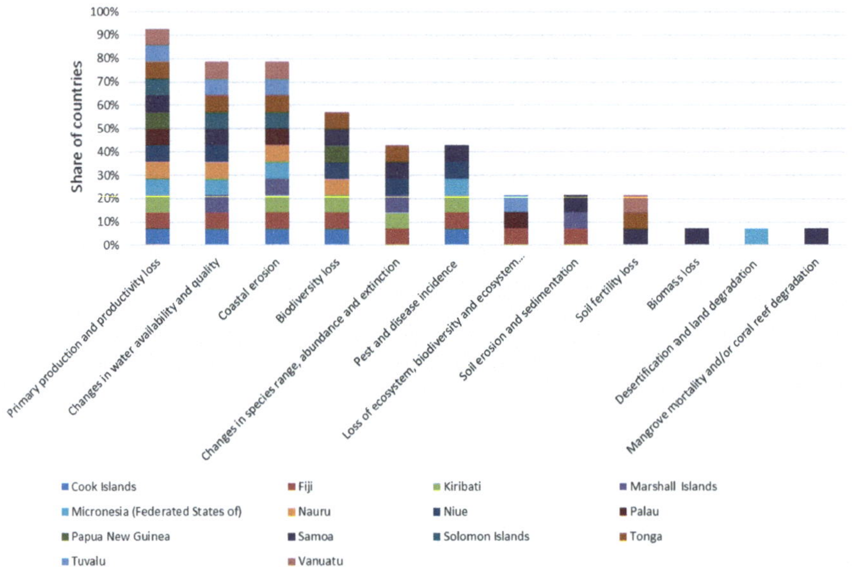

Fig. 3 Observed and/or projected climate-related impacts reported on ecosystem services in the Pacific, by type (share of countries)

others. Figure 3 illustrates the share of countries in the region that report observed and/or projected climate-related impacts in ecosystems, by ecosystem service impact category.

Climate-Related Impacts, Vulnerabilities and Risks in Social Systems

All countries in the region identify at least one observed and/or expected impact, vulnerability and risk induced by climate change in social systems. Overall, the majority of countries report health as social dimension at risk under climate change (93% of countries), followed by loss of productive infrastructure and assets, food insecurity and malnutrition and rural livelihoods and income loss (71% each), migration and displacement (57%) and gender and youth inequality (50%), amongst others. Figure 4 illustrates the share of countries in the region that report observed and/or expected climate-related impact, vulnerability and risk in social systems, by type.

Geography and topography are reported as the largest non-climatic stressors of vulnerability, followed by the economic dependence on agriculture and natural resources (79%), poverty and low levels of development (71%), undermining the adaptive capacity of people to respond to climatic shocks and stresses.

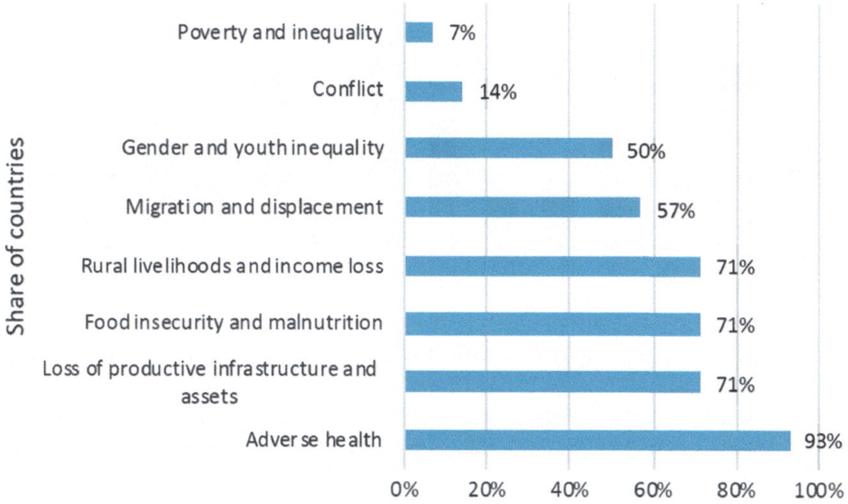

Fig. 4 Observed and/or projected climate-related impacts, vulnerabilities and risks reported in the Pacific, by type (share of countries)

Adaptation Components in Pacific NDCs

Including when a country makes reference to key adaptation plans in their ND, all countries in the Pacific communicated an adaptation component of which include the agriculture and land use sectors. The level of detail included in each country's adaptation component varies, as some countries detailed their adaptation visions, goals and measures, while other countries made reference to national adaptation and climate change plans. For the sake of this analysis, the agricultural adaptation component is differentiated in terms of ecosystems and social systems.

Amongst priority sectors and cross-sectoral priorities for adaptation, all countries identify water resources as a priority, followed by the agriculture sector in general (93% of countries) and oceans and coastal zones (86%), as well as fisheries and aquaculture, biodiversity and forestry (36%, respectively), amongst others. Figure 5 illustrates the types of ecosystem-related adaptation priority sectors and cross-sectoral priorities, by country.

Health represents the greatest cross-cutting adaptation priority in social systems amongst countries in the region (71% of countries), followed by Disaster Risk Reduction (DRR) (64%), food security and nutrition (57%), resilient infrastructure (50%) and gender equality (21%), amongst others. Figure 6 illustrates the types of cross-cutting adaptation priorities in social systems, by country.

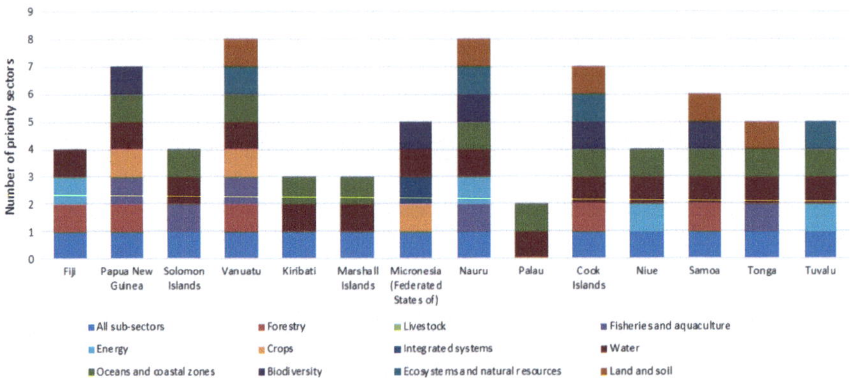

Fig. 5 Adaptation priority sectors and cross-sectoral priorities in ecosystems in the NDCs of Pacific countries

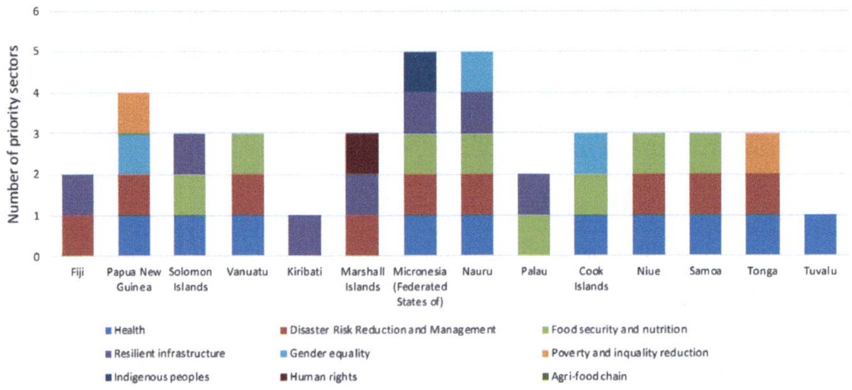

Fig. 6 Cross-cutting adaptation priorities in social systems in the NDCs of Pacific countries

Adaptation Measures in Ecosystems

In addition to prioritized sectors or cross-cutting priorities for adaptation, countries often include a set of policies or measures as part of their adaptation strategy. Overall, the majority of countries include at least one or more adaptation policy or measure in oceans and coastal zones, followed by agro-ecosystems and ecosystems in general.

Ninety-three percent of countries include at least one adaptation policy or measure in ocean and coastal zone ecosystems, of which the majority promote mangrove conservation and replanting (57% of countries), followed by coastal zone management (43%), biodiversity and ecosystem management (21%), flood management and land/soil management, restoration and rehabilitation (14%, respectively), amongst other measures. Figure 7 illustrates the share of countries in the region with one or

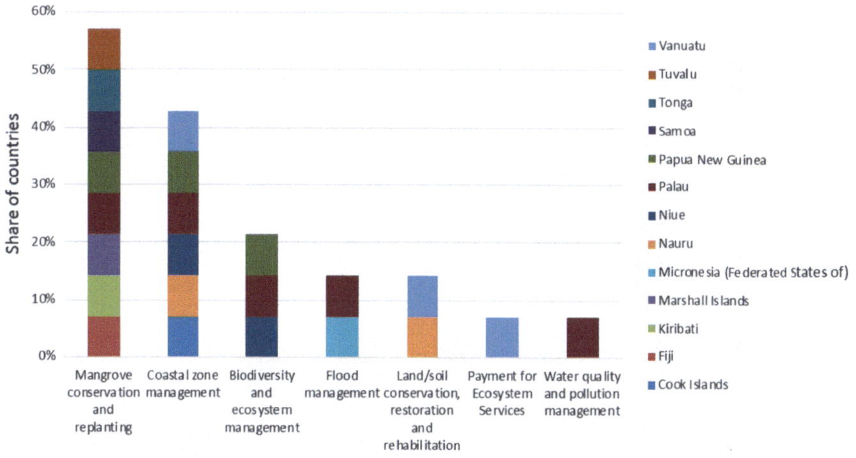

Fig. 7 Adaptation policies and measures in ocean and coastal zone ecosystems in the NDCs of Pacific countries, by type (share of countries)

more (to avoid bias of representation) adaptation policy or measure in ocean and coastal zone ecosystems, by management activity.

Eighty-six percent of countries include at least one adaptation policy or measure in agro-ecosystems. The majority of those countries promote adaptation in marine fisheries and aquaculture (71% of countries), followed by crops and agriculture in general (64%, respectively), forestry (57%), livestock (50%), freshwater aquaculture (14%) and integrated systems (7%). Figure 8 illustrates the share of countries in the

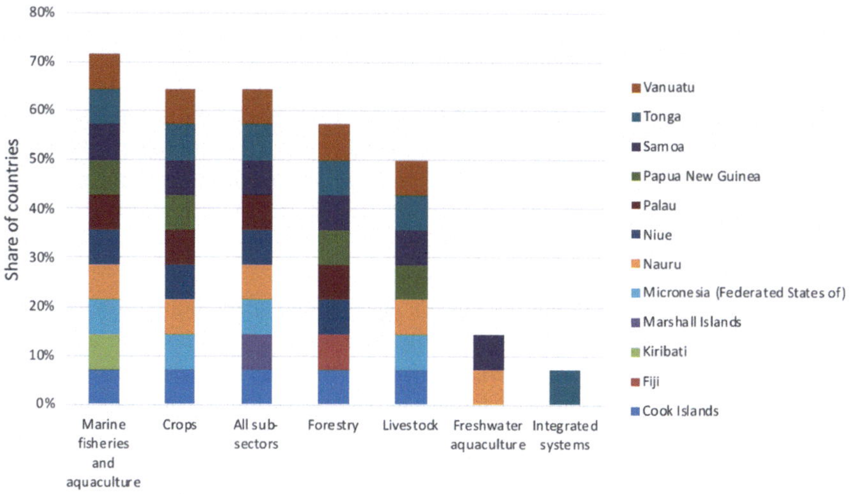

Fig. 8 Adaptation policies and measures in agro-ecosystems in the NDCs of Pacific countries, by sub-sector (share of countries)

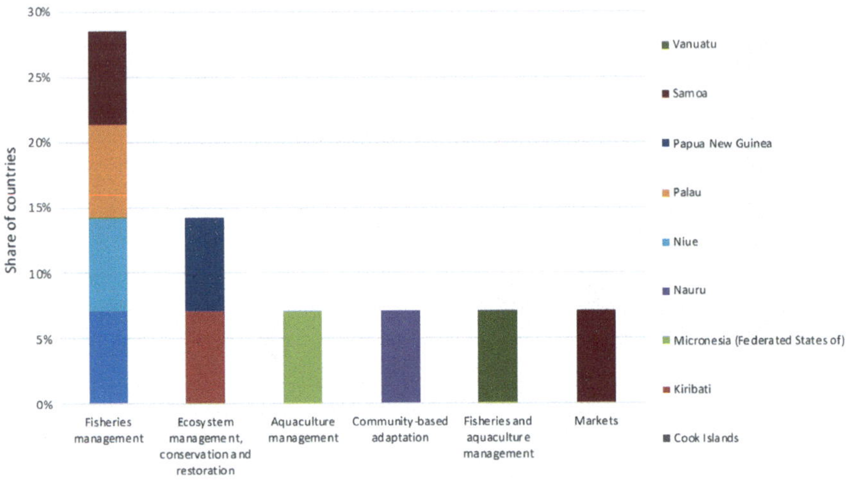

Fig. 9 Adaptation policies and measures in marine fisheries and aquaculture in the NDCs of Pacific countries, by type (share of countries)

region with one or more (to avoid bias of representation) adaptation policy or measure in agro-ecosystems, by sub-sector.

Seventy-one percent of countries include at least one adaptation policy or measure in marine fisheries and aquaculture. The majority of those countries promote fisheries management (29% of countries), followed by ecosystem management, conservation and restoration (14%), and equal shares of aquaculture management, community-based adaptation, market-based measures and fisheries and aquaculture management in general (7% each). Figure 9 illustrates the share of countries in the region with one or more (to avoid bias of representation) adaptation policy or measure in marine fisheries and aquaculture, by management activity.

Sixty-four percent of countries include at least one adaptation policy or measure in the crops sub-sector. The majority of those countries promote plant management (30% of countries), followed by water management and general crop management (22 and 18%, respectively), nutrient and on-farm soil management and pest and disease management (11% each), amongst others. Figure 10 illustrates the share of countries with one or more (to avoid bias of representation) adaptation policy or measure in the crops sub-sector, by management activity.

Fifty-seven percent of countries include at least one adaptation policy or measure in the forestry sub-sector. The majority of those countries promote afforestation/reforestation (36% of countries), followed by reducing deforestation and forest conservation (29%), reducing degradation and sustainable forest management (21%) and water management (7%), amongst others. Figure 11 illustrates the share of countries with one or more (to avoid bias of representation) adaptation policy or measure in the forestry sub-sector, by management activity.

Fifty percent of countries include at least one adaptation policy or measure in the livestock sub-sector. The majority of those countries promote animal breeding and

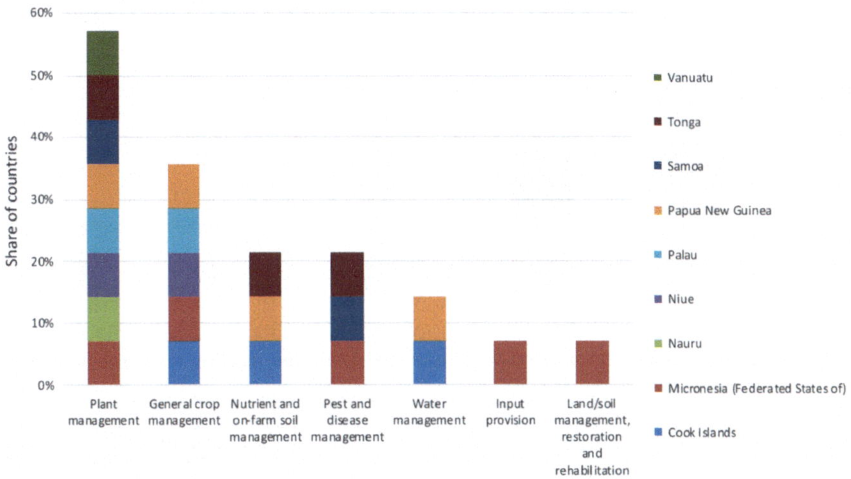

Fig. 10 Adaptation policies and measures in the crops sub-sector in the NDCs of Pacific countries, by type (share of countries)

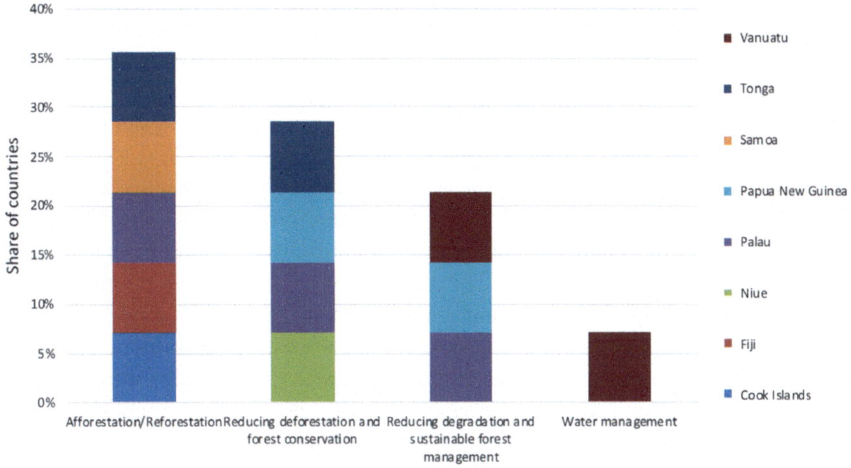

Fig. 11 Adaptation policies and measures in the forestry sub-sector in the NDCs of Pacific countries, by type (share of countries)

husbandry (20% of countries), followed by water and general livestock management (14% each), and manure and grassland management (7% each). Figure 12 illustrates the share of countries with one or more (to avoid bias of representation) adaptation policy or measure in the livestock sub-sector, by management activity.

Only two countries (Nauru and Samoa, 14% of countries) identify at least one adaptation policy or measure in freshwater aquaculture, while only one country

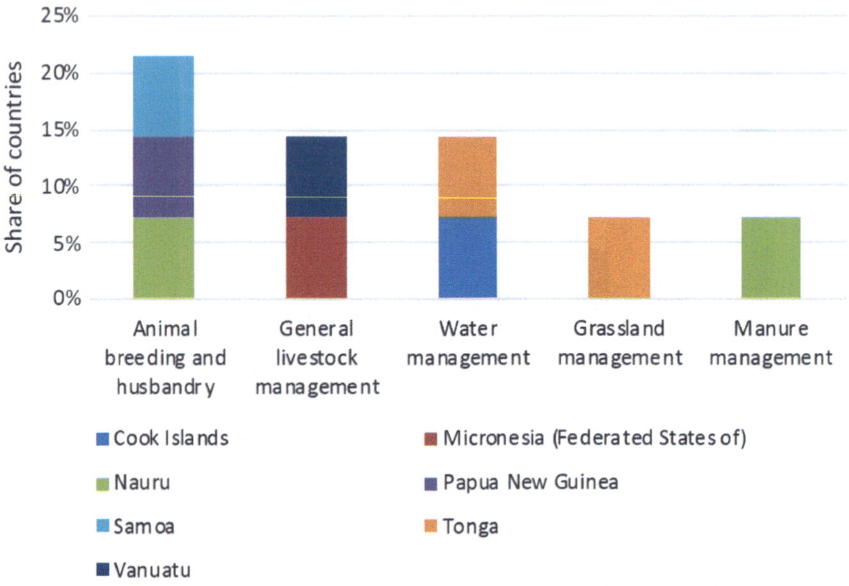

Fig. 12 Adaptation policies and measures in the livestock sub-sector in the NDCs of Pacific countries, by type (share of countries)

(Tonga, 7%) includes an adaptation policy or measure in integrated systems, namely agroforestry.

Natural resource use and management options are integrated within each of the approaches to adaptation identified above due to their cross-cutting nature. From a natural resource perspective, 86% of countries identify water resource use and management amongst adaption options. The majority of those countries promote water storage and harvesting (57% of countries), followed by integrated watershed management and sustainable use and management (29% each), availability and access, quality and pollution management and efficiency and use (21% each), amongst others. Figure 13 illustrates the share of countries with one or more (to avoid bias of representation) water-related adaptation policy or measure across all ecosystems, by resource use and management option.

Seventy-one percent of countries identify ecosystem and genetic resource use and management amongst adaption options. The majority of those countries promote pest and disease management (43% of countries), followed by the protection, conservation and restoration of biodiversity and ecosystems in general (36 and 29%, respectively) and payment for ecosystem services (7%). Figure 14 illustrates the share of countries with one or more (to avoid bias of representation) ecosystem and genetic resources-related adaptation policy or measure across all ecosystems, by resource use and management option.

Fifty-seven percent of countries with an adaptation component identify land resource use and management amongst adaption options. The majority of those

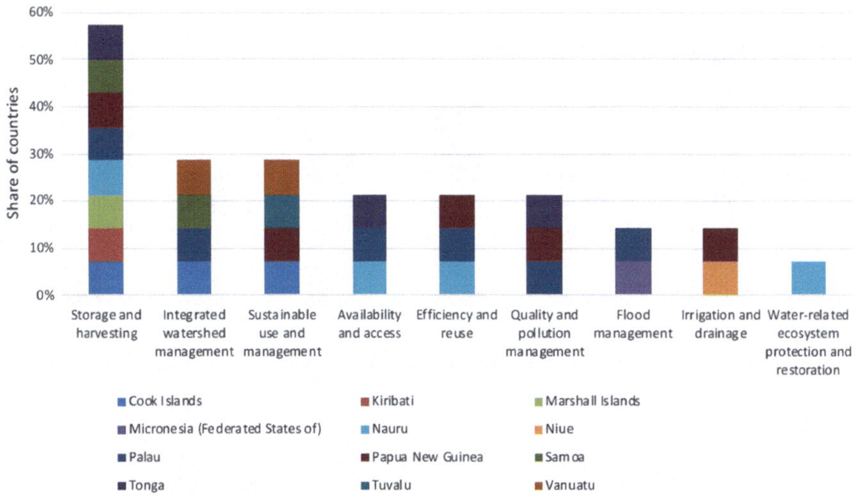

Fig. 13 Water-related adaptation policies and measures in the in the NDCs of Pacific countries, by type (share of countries)

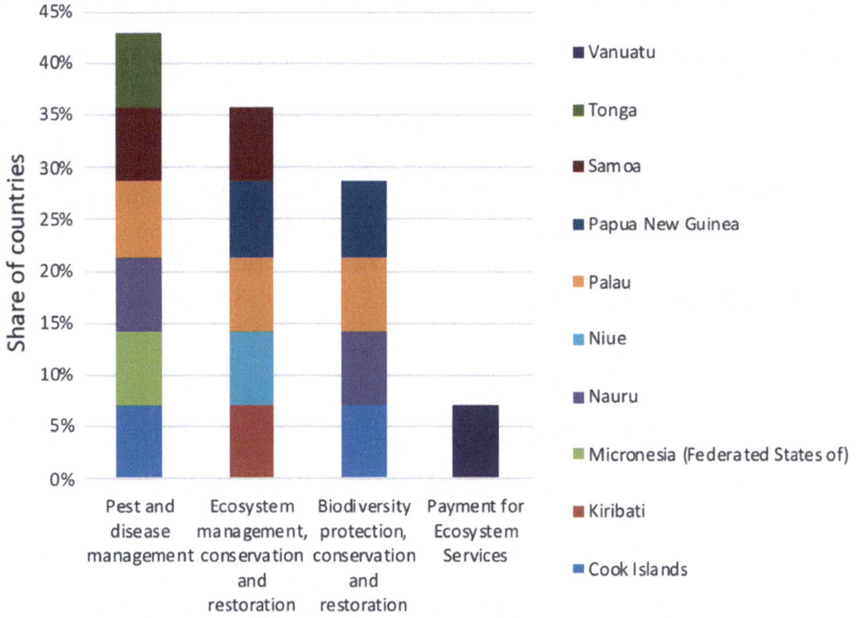

Fig. 14 Ecosystem and genetic resources-related adaptation policies and measures in the in the NDCs of Pacific countries, by type (share of countries)

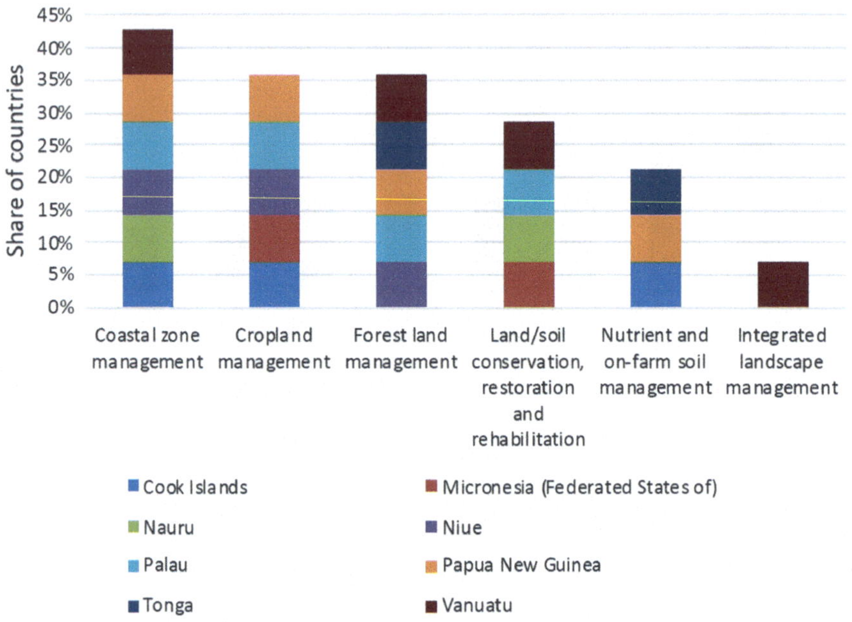

Fig. 15 Land-related adaptation policies and measures in the in the NDCs of Pacific countries, by type (share of countries)

countries promote coastal zone management (45% of countries), followed by forest and cropland management (36% each), land/soil conservation, restoration and rehabilitation (29%) and nutrient and on-farm soil management (21%), amongst others. Figure 15 illustrates the share of countries with one or more (to avoid bias of representation) land-related adaptation policy or measure across all ecosystems, by resource use and management option.

Adaptation Measures in Social Systems

For the sake of this analysis, adaptation measures in social systems are differentiated along three main pillars: socio-economics and well-being; knowledge and capacity; and institutions and governance. All countries in the Pacific identify at least one adaptation measure in social systems, primarily around the institutions and governance and socio-economics and well-being pillars (93% of countries each), followed by the knowledge and capacity pillar (86%).

Ninety-three percent of countries in the Pacific include at least one adaptation policy or measure related to institutions and governance. The majority of those countries

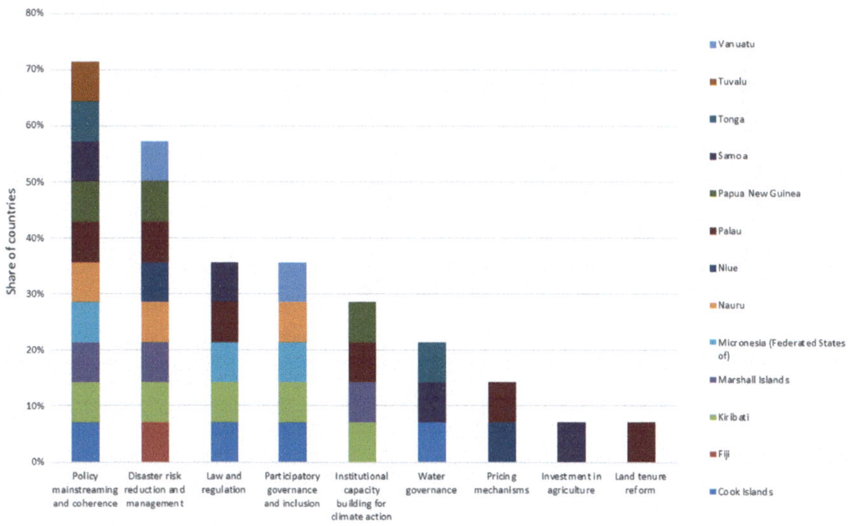

Fig. 16 Institutions and governance-related adaptation policies and measures in the NDCs of Pacific countries, by type (share of countries)

promote policy mainstreaming and coherence (71% of countries with policy or measure), followed by DRR and management (57%), participatory governance and inclusion (36%) and law and regulation reform (36%), and institutional capacity building for climate action (29%), amongst others. Figure 16 illustrates the share of countries with one or more (to avoid bias of representation) institutions and governance-related adaptation policy or measure, by intervention option.

Ninety-three countries in the Pacific include at least one adaptation policy or measure related to socio-economics and well-being. The majority of those countries promote health information and services (71% of countries), followed by resilient infrastructure (50%), social protection (29%), food security and nutrition, safe and responsible migration (21%) and resilience and adaptive capacity building (21%) and credit and insurance services and farmer cooperatives and services (14% each), amongst others. Figure 17 illustrates the share of countries with one or more (to avoid bias of representation) socio-economics and well-being-related adaptation policy or measure, by intervention option.

Eighty-six countries in the Pacific include at least one adaptation policy or measure related to knowledge and capacity. The majority of those countries promote hazard and vulnerability mapping (57% of countries), followed by awareness raising and education (50%) and early warning systems and climate information services (50%), research and development (R&D) (36%) and traditional knowledge (36%,), extension services for climate action (29%) and early warning systems (29%), amongst others. Figure 18 illustrates the share of countries with one or more (to avoid bias of representation) knowledge and capacity-related adaptation policy or measure, by intervention option.

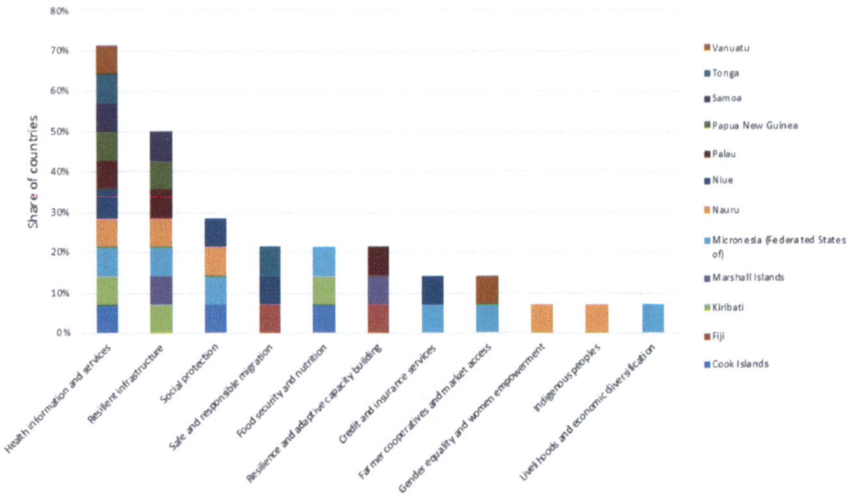

Fig. 17 Socio-economics and well-being related adaptation policies and measures in the NDCs of Pacific countries, by type (share of countries)

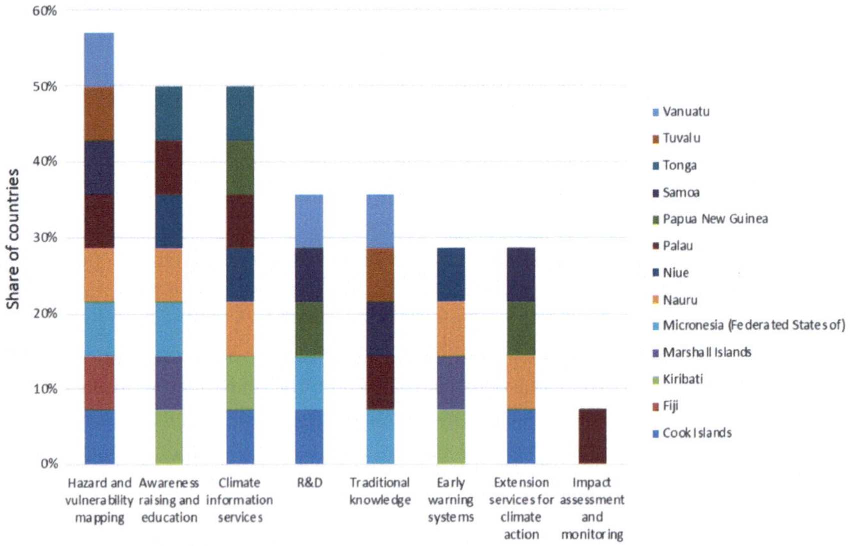

Fig. 18 Knowledge and capacity related adaptation policies and measures in the NDCs of Pacific countries, by type (share of countries)

Adaptation Gap and Opportunity Analysis

This section compares the major climate-related hazards, impacts, vulnerabilities and risks in ecosystems and social systems presented in the previous section against the adaptation measures found in the NDCs at either the ecosystem service level (for ecosystems) or social dimension (for social systems). The analysis aims to identify policy coverage gaps and, therefore, opportunities for enhancing adaptation options in the next round of NDCs.

"Policy coverage" refers to when at least one adaptation measure in a country's NDC aims to reduce vulnerability and/or increase adaptive capacity in relation to a given climate-related hazard, impact, vulnerability or risk reported, or "hotspot." The analysis is based on the methodological matrix and assessment framework that can be found in Crumpler et al. (2019). "Policy coverage" is quantified at the regional level as the share of countries with at least one adaptation policy or measure that addresses a given hotspot. A "policy coverage gap" refers to when there is an absence of at least one adaptation policy or measure that addresses a particular vulnerability hotspot. A policy coverage gap is the difference between the share of countries with a vulnerability hotspot and the share of countries with policy coverage. It should be noted that the analysis serves as a broad review of the coverage of adaptation priority sectors and measures mentioned in the NDC and not an assessment of their strength, which should be further assessed in terms of type (e.g. action, policy, project, programme or framework), scale, comprehensiveness and timeframe. The analysis, therefore, serves as an initial stocktaking of policy coverage and does not necessarily indicate policy effectiveness. Table 2 illustrates the range of policy coverage gaps and associated score.

In ecosystems, high to very high adaptation policy coverage gaps are found around climate-related losses in biological control services and ecosystem services regulating the moderation of extreme events, soil erosion and the maintenance of genetic diversity and abundance. Moderate gaps are observed in the relation to observed or projected losses in the provision of crops and fisheries, as well as ecosystem services supporting nutrient cycling and soil formation, and control against the increased invasion of pests and non-native species in agriculture. Table 3 illustrates the adaptation policy coverage gaps found around climate-related ecosystem hotspots, ordered from highest to lowest gap.

In social systems, high adaptation policy coverage gaps are found around climate-related migration and displacement. Moderate policy gaps are found in relation to

Table 2 Adaptation policy coverage gap scoring of NDC

Score	Policy coverage gap range (%)
Very high	61–100
High	31–60
Moderate	10–30
Low	0–9

Table 3 Ecosystem-related adaptation policy coverage gaps in Pacific NDCs

Climate-related ecosystem hotspot	Number of countries with hotspot (%)	Adaptation policy coverage gap
Reduced biological control services	21	Very high
Reduced moderation of extreme events services	29	High
Soil erosion	36	High
Reduced genetic diversity and abundance	43	High
Losses in fisheries provision	71	Moderate
Increased invasion by pests and non-native species in agriculture	29	Moderate
Reduced nutrient cycling and soil formation	71	Moderate
Losses in crops provision	57	Moderate

Table 4 Social system-related adaptation policy coverage gaps in Pacific NDCs

Climate-related social system hotspot	Number of countries with hotspot (%)	Policy gap
Migration and displacement	57	High
Gender and youth inequality	50	Moderate
Loss of productive infrastructure and assets	64	Moderate

observed or projected gender and youth inequality and loss of productive infrastructure. Table 4 illustrates the adaptation policy coverage gaps found around climate-related social system hotspots, ordered from highest to lowest gap.

Mitigation and Sustainable Development Co-benefits of Adaptation in the Pacific

Mitigation and adaptation in agriculture are closely interlinked through a web of feedbacks, synergies, and tradeoffs. Sustainable food and agriculture systems carry the greatest potential for generating synergies across climate change mitigation and adaptation efforts, as well as significant socio-economic and environmental co-benefits (FAO 2016a). In the Pacific, around 40% of countries explicitly reference the mitigation co-benefits of adaptation in their NDCs. Adaptation measures in ocean and

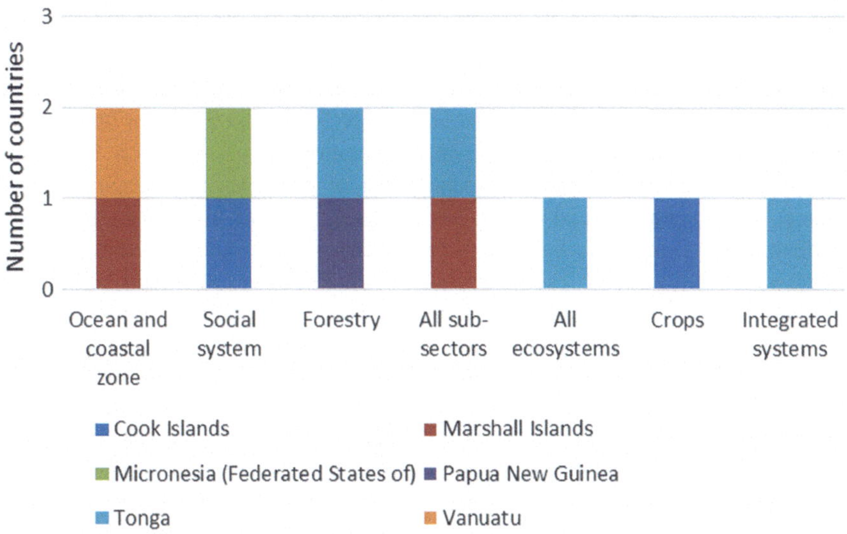

Fig. 19 Number of Pacific countries with explicit reference to the co-benefits of adaptation by land use/sub-sector in the NDCs

coastal zones, social systems and forestry are expected to generate the most mitigation co-benefits amongst adaptation measures in the agriculture and land use sectors, followed by crops and integrated systems. Out of those adaptation measures in ocean and coastal zones, mangrove conservation and replanting and land/soil conservation generate mitigation co-benefits. For instance, the Marshall Islands reference the capacity of mangroves to act as carbon sinks as well as protect water resources and human health. Out of those adaptation measures in social systems, disease management and awareness raising generate the majority of mitigation co-benefits. For instance, Micronesia stresses the benefits associated with raising awareness for the need for adaptation and mitigation, including shifting to renewable energy sources, reduced air pollution, consumption of local and more nutritious food and improved human health. Out of adaptation measures in forestry, reducing deforestation and sustainable forest management generate the majority of mitigation co-benefits. Figure 19 illustrates the number of countries with at least one adaptation measure with mitigation co-benefits explicitly referenced, by land use/sub-sector.

Barriers to Implementation and Support Needs

Article 9, 10 and 11 of the Paris Agreement reiterate the obligations of developed countries to support developing country efforts to build clean, climate-resilient futures through the provision of finance, technology and capacity-building support for climate change mitigation and adaptation. This section discusses the different types of support needs communicated by countries in the Pacific, as well as the barriers facing these nations to effectively put in place technologies and policies to achieve their climate goals and targets. Information from the NDCs was supplemented by a comprehensive review of country NCs and the TNAs to identify all support needs and potential barriers to implementation.

Overall, the majority of countries identify lack of technical capacities and human skills, economic and financial constraints and lack of proper institutions and organizations as the three main barriers to technology transfer and dissemination for climate action in the Pacific. Figure 20 illustrates the share of countries with barriers to climate action in the agriculture and land use sectors reported, by type.

Access to additional financial resources, capacity-building and technology transfer is the preamble to achieving many of the ambitious climate goals and targets. All countries in the region communicate either full or partial conditionality of NDC implementation to external financial support, but not all quantify the respective conditional and unconditional shares. Eighty six percent of countries communicate that NDC implementation is partly conditional to international financial support, while two countries (Samoa and Vanuatu) make their NDC totally conditional to it.

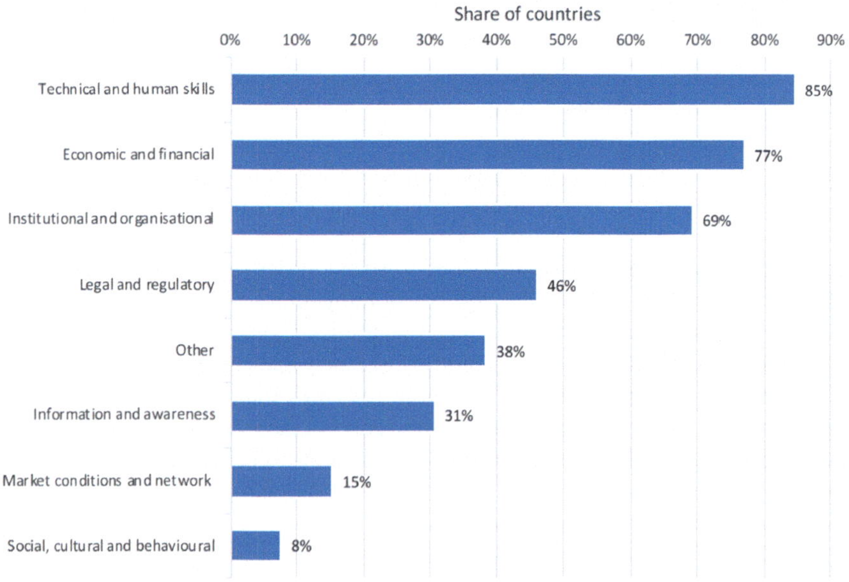

Fig. 20 Barriers to climate action in the agriculture and land use sectors reported in the Pacific, by barrier type (share of countries)

Conclusion

The incidence of climate-related hazards and natural disasters are on the rise in the Pacific (EM-DAT n.d.), threatening food security and nutrition through a cascading chain of impacts transferred from agroecosystems, along food value chains, to natural resource-dependent livelihoods (FAO 2018). Sea level rise, coastal erosion, water stress and storms are reported by the majority of countries in the Pacific as climate-related extremes and slow onset events, exacerbating the vulnerability of communities and households to other compounding social, economic and environmental stressors. The increasing intensity and frequency of climate extreme and variability is expected to result in the loss of productive infrastructure and assets, greater levels of food insecurity and malnutrition, negative impacts on incomes and rural livelihoods, wider gaps in gender and youth equality and increasing trends in migration and displacement, as reported by the majority of countries in the region. With some 486 million people still undernourished in Asia and the Pacific, and progress stagnated in all sub-regions, the increasing severity and incidence of weather extremes and climate-related disasters threaten to seriously burden food security and nutrition in the region, as well as challenge progress on poverty alleviation (FAO 2018).

Indeed, the agriculture and land use sectors feature prominently in the adaptation component found in the NDCs of countries in the Pacific. Around 90% of countries include adaptation polices or measures in ocean and coastal zones and in agro-ecosystems, particularly in the marine fisheries and crops sub-sectors. Almost all countries in the region also recognize the role of adaptation of institutions and governance, including measures promoting climate change policy mainstreaming and coherence, disaster risk reduction and management and health information and services. The opportunity to leverage mitigation and sustainable development co-benefits of adaptation in the agriculture and land use sectors is explicitly referenced by 40% of countries in the region. In particular, mangrove planting and conservation is associated with natural sinks for emission removals, while climate change awareness raising is associated with human health benefits.

However, high to very high adaptation policy coverage gaps are still found around climate-related losses in biological control services and ecosystem services regulating the moderation of extreme events, soil erosion and the maintenance of genetic diversity and abundance. In social systems, moderate to high adaptation policy coverage gaps are found in relation to climate-induced migration and displacement, gender and youth inequality and losses in productive infrastructure and assets.

By highlighting the gaps in the coverage of adaptation policies in the agriculture and land use sectors, as well as illustrating opportunities for enhancing adaptation ambitions in the next round of NDCs, this analysis can serve as an important roadmap for informing country programming and directing future investments in support of low-emission and climate-resilient agriculture and food systems in the region. Evidence suggests that an integrated approach to disaster risk reduction and management and climate change adaptation that promotes anticipatory, absorptive, adaptive and transformative capacity is the most effective in building the resilience to climate-related hazards and natural disasters of the community and ecosystem as a whole (FAO 2018). Ecosystem-based approaches also offer opportunities for leveraging mitigation and sustainable development co-benefits, enabling national governments to co-deliver on other international agendas such as the 2030 Agenda for Sustainable Development and the Sendai Framework for Disaster Risk Reduction. Unlocking the potential of transformational adaptation and resilience in the agriculture and land use sectors will require addressing the technical, economic and institutional barriers reported that are impeding progress on NDC implementation in the Pacific.

References

Asch RG, Cheung WWL, Reygondeau G (2018) Future marine ecosystem drivers, biodiversity, and fisheries maximum catch potential in Pacific Island countries and territories under climate change. Marine Policy 88:285–294. https://doi.org/10.1016/j.marpol.2017.08.015

Crumpler K, Meybeck A, Federici S, Salvatore M, Damen B, Dasgupta S, Wolf J, Bernoux M (2019) Assessing the role of agriculture and land use in Nationally Determined Contributions. Working paper no. 76. Environment and Natural Resources Management, FAO, Rome, 58pp. http://www.fao.org/3/CA5543EN/CA5543EN.pdf

Edmonds C, Noy I (2018) The economics of disaster risks and impacts in the Pacific. Disaster Prevent Manage 27(5):478–494. https://doi.org/10.1108/DPM-02-2018-0057

EM-DAT (n.d.) The international disaster database. In: Centre for research on the epidemiology of disasters (CRED). Brussels. https://www.emdat.be/. Accessed on 29 Oct 2019

FAO (2020) Regional analysis of the nationally determined contributions in the Pacific: gaps and opportunities in the agriculture and land use sectors. FAO, Rome

FAO (2016a) State of Food and Agriculture: Climate change, Agriculture and Food Security. FAO, Rome. 194 pp. http://www.fao.org/3/a-i6030e.pdf

FAO (2016b) The agriculture sectors in the intended nationally determined contributions: analysis. FAO, Rome. 92pp. http://www.fao.org/3/a-i5687e.pdf

FAO (2018) Asia and the Pacific regional overview of food security and nutrition 2018—accelerating progress towards the SDGs. FAO, Bangkok. 96pp. http://www.fao.org/3/CA0950EN/CA0950EN.pdf

Hallegatte S, Bangalore M, Bonzanigo L, Fay M, Kane T, Narloch U, Rozenberg J, Treguer D, Schilb A (2016) Shock waves: managing the impacts of climate change on poverty. World Bank Publications, Washington, DC

Hanich Q, Wabnitz CC, Ota Y, Amos M, Donato-Hunt C, Hunt A (2018) Small-scale fisheries under climate change in the Pacific Island region. Marine Policy 88:279–284. https://doi.org/10.1016/j.marpol.2017.11.011

Intergovernmental Panel on Climate Change (IPCC) (2014) Climate change 2014: impacts, adaptation, and vulnerability. Part B: regional aspects. Contribution of working group II to the fifth assessment report of the Intergovernmental Panel on Climate Change. Cambridge, UK, and New York, Cambridge University Press, 688pp. https://www.ipcc.ch/report/ar5/wg2/

IPCC (2018) Global warming of 1.5 °C. An IPCC special report on the impacts of global warming of 1.5 °C above pre-industrial levels and related global greenhouse gas emission pathways, in the context of strengthening the global response to the threat of climate change, sustainable development, and efforts to eradicate. IPCC (in press). https://www.ipcc.ch/sr15/

IPCC (2019a) Summary for policymakers. In: IPCC special report on the ocean and cryosphere in a changing climate. IPCC (in press). https://report.ipcc.ch/srocc/pdf/SROCC_SPM_Approved.pdf

IPCC (2019b) Summary for policymakers. In: Climate change and land: an IPCC special report on climate change, desertification, land degradation, sustainable land management, food security, and greenhouse gas fluxes in terrestrial ecosystems. IPCC (in press). https://www.ipcc.ch/report/srccl/

United Nations Statistics Division (UNSD) (n.d.) Standard country or area codes for statistical use (M49). In: United Nations. https://unstats.un.org/unsd/methodology/m49/. Accessed on 28 Oct 2019

Place Attachment and Cultural Barriers to Climate Change Induced Relocation: Lessons from Vunisavisavi Village, Vanua Levu, Fiji

Priyatma Singh, Dhrishna Charan, Manpreet Kaur, Kelera Railoa and Ravneel Chand

Abstract Relocation from coastal areas is a huge challenge for communities vulnerable to the impacts of climate change induced inundation. This study focuses on Vunisavisavi Village in Fiji, where severe coastal erosion and frequent inundation events have increased to such an extent that relocation is the only feasible option remaining. This paper explores the social and cultural challenges faced by Vunisavisavi villagers in relocation, with an emphasis on the extent to which place attachment acts as a barrier for relocation. The paper summarizes the findings from individual and focus group interviews of Vunisavisavi villagers. The research findings provide an insight into the existing adaptation patterns of the villagers and recommends an early intervention in assessing the vulnerability of communities to ensure that best adaptation strategies are implemented.

Keywords Climate change · Place attachment · Relocation · Cultural barriers · Ecosystem-based adaptation

---

P. Singh (✉) · D. Charan · R. Chand
School of Science and Technology, The University of Fiji, Saweni, Lautoka, Fiji
e-mail: priyatmas@unifiji.ac.fj

D. Charan
e-mail: dhrishnac@unifiji.ac.fj

R. Chand
e-mail: ravneelc@unifiji.ac.fj

M. Kaur
School of Humanities and Arts, The University of Fiji, Saweni, Lautoka, Fiji
e-mail: manpreetk@unifiji.ac.fj

K. Railoa
Technical and Validation Team, Douglas Pharmacy, Nadi, Fiji
e-mail: kelerar@douglas.com.fj

W. Leal Filho (ed.), *Managing Climate Change Adaptation in the Pacific Region*, Climate Change Management, https://doi.org/10.1007/978-3-030-40552-6_2

Introduction

The Pacific island nations have experienced periodic changes in climate that have affected the well-being and economy of the small island nations for decades (Boege 2011). The projected changes in climatic conditions pose threat to the livelihood of the Pacific Islanders, particularly those in the low-lying coral islands (Nunn 2012 in Charan et al. 2016). There is no denying that the livelihoods of most Pacific Islanders is being severely affected due to rising sea levels, increased coastal erosion, inundation, flash flooding, cyclones and salinization of coastal aquifers (Ferris et al. 2011). While the Pacific people have been adapting to the changing climate, further adaptation efforts will be required as the effects of climate change intensify. Over the past few years, numerous research projects have sought to analyze and document the links between climate change and human migration but only a few of these have attempted to explore the dimensions of planned relocation made obligatory due to climate change (McAdam and Ferris 2015).

The inevitable need to familiarize to climate change adversities has intensified over the last few decades as climate change continues to remain a priority in discussions in many island nations in recent discourses (Barnett and Campbell 2010). The thermal expansion of sea water coupled with the melting of land ice is threatening to increase sea levels to limits which will render most low-lying atolls inhabitable. Since 1993, Fiji has recorded a 6 mm (0.2 in.) increase in its sea level per year, larger than the global average (COP 23, Fiji 2017). The rise in sea level and the resulting sea water intrusion that stems from the increased ferocity of coastal floods have made portions of the island uninhabitable. Although, 6 mm may seem as an inconsequential sum, its ramifications on island nations are colossal.

According to IPCC's Fifth Assessment Report, sea-level rise will significantly affect islands where communities and infrastructure are located in coastal zones with limited possibility to relocate inland (IPCC 2014). This is evident in many Pacific island communities where the increasing impacts of climate change is causing unprecedented adverse effects to the coastal ecosystem, which has left many of these coastal communities exposed to the threats of flooding and inundation. For Pacific island coastal communities, retreating slightly inland due to increased outreach of ocean waves has been an ongoing adaptive practice in the past, but the recent impacts of the rapidly changing climate is forcing some of these communities to seek land much further inland, away from the ever increasing reach of the ocean waves. The onset of the disastrous impacts of climate change, and permanent relocation further inland seems inevitable (Tronquet 2015).

For the iTaukei (indigenous Fijians) communities, land is seen as an ancestral trust passed down through successive generations, thus resulting in the formation of strong cultural links between people and their land. Land in Fiji is categorized into three main types namely: freehold land, state land and native land. Out of these, about 9% of the land is freehold, 3% belongs to the state, and the remaining 88% is native land (Shah 2017). Land in iTaukei communities is communally owned by the *mataqali*[1] with any decisions regarding land use being presided over by the traditional leaders with collective input from the clan. Fiji's native land is also administered by the iTaukei Land Trust Board (formerly known as the Native Land Trust Board) who are the authorized body assigned to oversee the matters concerning native land use.

This paper explores the extent to which land ownership, place attachment and culture affect population movements in Fiji, exemplified by the case of Vunisavisavi Village. It builds on the notion that place attachment and community structures are essential elements to sustainable relocation. The two dimensions of place attachment, namely place identity and place dependence have a large bearing on the extent of connection an individual has with the land (Low 1992). According to Stokols and Schumaker (as cited in Pretty et al. 2003), place dependence reflects the functional attachment of a setting in achieving the specific goals of the individual. Prohansky (as cited in Kyle et al. 2004) equates place identity to the cognitive connection of the self with the setting, while Kals and Maes characterize it as an emotional bond which is developed between an individual and a place (2002). For the case of Vunisavisavi Village, place attachment is explored as a social bonding developed through shared experiences in the village site. An equally pivotal aspect is the cultural connection to land and ocean that is deeply ingrained in the people of Vunisavisavi. Culture is important to understand the way in which people identify risks and react to adversities (Adger et al. 2012). In a Fijian context, the notion of culture is essential to understanding the causes and meaning of human responses to climate change.

The main aim of this paper is to explore the role of place attachment and cultural obligations in the Vunisavisavi Village relocation. The paper also examines few other cases of recent full scale and partial relocations in Fiji to understand the processes through which the traditional customs, shared institutions and structures were reconciled in past relocations. Financial limitations as a barrier to relocation in Fiji is briefly considered. The paper also includes a decription of community based adaptation (CBA) practices and traditional practices that have enabled the people of Vunisavisavi to adapt to their changing environment and explores options for Ecosystem-based Adaptation (EbA) practices that could improve the overall adaptation strategy for the Vunisavisavi community.

The Vunisavisavi Community

The study site of Vunisavisavi is a village of approximately 200 inhabitants (25 households), which is located in the South Coast of Savusavu Town in Cakaudrove Province, Vanua Levu, the second largest island in Fiji (Figs. 1 and 2). The village, which is an hour's drive from Savusavu Town, is situated a few meters from the sea. The main source of income for the villagers is from *Yaqona*[2] (*Piper methysticum*) farming and fishing. The village suffers from frequent inundation, particularly from

[1] Traditional iTaukei land owning unit or clan.

[2] Referred to as kava and is strongly associated with the Fijian culture and traditional ceremony.

Fig. 1 Map of Fiji. Adapted from Fiji Google Maps

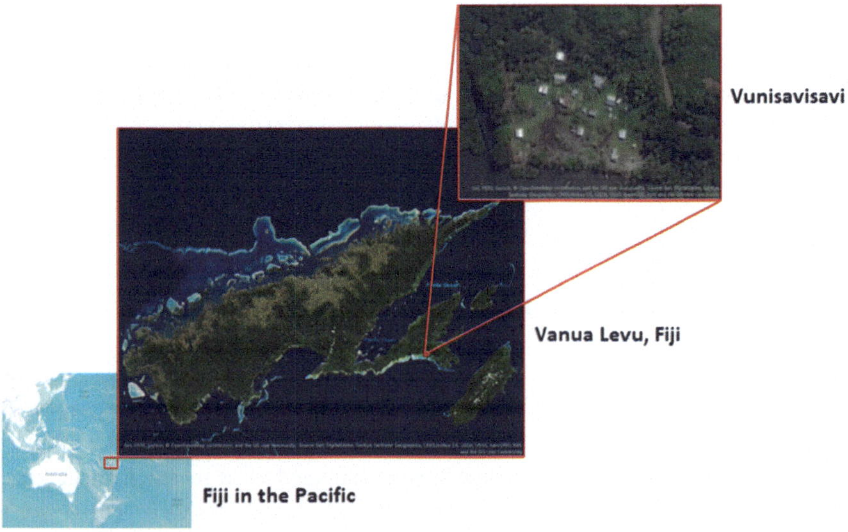

Fig. 2 Locality map showing Vunisavisavi Village, Savusavu, Fiji

the influence of king tides. It is poignant to observe that the resources the villagers are heavily reliant on are repeatedly impinged by flood waters. The quality of soil has deteriorated due to salt water intrusion, making it difficult for the villagers to grow crops in their *teitei.*[3]

The villagers have been experiencing the recurrent issue of flooding for the last thirty years, but recently the flooding events have intensified in frequency. The constant inundation has resulted in substantial damage to the village site with a shriveled sea front containing only a few surviving trees and small patches of grass (Fig. 3). There are small mangrove strands near the village and some scattered vegetation along the shoreline. In addition, the streams which were once a source of drinking water is now salty due to inundation. Even the village septic tanks fill up with seawater during high tide. During the period of the field visit, the village was plagued by a prolonged dry spell which caused the households to be completely dependent on government water trucks to fill up their tanks for the purpose of drinking, cooking

Fig. 3 Soil quality has deteriorated with few trees and patches of grass remaining

[3]Fijian term for farms.

and bathing. Other essential infrastructure such as access road, main road and the village drain are exposed and vulnerable to flooding and extreme weather events.

The village is a site for the USAID Coastal Community Adaptation Project (C-CAP). Through this project, four new houses have been constructed and four families successfully relocated. The houses moved a few meters inland from their previous location and are still located in the same vicinity. Indisputably, most of the remaining homes in Vunisavisavi need to be relocated. However, there are several barriers that need attention redress prior to facilitating complete relocation. During the field visitation, it was revealed by the villagers that the village is located on the original site of the paramount Chief's house and many villagers feel duty bound to stay and watch over the land, fearing customary punishment will follow if they leave. This unyielding belief is rooted in Vunisavisavi settlement, making it difficult to convince the villagers to relocate.

Method

The sentiments surrounding the notion of relocation is complex and as such a participatory research approach was adopted to gain insights into the way the Vunisavisavi villagers perceive relocation. This research examines the prospect of relocating Vunisavisavi Village and considers the challenges faced by the villagers in terms of their attachment to their land and cultural beliefs that is deeply rooted in the village traditions. The study utilized literature and secondary data sources to recourse information on the process of relocation, which provided the initial research direction. This was accomplished through means of a literature review, which focused on analyzing scholarly literature on forced relocation and recent and past examples of climate change induced relocation in Fiji.

Semi-structured interviews were conducted as a means of obtaining qualitative data from the villagers and key informants. The interviews were conducted over a period of two weeks in July, 2018. The study gathered information from 20 individual interviews and 2 focus group interviews. Individual interviewees included 16 villagers from Vunisavisavi and 4 administrators from the Cakaudrove Provincial Council. Equal number of men and women were interviewed. The participants ranged from the age of 20 years to 70 years. There were two focus groups. For ease of communication, one of the focus groups consisted of 5 women and the other comprised of 5 men. A questionnaire guide was used for the interviews, where English and iTaukei languages featured as a medium of communication, focusing on eliciting information in relation to climate change relocation and adaptation measures for Vunisavisavi settlement.

Limitations of the Research

Due to funding and time limitations, the researchers were unable to carry out surveys, however, qualitative research methods was considered appropriate due to the nature of the research. While, this meant that the study did not yield statistically significant results, it provided details that assisted the researchers to comprehend the significance the participants placed on the matter under investigation.

Place Attachment and Its Implications on Relocation

The role of place attachment in climate induced relocation warrants deliberation. Place attachement revolves around emotions, local knowledge and customary behaviour (Chow and Healy 2008). The notion of attachment adjoins to individual and community well-being and value of life and is extensively used as a constituent in assessing community sustainability. Thus, place attachment is an important factor for climate change adaptation in regions where livelihoods of people get affected as impacts of climate changes are increasingly manifested (McAdam and Ferris 2015). Relocation from their land, either voluntarily or involuntarily, has emotional and psychological implications for its people (Twigger-Ross and Uzzel 1996). In the case of Vunisavisavi Village, the examination of place attachment as an emotional bond has shed light on the distress and grief expressed by those who are forced to relocate. In order to completely understand this, one must first appreciate the value that indigenous communities give to their land in Fiji.

Land is more than just a means of earning livelihoods for an indigenous Fijian, it is strongly linked to traditional and emotional well-being. The village kinsmen of Vunisavisavi recognize the traditional significance of the site on which the village is situated. The Assistant Village Headman explains this as "our elders were sent here from Somosomo, Taveuni to occupy this site as this is where the first *Tui Cakau*,[4] Ro Kevu, was installed as paramount Chief of Cakaudrove Province" (Vunisavisavi Village Interviewee, personal communication, 20 July 2018, Savusavu, Fiji). According to the majority of the villagers, the current village site has deep significance due to their lineage. Naidu and Reddy (2004) capture the intrinsic relationship one shares with the land aptly describing the concept of the *vanua*[5] as not just a detached entity but as "inclusive of the land with its flora and fauna, rivers and adjacent seas, the people (the ancestors, those living and those yet to be born) and their customs, norms, beliefs, social organization (traditional hierarchy or *tutu vakavanua*[6]) and sacred sites". According to the Cakaudrove Provincial Council, it is these sentiments that

---

[4]The Tui Cakau is the Paramount Chief of Cakaudrove Province in Fiji.

[5]Vanua is the Fijian term used to refer to the physical land, home or village.

[6]Tutu vakavanua is the Fijian term referring to social organization or traditional hierarchy.

continue to evoke strong emotional attachments to the *vanua* and more particularly to a certain place.

One of the women participants expressed her family's unwillingness to demolish their home and relocate inland because of the memories attached to the house and the village site. As stated by her, "This house was built by my grandfather way back when chainsaws were not available so they had to use an axe to fetch enough trees for timber" (Vunisavisavi Village Interviewee, personal communication, 21 July 2018, Savusavu, Fiji). She explained that it is difficult for her family to leave their home, a place where they were born and raised and for which their father and elders worked tirelessly to construct. An elderly woman, who spent most of her time at the village, stated, "We would rather let nature come and take its course after which we will then move" (Vunisavisavi Village Interviewee, personal communication, 22 July 2018, Savusavu, Fiji). The idea of abandoning their place of birth and moving elsewhere even if it is still within the land boundaries of the same *mataqali*[7] is quite stressful for the aged members of the village, given the memories and emotional attachments that they have with the site.

The young participants, mostly in their twenties were open to discussions and able to understand the need to relocate in the face of adversity. Nevertheless, the kinship ties and communal lifestyle make it challenging for even the young people to completely embrace the notion of relocation. Attachment to a place may be closely linked to a sense of belonging to a community and this is quite apparent in the Fijian context, where many communities still live in villages governed by chiefly systems. These familial units with strong attachment to their community are often unwilling to relocate because they are reluctant to leave behind their social and emotional support groups and adapt to a new community.

Cultural Barriers and Traditional Obligations

The multifaceted nature of climate change is essential to consider when developing strategies for adaptation. For example, while it is important to focus on environmental, economic and political aspects of climate change, it is equally vital to emphasize the cultural and social norms that may increase or reduce the exposure to climate-related risks. This indicates that climate change is a cultural challenge as well, especially considering the diverse ways in which culture is interpreted by various communities. In a Fijian society, the notion of the *vanua* is acknowledged and used as a powerful instrument in securing conformity and forging solidarity. The members of the community play a unique role in fulfilling their duties to the paramount chief and ultimately the *vanua*.

[7]Mataqali is the Fijian term referring to a Fijian clan/tribe or landowning unit.

Bound by traditional obligation to the *Tui Cakau*, the people of Vunisavisavi have been bestowed stewardship over the *lalagavesi.*[8] One of the elderly participants, from a chiefly family stated, "It is our traditional duty to the *Tui Cakau* to remain here and watch over the *lalagavesi* and *yavu*" (Vunisavisavi Village Interviewee, personal communication, 21 July 2018, Savusavu, Fiji). The villagers strongly believe in chastisement that may befall them at the new site should they leave the original site. Almost all of the participants mentioned that the newly married couples are encouraged to build new homes away from the current site and the households that currently exist are to remain and fulfill their obligation to the *vanua* until the village becomes completely inundated. Assistant Village Headman stated, "The elderly will remain here until the homes are completely destroyed". We know we will face *sau*[9] if we move now so we do not want to take that risk" (Vunisavisavi Village Interviewee, personal communication, 20 July 2018, Savusavu, Fiji). Hidalgo and Hernandez (2001) stated that when a community relocates from their place of birth—a place they value so much, their cultures may be weakened, and in some cases endangered as well. According to about 90% of the interviewees there is absolutely no replacement for, or sufficient reparation for lost sites of meaning.

One of the participants, who occupies the newly constructed home, revealed that they only moved because their former house was completely inundated during king tide and frequent coastal floods. According to the village spokesperson, an elderly male, the villagers understand and respect the power (*sau*[10]) or blessing (*mana*[11]) entrusted to the Chief's. The spokesperson mentioned that, "even when the tides come in and village is flooded, the residence of the former *Tui Cakau,* remains above the water". The spokesperson further described it as a miracle and stated, "It's as if the sea understands the significance of the *lalagavesi* and refuses to touch it" (Vunisavisavi Village Interviewee, personal communication, 24 July 2018, Savusavu, Fiji).

In an iTaukei community, culture plays an integral role in shaping the Fijian way of life and ultimately all major decision making processes. The Fijian culture encompasses the *vanua*, traditional roles as a clan or *mataqali* and as an individual these define or give identity to one as *iTaukei.* Cultural and traditional obligations to Chiefs and the *vanua* are acknowledged barriers to relocation across many vulnerable Fijian communities and unless these are integrated at the onset of any relocation effort, there is very little guarantee of success. The impacts of climate change experienced by the people of Vunisavisavi are significant and even though relocation in future seems inevitable, the members of the community are in a quandary over their treasured land and the need to move to the new site as ramifications of climate change become palpable. The members realize the seriousness of the situation but their social and cultural values prevent them from relocating.

---

[8]Lalagavesi is the Fijian term referring to the foundation of the former and first Tui Cakau's chiefly residence.

[9]Sau is the Fijian term used to refer to customary punishment and can also be used to mean 'power'. In this instance it means customary punishment.

[10]Sau in this instance is used to refer to power or chiefly might.

[11]Mana is the Fijian term used to refer to heavenly blessings.

Funding Constraints

The study indicated that financial constraint is also one of the biggest hurdles for climate induced relocation. The lack of governance and funding mechanisms create countless challenges for villagers in need of relocation. For example, the focus group discussions revealed that some families whose houses are close to the shores and frequently flooded might be willing to relocate but they do not have the funds to do so. Gharbaoui and Blocher (2016) define planned relocation as premeditated movement of a population in consultation with the community being relocated which is planned and supported by the government, its partners, or a regional entity. However, Edwards (2013a, b) reports that national governments usually do not have the capacity to offer displaced people financial support. This is especially applicable to small island developing states that face numerous political and economic challenges.

Campbell (2010) argues that the cost of climate induced relocation should not be borne by the community or their national governments and any costs incurred should be compensated via international adaptation financing. In a recent case of the Vunidogoloa Village relocation, the Fijian government was able to sponsor up to two thirds of the cost of the entire relocation, which is small feat in itself for a small island state which is still in its developing stages. In 2019, the Fijian government launched Fiji's Climate Relocation and Displaced Peoples Trust Fund for communities and infrastrature. A certain percentage of the money will be deducted from the Environment and Climate Adaptation levy to finance the fund. However, this may not be sufficient to cater for the increasing number of villages in need of relocation.

It is clear that international source of funding from donor agencies will be needed to support Vunisavisavi and other future relocations. However, developing nations like Fiji usually face limitations in accessing the international climate fund, mostly due to its intricate application and accreditation procedures. In response to these issues of accessibility raised by developing nations in the past few years, major funding agencies such as the Green Climate Fund are now improving their strategic planning to be more responsive to the needs and priorities of developing countries' which include enhancing direct access to funds by these nations (Green Climate Fund 2017).

Overview of Past Climate Induced Relocation Efforts in Fiji

The first successful climate induced community relocation was that of Vunidogoloa Village, Vanua Levu (Fig. 4). The Relocation Project was supported and largely funded by the Fijian government as encroaching seas due to rising sea levels were making the former village site uninhabitable (Charan et al. 2016). The Vunidogoloa villagers made several attempts to adapt by building seawall, raising houses on piles and moving slightly inland, but eventually decided to move further inland as it became extremely challenging to sustain livelihoods. The new site was within the same

Fig. 4 Vunidogoloa Village at new site

customary boundaries of the community, so the land ownership issues were avoided. However, the process of relocation was an emotionally disturbing experience for the villagers who had strong connections to former village site coupled with fond memories of the place.

Subsequent to the successful relocation of Vunidogoloa, a small coastal community of Narikoso on Ono Island was the next to relocate as sea level rise had severely inundated the coastal area. The sea wall which was constructed over 50 years ago gradually eroded and did not offer much protection against the storm surges. The relocation of the Narikoso community started in the year 2012 and since then it faced a myriad of issues, especially associated with the unavailability of financial resources and villagers' resistance to relocation (Bertana 2019). Additionally, 10 houses earmarked for relocation belonged to a *mataqali* which has no land to relocate to, which further complicated the relocation process (Bertana 2019).

The coastal village of Solodamu on Fiji's third largest island, Kadavu is yet another example of relocation due to climate change (Cagilaba 2005). The Solodamu villagers had to negotiate access to the new site with a neighboring village. Following traditional negotiations between the two village chiefs, the Solodamu community members were able to move to the new site—a hilly area 2 km inland (Rokocoko 2006). Gradually, social tensions began to grow between the new settlers and the host community. In such cases the protection of the host community's land rights and appropriate compensation for the displaced community needs to be considered and discussed among all stakeholders before relocation.

A more sudden relocation compared to the previous three cases, is the Denimanu relocation which was completed in 2013. The Denimanu Village located on the Yadua Island was a planned partial relocation which took place after cyclone Evan in December 2012 destroyed around 19 houses in the village. The relocation which was supported by the government, involved relocating these 19 houses to a closely related *mataqali* land, some 500 m away from the original site (Piggott-McKellar et al. 2019). As such the relocated villagers did not experience land issues like in the case of the Narikoso and Solodamu village.

The Fijian Government has developed a National Relocation Guideline to support and steer relocation efforts at the local level and is working towards formulating a

National Relocation Policy. Such a policy must circumspectly address social and cultural issues which are the foundation of traditional Fijian structures. There are numerous other issues to take into consideration while developing the relocation policy. These entail: new sources of income and livelihood, environmental impact assessment at the new site, and consistent monitoring and evaluation to assess the livelihoods of people at the new location. An extensive consultation and deliberation is necessary with the relevant stakeholders before Fiji's National Relocation Policy is made available. The *iTaukei* land and socio-cultural elements are an integral component of most communities in Fiji, and these must form the basis of national relocation framework. It is also essential to consider funding issues as several communities might be able to overcome cultural barriers to relocation, but may be constrained by financial issues.

The Vunisavisavi relocation presents a unique case as traditional obligation and place attachment forms an important hurdle that may need to be overcome through successive traditional consultations, particularly between the communities at stake, traditional chiefs and representatives of government ministries. In the meantime, it is imperative to continue exploring community based adaptation (CBA), traditional practices and ecosystem based adaptation (EbA) that have enabled people of Vunisavisavi to increase their community resilience.

Community Based Adaptation

The Pacific region consists of a number of small island nations that are entrenched in rich culture and traditions, which have both been a limitation as well as an opportunity for adaptation (Barnett 2001). In the case of Vunisavisavi Village, cultural obligations were observed to impede relocation. However, there are robust knowledge-practice belief systems that have allowed communities to adapt to their surrounding environment in order to maintain sustainable livelihoods over long periods of time (McNamara and Prasad 2014). Berkes and Jolly (2002) argue that although climate change affects the livelihood and subsistence activities, there are several ways that communities can adapt to the changing climate by varying their coping strategies.

A community's adaptation to climate change is based on a number of factors, some of which extend beyond technical and financial prerequisites. It is important for communities to demonstrate their adaptive capacity by restructuring their strategies as per their social response options (Berkes and Jolly 2002). For example, community members fully understand their local environment, and as such have the capacity to reduce the impacts of climate change by utilizing their local knowledge of the environment and natural resources (Eisenack et al. 2014). The people of Vunisavisavi are cognizant of the impacts of climate change and have been collectively addressing the challenges for over 20 years. For example, they have shifted their plantations further inland as the agricultural land close to the village is not suitable for growing crops due to saltwater inundation. The stream closer to the village is also affected by seawater intrusion. The community members are storing freshwater supplied by Government as their own water supply from a stream in the highlands is often disrupted.

Although, the iTaukei communities normally have a division of labor based on gender, the challenges brought about by the changing climate are inspiring more women to share responsibilities in planting of *Yaqona*, especially with weeding, sorting and sometimes even marketing. The support of women in CBA initiatives is commendable and requires further research to understand their contribution to community resilience. The Fijian government successfully commissioned solar panels for homes in Vunisavisavi in June 2017. Introduction of solar electricity has significantly increased the adaptive capacity of the villagers by improving their quality of life ensuing that they have access to a clean and reliable source of energy. The villagers no longer depend on diesel generators. Instead, they are able to invest their savings in other projects that increase their livelihood.

Integrating Traditional Practices in Adaptation Strategies

Traditional knowledge is consistent with the adaptive capacity of community members and includes an integrated approach to resource management (Berkes et al. 2000; McNamara and Prasad 2014). Janif et al. (2016: 7) states that "Among the components of resilience are traditional environmental knowledge, community unity, and respected decision making processes". Indigenous customary knowledge is recognized as an essential factor in improving the adaptive capacity of villagers. According to an elderly participant, "When hornets begin to build their nests very close to the ground, it is a clear indication that a tropical cyclone is approaching" (Vunisavisavi Village Interviewee, personal communication, 23 July 2018, Savusavu, Fiji). This prompts villagers to begin preparation for alleviating the impacts of natural hazards. People of Vunisavisavi, particularly women exhibit adequate knowledge of food preservation techniques. They often seek assistance from men to create earth oven pits to cook fish and root crops. Food made in this way can be conserved for over a week and is especially useful in times of food shortages post cyclone.

Villagers also smoke fish, which can be preserved for up to one week. Similar findings are shared by McNamara and Prasad (2014) in their case study of three villages in Fiji namely; Naselesele, Qeleni and Yanuca. The study which documented and synthesized local and traditional experiences and knowledge used in preparation for extreme weather events revealed that food crops of traditional iTaukei resilient staple foods such as *taro,*[12] *yam,*[13] *cassava,*[14] *breadfruit* and *kumala*[15] are often valuable in times of extreme climate events such as drought and cyclone. The traditional practices also overlap with resource sustainability. As explained by one of the participants,

[12]Taro—*Colocasia Esculenta.*

[13]Yam—common name for some plant species in the genus *Dioscorea* that form edible tubers.

[14]Cassava *Manihot Esculenta.*

[15]Kumala—Sweet Potato (*Ipomoea batatas).*

"as part of cultural obligation, as much as 50% of the fishing grounds are closed following the death of village elderly and reopened after 3–6 months" (Vunisavisavi Village Interviewee, personal communication, 22 July 2018, Savusavu, Fiji). Preservation and utilization of medicinal flora for traditional treatments is also a common practice in iTaukei communities and is an effective way of assisting communities cope with the adverse impacts of climate change, particularly, infectious diseases that increase during extreme weather events.

Ecosystem Based Adaptation

Coastal ecosystems offer various services and opportunities for communities and this interdependence if managed properly can increase a community's resilience (Jones et al. 2012). Ecosystem Based Adaptation (EbA) is a relatively new strategy used in disaster risk reduction and climate change adaptation, which utilizes the most accessible asset; the natural system, to adapt to the impacts of climate variability (Munang et al. 2009). As described by the Convention on Biological Diversity (2009), EbA is the "use of biodiversity and ecosystem services as part of an overall adaptation strategy to help people to adapt to the adverse effects of climate change". Natural ecosystem provides services that reduces the impact of natural hazards, improves resource management, increases food security and enhances livelihood diversification (IPPC 2014).

In the case of Vunisavisavi, socio-cultural and financial barriers impede effective relocation, thus increasing exposure to risks posed by natural hazards. However, majority of the participants expressed some level of understanding of the EbA approach that can increase community resilience. For instance, one of the participants showed prior knowledge about the importance of mangrove rehabilitation activities to strengthen shoreline and prevent coastal degradation. Mangrove ecosystems can dissipate the energy of tropical cyclones along exposed coastlines. Mangrove forest cover is not only an essential tool for adaptation, but also contributes to climate change mitigation due to its high carbon sequestration value. Alongi (2012), argue that coastal vegetation such as mangrove forests are a valuable resource and has remarkable capacity to offer coastal protection, protect biodiversity and regulate climatic processes. The existing coastline along Vunisavisavi has small strands of mangroves, which does not deliver optimal services due to its patchy distribution. Thus, a concerted effort by community members and external stakeholders is essential to induce a community steered mangrove restoration program.

The use of coir logs in Fiji has recently garnered some consideration, particularly because it is a cost effective approach to increase community's resilience. Hashim et al. (2010) describes coir logs as compressed cylindrical structures made from natural biodegradable coconut fiber, sheathed in polypropylene netting. Generally, coir logs are implanted along with vetiver grass and local shrubs to ensure sustainable waterways and riverbanks. The coir logs function as a buffer against wave action and protect vegetation beyond by deeply anchoring plant root system into the soil. Coir logs also help to stabilize soil which supports coastal infrastructure. World Wide Fund for Nature (WWF)—Pacific has implemented coir-logs in some villages in Fiji as part of their USAID funded project aimed at reinforcing local institutional structures and resource management for communities in Fiji (WWF-Pacific 2016). This technique is fairly new for Fiji and warrants further observation and evaluation to gather its effectiveness in sustaining coastal systems.

Resilient ecosystems are important to support communities' efforts to sustain livelihood and adapt to the changing climate. Berkes (2004, p. 623) argue that "it has become increasingly important to incorporate the dynamic interactions between societies and natural systems in addressing environmental issues". Consequently, the emergence of social-ecological mechanisms may be relevant for several small coastal communities trying to adapt to the impacts of climate change. Social mechanisms and community dynamics play a crucial role in the integration of ecological and community based approach towards adaptation (Berkes et al. 2000). Community cohesion observed in iTaukei communities can play a pivotal role in implementation of EbA strategies. Ecosystem based approach to adaptation entails capacity building of community members and may be difficult to execute if there is a lack of collaboration between communities, institutions and government agencies (Jones et al. 2012).

Conclusion

The complexities surrounding relocation often places pressure on the communities' that are repeatedly challenged by rising sea level. This case study finds that there are numerous obstacles associated with the relocation of Vunisavisavi villagers to a new site. For the people of Vunisavisavi, the need to relocate is overpowered by strong bonding to the original village site. The study highlights the importance of considering the cultural and traditional dimensions associated with relocation. We find that these cultural factors and social norms act as barriers to the Vunisavisavi community relocation. It is also evident from the research that financial challenges faced by villagers contribute to the problem. The Fijian government continues in its pursuit to relocate the coastal communities, needless to say, financial restrictions also affect them. The role of government is fundamental, especially when it comes to designing, supporting and implementing planned relocation. It is thus, imperative for the government to invest conscientiously in preparing a detailed and systematic relocation policy that takes into consideration social, economic and ecological factors. Additionally, although the context is important, the critical factors of culture, finances and land come up in other such relocations in Fiji and so the findings of this study has important generalizations that can be used to understand and plan for most small island community relocations in the Pacific. We also find that the traditional practices and ecosystem based adaptation has potential to increase the adaptive capacity of the Vunisavisavi villagers. It is therefore, imperative for communities to take ownership of their community adaptation plans and learn to adapt as and when required through various social-ecological systems.

References

Adger WN, Barnett J, Brown K, Marshall N, O'Brien K (2012) Cultural dimensions of climate change impacts and adaptation. In: Nature climate change, vol 3. Macmillan, pp 112–117. www.nature.com/natureclimatechange. Accessed on 7 Sept 2017

Alongi DM (2012) Carbon sequestration in mangrove forests. Carbon Manage 3(3):313–322

Barnett J (2001) Adapting to climate change in Pacific Island countries: the problem of uncertainty. World Dev 29(6):977–993

Barnett J, Campbell J (2010) Climate change and small island states: power, knowledge, and the South Pacific. Earthscan, London

Berkes F (2004) Rethinking community-based conservation. Conserv Biol 18(3):621–630

Berkes F, Jolly D (2002) Adapting to climate change: social-ecological resilience in a Canadian western Arctic community. Conserv Ecol 5(2):18

Berkes F, Colding J, Folke C (2000) Rediscovery of traditional ecological knowledge as adaptive management. Ecol Appl 10(5):251–1262

Bertana A (2019) Relocation as an adaptation to sea-level rise: valuable lessons from the Narikoso Village relocation project in Fiji. Case Stud Environ

Boege V (2011) Challenges and pitfalls of resettlement measures: experiences in the Pacific region. Centre on Migration Citizenship and Development Workers Paper Series, Bielefeld

Cagilaba V (2005. Fight or flight? Resilience and vulnerability in rural Fiji. Department of Geography, University of Waikato, Hamilton, p 123

Campbell J (2010) Climate change and population movement in Pacific Island Countries. Climate Change Migr 29–50

Charan D, Kaur M, Singh P (2016) 'Customary land and climate change induced relocation—a case study of Vunidogoloa Village, Vanua Levu, Fiji', In: Leal Filho W (ed) Chapter 2 in Climate change adaptation in Pacific Countries, Climate Change Management, pp 19–33. https://doi.org/10.1007/978-3-319-50094-2_2

Chow K, Healy M (2008) Place attachment and place identity: First year undergraduates making the transition from home to university. J Environ Physcol 28(4):362–372

Convention on Biological Diversity (2009) connecting biodiversity and climate change mitigation and adaptation: report of the second ad hoc technical expert group on biodiversity and climate change. CBD technical series no. 41. Secretariat of the CBD, Montreal, QC, Canada, p 126

Edwards J (2013a) Climate-induced relocation: a first for Fiji. http://www.methodist.org.uk/media/757401/wcr-pcc-relocation-policy-fiji-jan13.pdf. Accessed on 7 Sept 2017

Edwards J (2013b) Financing the future wave of relocations. http://www.methodist.org.uk/media/757401/wcr-pcc-relocation-policy-fiji-jan13.pdf. Accessed on 5 Dec 2015

Eisenack K, Moser SC, Hoffmann E, Klein RJ, Oberlack C, Pechan A, Rotter M, Termeer CJ (2014) Explaining and overcoming barriers to climate change adaptation. Nat Clim Change 4(10):867

Ferris E, Cernea MM, Petz D (2011) On the front line of climate change and displacement: learning from and with Pacific island countries. In: Proceedings of the international disaster and risk conference, 1st edn. Brookings Institution, New York

Gharbaoui D, Blocher J (2016). The reason land matters: relocation as adaptation to climate change in Fiji Islands. In: Migration, risk management and climate change: evidence and policy responses. Springer, pp 149–173

Government of Fiji COP 23 (2017) How Fiji is affected by climate change. https://cop23.com.fj. Accessed on 7 Sept 2017

Green Climate Fund (2017) About the fund. http://www.greenclimate.fund/who-we-are/about-the-fund/governance. Accessed on 3 July 2017

Hashim R, Kamali B, Tamin NM, Zakaria R (2010) An integrated approach to coastal rehabilitation: mangrove restoration in Sungai Haji Dorani, Malaysia. Estuar Coast Shelf Sci 86(1):118–124

Hidalgo MC, Hernandez B (2001) Place attachment: Conceptual and empirical questions. J Environ Psychol 21(3):273–281

Intergovernmental Panel on Climate Change (IPCC) (2014) Chapter 29: Small islands. http://ipcc-wg2.gov/AR5/images/uploads/WGIIAR5-Chap29_FGDall.pdf. Accessed on 25 Feb 2017

Janif SZ, Nunn PD, Geraghty P, Aalbersberg W, Thomas FR, Camailakeba M (2016) Value of traditional oral narratives in building climate-change resilience: insights from rural communities in Fiji. Ecol Soc 21(2):7. http://dx.doi.org/10.5751/ES-08100-210207. Accessed on 5 Dec 2015

Jones HP, Hole DG, Zavaleta ES (2012) Harnessing nature to help people adapt to climate change. Nat Clim Change 2(7):504

Kals E, Maes J (2002) Sustainable development and emotions. In: Schmuck P, Schultz W (eds) Psychology of sustainable development. Kluwer, Norwell, MA, pp 97–122

Kyle G, Graefe A, Manning R, Bacon J (2004) Effects of place attachment on users' perceptions of social and environmental conditions in a natural setting. J Environ Psychol 24(2):213–225

Low SM (1992) Symbolic ties that bind. In: Altman I, Low SM (eds) Place attachment. Plenum Press, New York. pp 165–185. https://link.springer.com/chapter/10.1007/978-1-4684-8753-4_8. Accessed on 8 Sept 2017

McAdam J, Ferris E (2015) Planned relocations in the context of climate change: unpacking the legal and conceptual issues. Cambridge J Int Comparative Law 4(1)

McNamara KE, Prasad SS (2014) Coping with extreme weather: communities in Fiji and Vanuatu share their experiences and knowledge. Clim Change 123(2):121–132

Munang R, Rivington M, Liu J, Thiaw I (2009) Ecosystem management: part of the climate change solution. ENEP Res Brief. http://www.macaulay.ac.uk/copenhagen/documents/UNEP-CC-EM-5-page-brief.pdf. Accessed on 8 Aug 2017

Naidu V, Reddy M (2002) ALTA and expiring land leases: Fijian farmers' perception of their future. Ford foundation. https://www.academia.edu/8612525/ALTA_and_expiring_land_leases_Fijian_farmers_perceptions_of_their_future. Assessed on 8 September 2017

Piggott-McKellar AE, McNamara KE, Nunn PD, Sekinini ST (2019) Moving people in a changing climate: lessons from two case studies in Fiji. Soc Sci 8(5):133

Pretty GH, Chipuer HM, Bramston P (2003) Sense of place amongst adolescents and adults in two rural Australian towns: The discriminating features of place attachment, sense of community and place dependence in relation to place identity. J Environ Psychol 23(3):273–287

Rokocoko V (2006) Fight and flight: challenges faced by two Fijian villages adapting to climate change. In: Oceanic waves, vol 8, no. 4. University of South Pacific, Suva, Fiji

Shah K (2017) Facilitating property developments in the Fiji Islands: the legal mechanisms a review of ALTA and NLTA: reforms. In: Paper presented at the tenth annual conference Pacific Rim Real Estate Society (PRESS), 25–28 Jan 2004, Bangkok, Thailand, pp 1–9. http://www.prres.net/Papers/Shah_Facilitating_Property_Developments_In_The_Fiji_Islands.pdf. Accessed on 8 Sept 2017

Tronquet C (2015) From Vunidogoloa to Kenani: an insight into successful relocation. State Environ Migr 2015:128

Twigger-Ross CL, Uzzell DL (1996) Place and identity processes. J Environ Psychol 16:205–220. http://www.wsl.ch/info/mitarbeitende/hunziker/teaching/download_mat/07-1_Twigger-Ross_Uzzell.pdf. Accessed on 16 Sept 2017

WWF-Pacific (2016) Biodegradable erosion control devices—a new frontier for WWF-Pacific. http://wwf.panda.org/?273110/Biodegradable-erosion-control-devices—a-new-frontier-for-WWF-Pacific. Accessed on 20 July 2017

"Adaptation in Small Islands: Research Themes and Gaps"

Tony Weir

Abstract This paper classifies the literature relating to adaptation in small island developing states (SIDS), especially in the Pacific Islands, with a view to identifying gaps where further research could facilitate useful action. The main themes emerging are: (1) Social issues. (2) sectoral impacts of climate change. (3) community-based adaptation (on which most studies are only in informal literature). (4) relocation of communities, both internationally (widely studied though little acted on as yet) or in-country (an emerging response in the Pacific but comparatively little studied). (5) financing at various levels for adaptation (far outweighed by financing for mitigation), (6) islander perceptions of climate change and their information sources. Researchers based in the islands and regional organisations have an important role in recognising these issues and in developing the local skills base needed to deal with them. The Paris Agreement of 2015 is a positive (but as yet inadequate) step towards the international action on climate change that small island developing states need. It would be particularly useful for researchers to document more cases of successes and failures (and the reasons for them) in community-based adaptation, community relocation (especially in-country relocation), and adaptation (or lack of it) in looser peri-urban communities. Researchers can also greatly assist the positions of the islands' negotiators by documenting the economic and social costs incurred by current and projected climate change, the effectiveness (or otherwise) of the financing mechanisms, and the extent to which 'loss and damage' is being incurred.

Keywords Pacific islands · Community relocation · Climate change · Adaptation · Research gaps

Introduction

The United Nations Framework Convention on Climate Change (UNFCCC 1992) specifically recognises that among the countries particularly vulnerable to climate

T. Weir (✉)
Fenner School of Environment and Society, Australian National University, Canberra, ACT 2601, Australia
e-mail: tony.weir@anu.edu.au

W. Leal Filho (ed.), *Managing Climate Change Adaptation in the Pacific Region*, Climate Change Management, https://doi.org/10.1007/978-3-030-40552-6_3

change are ‘low lying and other small island countries’. This obviously includes the Pacific Island countries (PICs). The PICs are also ‘developing countries’, and as such are widely expected to bear a disproportionate burden from climate change, because of their inability to make the necessary adjustments without help. Thus a major environmental change of concern to SIDS is climate change, both as it is emerging now and is projected to do so in future.

Accordingly, this paper classifies the literature relating to adaptation in small island developing states (SIDS), with a view to identifying gaps where further research could facilitate useful action, especially in the Pacific Islands.

Among the common themes identified in this paper are:

(1) social issues,
(2) sectoral impacts of climate change,
(3) community-based adaptation,
(4) relocation of communities, both internationally and in-country,
(5) financing of adaptation.
(6) islander perceptions and information sources.

The paper continues with two more sections relating to ‘what next?’:

(7) The Paris Agreement and its follow-up:
(8) Conclusions: implications for research and policy.

Thus this paper relates directly to most of the themes of this conference (*Pacific Adapt* 2019), especially:

- Implementing climate change adaptation in rural areas and communities [themes (3) and (4)] above;
- Climate change resilience and hazards [themes (1) and (2)];
- Financing climate change adaptation [theme (5)];
- Information, communication, education and training on climate change [theme (6)].

Although relocation and migration forced by environmental change (theme 4 of this paper) is certainly related to ‘adaptation’, it is important to note that many in the Pacific view it as an impact and thus deserving of compensation for ‘loss and damage’, as island governments seek in the continuing international negotiations following the Paris Agreement of 2015 (see later section of this paper).

Of course it is not claimed that these themes cover all aspects of adaptation in SIDS. *The Fifth Assessment Report* of the Intergovernmental Panel on Climate Change (IPCC) includes a fuller review of impacts and adaptation in SIDS, which highlights also their vulnerability to climate change, particularly as an enhancement of the stresses they already face (Nurse et al. 2014), as does (even more strongly) the later IPCC special report on a 1.5° target (IPCC 2018) and the review by Schleussner et al. (2018).

Nor is it claimed that this paper is a ‘systematic’ review, with actual counts of published research papers under each theme, nor even that it cites all relevant peer-reviewed literature. Rather it is an update and widening of an earlier, more restricted, review of themes (Weir and Pittock 2017). Consequently the emphasis here is on papers published and work done since about 2015, supplemented by some background from the author's own experience over several decades of the Pacific Islands and climate issues.

Social Issues

The IPCC review of the vulnerability of SIDS to climate change highlights that small islands will face as soon as 2030 a high risk of loss of much of their biodiversity, fish abundance and of the coastal protection afforded by coral reefs (Nurse et al. 2014, Table 29.4). That review further projects that these risks will become very high later this century even if the rise in global mean surface temperature (GMST) is held below 2 °C. Even more widely publicised in popular media is the similarly very high risk that will occur to low-lying coastal areas from the interaction of high-water events, such as storms and king-tides, with rising global mean sea level (which is a direct physical consequence of the rise in GMST). Indeed all these risks are already occurring in many SIDS at levels perceptible to the islanders (IPCC 2019).

These and other associated risks constitute grave threats to the social fabric and, in some cases, the continuing existence of small island states, as they threaten the sustainability of food security, livelihoods from tourism and fisheries, and the continuing habitability of the coastal villages and towns in which most of the population lives.

The article by Weir, Dovey and Orcherton (2017) presents an overview of such socio-economic aspects, drawing on case studies from a wide range of Pacific Islands. They point out that, although most of these issues occur in other SIDS in some form or other, one distinctive aspect of the Pacific Islands is the continuing importance there of traditional environmental knowledge and of traditional social networks, both of which contribute to the continuing resilience of Pacific Island communities (villages) in the face of current climatic hazards such as tropical cyclones, but may be less effective against the unprecedented steady rise in sea levels.

Notwithstanding popular perceptions and historically-oriented policies that focus on rural populations, in many SIDS around half of the population now lives in urban and peri-urban areas mostly in and around the national capital (Connell 2013). Many national governments are having difficulty in grappling with this new reality, though they can hardly be unaware of this rapid urbanisation which includes large informal settlements of poorer people, often in areas vulnerable to climatic hazards, especially flooding (Orcherton et al. 2017). But in terms of adaptation and disaster risk management, these looser 'communities' may require different policies and programs from those in rural areas, with more direct involvement by their national governments.

Sectoral Impacts of Climate Change

No human settlement can be sustained without secure access to food and drinkable water. Consequently both of these have been widely recognised as human rights and in the Millennium Development Goals set by the United Nations in 2000 and the Sustainable Development Goals of 2015 that replace them [see http://www.un.org/millenniumgoals/ and http://www.un.org/sustainabledevelopment/]. Both of these securities are affected by global environmental changes. These issues are particularly acute for SIDS, whose resilience in the face of such changes is often less than that of other countries, not least because of socio-economic factors.

The research literature contains numerous articles by specialists in various sectors examining the impact on small islands of current and/or projected climate change on 'their' sector, for example on agriculture (Campbell 2015; Wairiu 2017), water supply (Holding and Allen 2016), biodiversity (Taylor and Kumar 2016), fisheries (Valmonte-Santos et al. 2016), and tourism (Pratt and Harrison 2015).

Coral bleaching due to global warming has become increasingly widespread and frequent world-wide, including in the Pacific (Hughes et al. 2018). As noted by IPCC (2014b: Assessment Box SPM-2) this results in reduced biodiversity, fisheries abundance, and coastal protection, and thus impacts on all of food security, habitability of coastal settlements, and tourism.

Thomas and Benjamin (2018) note that few if any SIDS have mechanisms in place to document the extent to which such present and future impacts constitute 'loss and damage' under the Paris Agreement. Such documentation would ideally include both clear attribution to climate change and economic costs.

Community-Based Adaptation

In most developing countries, including SIDS, a large proportion of the population continues to live in rural communities (villages). In such countries, effective adaptation to climate variability and climate change needs to have a firm base in these communities and their quest for sustainable development on their terms (Barnett and Campbell 2010; Nunn 2009).

In the Pacific, several traditional measures to enhance resilience are still widely practiced, e.g. biodiverse gardens including some long-lasting crops that are held back for emergencies (Fig. 1a). Social capital, in the form of a coherent community willing to all pitch into help those affected by disasters, also remains fortunately widespread.

Nevertheless, community-based adaptation (CBA) has been under-represented in the formal academic literature, as found by IPCC in its successive Assessment Reports (Nurse et al. 2014; Mimura et al. 2007) and by Schipper et al. (2014). The useful systematic review by Klöck and Nunn (2019) of academic (peer-reviewed)

(a) (b)

Fig. 1 Examples of community-based adaptation. **a** A biodiverse garden at Teouma in Vanuatu. **b** Planting long-rooted vetiver grass to lessen river bank erosion at Buretu in Fiji

literature on adaptation in SIDS between 2000 and 2016 found only about 10 such papers, most of which were about projects in the Pacific.

This conference (*Pacific Adapt 2019*) and its predecessor in 2016 (Leal Filho 2017) thus add substantially to the small body of peer-reviewed literature on adaptation in the Pacific.

Nevertheless, much of the relevant practice often has to be learnt afresh, though some is passed on verbally or recorded in informal sites aimed 'community development', such as *Participatory Learning and Action* (the numerous issues of which are freely downloadable from www.planotes.org) and WikiADAPT (online at http://www.weadapt.org/person/) and magazines such as *Tiempo*. Of such grey literature, Piggott-McKellar et al. (2019) give a useful review, though partial by its very nature. Like Klöck and Nunn (2019), they found relatively few reports about the Pacific Islands. Postgraduate teaching at the University of the South Pacific has also made a significant contribution to spreading such knowledge around the region.

Limalevu et al. (2010) list some lessons learnt about CBA from long experience in the Pacific Islands. One key lesson is that 'fly-in, fly-out' projects almost always fail to bring sustainable benefits. In island cultures, patience is rewarded and haste is not. Time spent early in listening to community concerns and helping the whole community to think through feasible adaptation options ensures their engagement in implementation in the medium term, but needs to be coupled with an appropriate (small but non-zero) level of external resources and sustained follow-up 'adaptive monitoring'. A good starting point is the community's perception of the climatic hazards that affect them already and a discussion of how they have acted in the past, and can do now, to lessen the impact of those all-too-familiar hazards, such as tropical cyclones or droughts. It is also important to share information about climate change and adaptation (what worked and what did not) with other communities, who have similar issues to deal with. Success can be gauged—and indeed achieved—only with long-term 'adaptive monitoring' and evaluation, beginning as early as possible and not left to the 'end' of the project. 'Adaptive monitoring' means if something's visibly not working, doing something to fix it; this implies a need to allocate some resources, including technical assistance, beyond the nominal 'end' of project.

These lessons are reinforced by the experience of Cambers et al. (2017) with the EU-funded GCCA—PSIS program of 2011–2015, which ran 'demonstration' CBA projects in nine Pacific Island countries, with the majority of these projects being in remote outer islands. The Pacific Adaptive Capacity Analysis Framework presented by Warrick et al. (2017) likewise draws on the experience of many of the agencies working in CBA in the Pacific, and has been extensively field-tested, and can therefore be expected to be the basis of many future projects. Among those agencies is the University of the South Pacific (USP), which ran a regional adaptation program, complementary to the PSIS but also funded by the EU-GCCA. The USP program included (as one would expect) a strong emphasis on education and training as well as pilot projects and has been succeeded by several other similar programs funded by a range of donors (see https://pace.usp.ac.fj/projects/). All these projects find that social capital, which remains strong in most parts of the Pacific, is a significant factor in adaptive capacity, as confirmed by the work of Hagedoorn et al. (2019) in Micronesia. Note how the villagers are working together to protect their riverbank from erosion in Fig. 1b.

Although building a sea wall is often one of the first adaptation measures to be thought of by a vulnerable coastal village, it is well known to coastal engineers that such sea walls are usually ineffective for more than a year or two, due to poor design and construction (Nunn 2009; Betzold and Mohamed 2017; Donner and Webber 2014).

Nunn et al. (2017) present case studies from coastal communities in the Federated States of Micronesia, which illustrate the range of contemporary responses and the tensions between culturally-grounded responses and those imposed from outside. Gau island in Fiji was overwhelmed by super-strong Cyclone Winston in February 2016, but has since made a considerable recovery, though this recovery required its social capital reserves and resilience (Veitayaki 2006; Remling and Veitayaki 2016), to be complemented by significant technical and financial assistance for immediate disaster recovery.

NGOs, notably the national Red Cross Societies through their networks of trained in-village volunteers and distributed emergency stores, have a significant role in CBA, at least so far as it overlaps with preparation for sudden-onset climate hazards such as cyclones. Drawing on their cross-border links they are also important players in relocations, both planned and sudden (IFRC 2018). So too are churches for similar reasons (Weir and Virani 2011; Edwards 2013), and also because they are particularly trusted sources of information (see Section "Islander Perceptions and Information Sources" below).

Relocation of Communities[1]

Three Levels of Relocation

Campbell and Warrick (2014) in their review for the 2014 UN conference on SIDS define community relocation in general terms:

> *Community relocation* means the forced permanent or long-term movement of an entire community (or a significant part of it) from one location to another.

They note that this term is generally used in the literature in two ways: (a) in a positive sense, to describe the planned relocation of unified communities that maintain structure and function; (b) sudden or forced displacement, which often indicates the fragmentation of communities.[2]

It is useful to distinguish between three levels of community relocation (Gharbaoui and Blocher 2017; Weir et al. 2018). In order of cultural, social, and political difficulty, they are (a) relocation within the community's own (village) land; (b) relocation elsewhere within the same country, and (c) relocation (emigration) to another country.

The cultural difficulties from the last two of these arise principally because of islanders' cultural and economic link to their traditional land (Curry et al. 2015). Most land in the independent Pacific Islands, especially that outside the capital, is under indigenous customary ownership; in effect the land belongs to a collective of owners in a local village (Crocombe 2008, Chap. 11). The degree to which ownership this is formalised in the national legal system varies; it is greatest in Fiji, where the individuals in the landowning unit (*mataqali*) are named.

The bond between indigenous people and their land, expressed in the concept called *vanua* in Fijian, is important throughout the Pacific. One Fijian anthropologist (Ravuvu 1983, p. 70) summarises the concept:

> [*Vanua* means not only] the land area one is identified with, and the vegetation, animal life and other objects on it, but it also includes the social and cultural system - the people, their tradition and customs, beliefs and values, and the various other institutions established for the sake of achieving harmony, solidarity and prosperity within a particular social context.

See also Batibasaqa et al. (1999), Nabobo-Baba (2010, p. 10), Long (2018).

(a) *Relocation within the community's own land*

Local relocation (sometimes referred to as 'managed retreat', especially in the disaster risk literature) is culturally relatively easy compared to relocations further afield, as the land required for relocation is already held within the community concerned.

---

[1] *Communities* is the generic term used in the Fiji draft guidelines on relocation to describe: villages, formal settlements, informal (squatter) settlements, and sub-communities within larger urban areas.

[2] Note that there is a variety of other definitions in the literature. In particular the Fiji *Planned relocation guidelines* define community relocation as "The voluntary, planned and coordinated movement of climate-displaced persons within [a State] to suitable locations, away from risk-prone areas, where they can enjoy the full spectrum of rights including housing, land and property rights and all other livelihood and related rights."

Such relocations have been one of the traditional responses to climate extremes. For example, the community of Biausevu in Fiji relocated four times between 1880 and 1983 (Campbell et al. 2005). Gharbaoui and Blocher (2017) discuss this and two other such relocations in Fiji, all of which occurred more than 50 years ago. Although it was more immediate environmental factors (e.g. flooding or cyclone damage) that drove these relocations, these relocations have all greatly reduced community vulnerability to sea level rise and climate change.

One of the most documented recent cases is the village of Vunidogoloa, which in 2014 was one of the first villages in Fiji to receive substantial government assistance to relocate about 2 km inland away from its coastal site on Vanua Levu, which had long been troubled by inundation, salt intrusion and coastal erosion despite failed seawalls (McNamara and Jacot des Combes 2015; Charan et al. 2017). This assistance took the form of help with planning, levelling of land on the new site, and finance. But even though the new and old sites are both owned by the same *mataqali*, some villagers (especially older ones) were reluctant to leave the very land of their birth that they perceive as defining them and solidifies their identity; indeed such objections had inhibited internal discussions of relocation for over 40 years! (Charan et al. 2017).

A more critical study of the relocation of Vunidogoloa by Rika (2017) points out that that although the Fiji government spent about US$250,000 on the relocation about half of this was wasted on premature earthworks, on top of which the village contributed some US$130,000 from the sale of much of their native forest to local loggers in return for material with which the new houses were built to state-approved plans. Rika also points to the important facilitating role played by the Methodist church, especially in overcoming the villagers' spiritual objections to the move.

Following consultations beginning in 2012 and with the then emerging Vunidogoloa experience in mind, the Fiji government issued draft guidelines, which outlined a consistent procedure for communities to follow in approaching government for financial or technical assistance for relocation. The extensive damage caused by Tropical Cyclone Winston in February 2016 injected a sense of urgency into these considerations, with more definitive working guidelines released in 2018 (GOF 2018; see Fig. 2). Also a series of vulnerability assessments led to the government identifying 42 villages as deserving of assistance for relocation. The Fiji government estimated that this work (and other infrastructural adaptations) would cost more than F$100m (GOF 2017). Fiji's own financial and technical resources available for this work remain well short of this amount, although the Fiji Prime Minister tabled the consolidated vulnerability assessment at COP-23 (of which he was chair) in December 2017 in an effort to seek such funding. Consequently not much more than the most urgent rehabilitation (e.g. of villages on Koro island which were almost totally destroyed) has yet taken place (Gard and Veitayaki 2017).

Many other vulnerable villages, such as Nabukadra in Ra province studied by Nichols (2019), have had to fend for themselves after Cyclone Winston, which has usually precluded relocation and resulted in houses being rebuilt in the vulnerable

Fig. 2 Front cover of the Fiji guidelines on planned relocation, showing part of one of the many coastal villages vulnerable to sea level rise and storm surges (GOF 2018)

location. That case study also illustrates the complications that can arise if the householders suffering most from coastal erosion want to move to land held by a different sub-community (*mataqali*) within the village.

In the Solomon Islands, Albert et al. (2018) report on two villages where the inhabitants were forced to relocate by coastal erosion and flooding, in the absence of any government assistance. In both cases, relocation was essentially piecemeal, with each household relocating to another site where that household had rights to land. Because of the fragmented nature of these land parcels, the result was that some households moved several km, and community solidarity declined markedly.

(b) *Relocation elsewhere in the same country*

This is culturally harder than (a) because of traditional attachment to land, held in customary tenure. In most (all?) PICs, this 'attachment' traditionally includes a person's sense of being an inherent part of the land and its environment in their place of origin (the *vanua*), along with its traditional social structures.

Since most Pacific Islanders (except in Papua New Guinea) live on the coast, many of those in non-atoll countries, particularly those now living right on the coast or in low-lying river deltas, may also be forced to move, though most likely to other parts of the same country.

Campbell (2014) rightly points to the 'essential link' between Pacific Islands people and their land as a key issue that poses major problems not only for those forced to leave but also for communities within the region that may be required to give up land for relocatees.

Nevertheless, the past few decades has seen much 'unforced' internal migration within each island country, most of it towards the urban centre(s) driven by young people seeking better education and income-earning opportunities (Connell 2013). [For example, according to the 2007 census, about one third of the population of Fiji now lives in the greater Suva area.] This movement is referred to as 'migration', as distinct from relocation, as it is individuals or families who move rather than a whole community at once. However, there are cases where over time one family from a village follows another from the same village in a 'chain migration', with the effect that a particular peri-urban informal settlement becomes a relocated community.

An example is 'Lord Howe' settlement near Honiara (Solomon Islands), in which almost everyone can be traced back to Lord Howe Island (now called Ontong Java), which is one of the largest atolls in the Pacific with only limited agricultural land and vulnerable to SLR.[3] Although migration from Ontong Java goes back at least 50 years, more recent recognition of climate change and its impacts has forced consideration of planned relocation of all or most of the population of that island to elsewhere in Solomon Islands (Bonie 2010; Rasmussen et al. 2011). This and other cases have driven the Solomon Islands government, like Fiji, to attempt to formulate guidelines for planned community relocations (Filia 2018; Wilson 2013).

[3] 'Traced back' allows for people born near Honiara, but whose parents or [great-]grandparents moved from Ontong Java.

As usual in the Pacific islands, finding suitable land onto which to move is a major issue (Monson and Fitzpatrick 2016). This has certainly held back another proposed large-scale relocation in Solomon Islands, that of Taro, the government capital of Choiseul province, for the past 20 years (Albert et al. 2018).

Note that a literature survey found 86 reported cases of forced relocation between 1920 and 2005, of which only about half were driven by natural hazards such as coastal erosion or disasters such as cyclones or volcanos (Campbell et al. 2005). Most of the other relocations were driven by armed conflict or major projects such as mines, new airports, or nuclear testing—the most famous example being the relocation of the Bikini islanders to a particular *motu* (islet) on Majuro atoll, the government centre for the Marshall islands. Many of these historic relocations were easier than they would be today because of the rapid increase of population in some PICs, which made it easier in the past to get landholder agreement to release a site (Weir and Virani 2011).

(c) *International migration*

The sea level rises projected by IPCC (2013) would swamp substantial parts of many coral atolls, in which most land is less than two metres above mean sea level. In fact an atoll could become uninhabitable as soon as 2030, because once sea level rises to the stage that saltwater inundation at higher tides occurs every few months instead of every few years, then the underground fresh water lens, on which crops and most drinking water supplies depend, will remain salty (Weir and Virani 2011; Storlazzi et al. 2015).

Consequently policy makers in atoll countries, such as Kiribati, Tuvalu and the Maldives are reluctantly having to consider the possibility of substantial portions of the population of atoll countries being forced to migrate (Smith and McNamara 2015; Nurse et al. 2014, Sect. 29.6.2.4). In turn, some island leaders (e.g. Tong 2015) are forcing these issues to the attention of the developed countries, who they see as responsible for this threat to their continued sustainability and who would be the receiving countries for such migrants.[4]

A flood of scholarly articles continues to examine the probability and implications of such migration, including convenient collections of such articles (Burson 2010; McAdam 2010; Leckie et al. 2012; McLeman and Gemenne 2018) and special issues of such journals as *Forced Migration Review, Population and Environment,* and *Climatic Change*.

Among recent articles, some are more optimistic than others. Farbotko et al. (2018) are among those who argue that such migration is a form of adaptation, and that mobility can contribute to both climate change adaptation and human development, particularly in vulnerable areas where in situ adaptation options have been exhausted. Barnett (2017) is exceptionally optimistic in arguing that "for as long as the science of loss remains uncertain, and the limits to adaptation are unknown,

[4]The current President of Kiribati, Mr Taneti Maamau, prefers not to consider mass migration, and looks instead to ambitious engineering adaptations (for which funding is as yet uncertain) to protect the people and culture of Kiribati (Tong and Rytz 2018).

forced migration cannot be taken as a matter of fact and could possibly be averted through emission reductions and a vastly improved and significantly more creative approach to adaptation".

Encouragingly, at the Second Pacific Climate Change conference in Wellington in February 2018, New Zealand's Minister for Pacific Peoples, 'Aupito Tofae Su'a William Sio, said New Zealand must have policies in place to deal with the possibility of climate induced migration from the Pacific Islands. However, such policies are not yet in place. Cass (2018) identifies some of the issues that have to be addressed before public acceptance is likely: public perceptions, the law, maintaining Island culture and identity and changing the existing media narrative on social change.

Less optimistic are Constable (2017) and Curtain and Dornan (2019), who examine both options for migration to neighbouring Pacific Island countries and for migration to developed Pacific Rim countries, particularly Australia, New Zealand and the United States. Although citizens of the Marshall Islands legally have free access to the United States under the Compact of Free Association (at least until 2023 when the current Compact expires), social, cultural and economic factors currently limit such migration. Weber (2017) notes that both multilateral trade agreements and regional trade and seasonal labour agreements between Pacific Island countries and Australia and New Zealand have potential to include labour mobility, but this aspect has proven to be politically very sensitive, especially in the receiving countries, and therefore such agreements do not currently offer much, if any, assistance to those who may be forced to migrate. The 2018 UN Global Compact on Migration is unlikely to improve access, as it is (i) non-binding, and (ii) opposed by many of the possible receiving countries (Espinoza et al. 2018). Nor has much agreed action flowed from the recommendations of the UNFCCC task force on displacement. A review by Obokata et al. (2014) points out that although there are "significant levels of international migration from highly vulnerable Pacific states like Tuvalu, at present, little of it can be directly attributed to environmental factors as compared with economic factors or traditions of mobility". It remains to be seen how long this will remain the case.

For migration from SIDS more generally, I agree with Weber and with former President Tong of Kiribati that now is the time to equip people with skills that enable them to positively contribute to the societies that are supposed to receive them, and that it is equally important that there is an islander community in the receiving community that sees it as natural to help out and welcome those who need support in finding and settling into new homes. There are certainly substantial and long-established populations of people of island origin in the former colonial powers, especially Britain and France, and also in New Zealand and (to a lesser extent) Australia. However the migration laws of all those countries have tightened since the main wave of islander migration (Connell 2013), so some of those of island ancestry are now in the third generation and may not consider themselves as 'islanders'.

Financing Adaptation

National Governments

In practice, almost all funding for adaptation measures in SIDS comes from developed countries, with relatively little coming from their own sparse national budgets (Nunn 2013). Through to 2007 at least, my observation suggests that such of this funding that reaches local communities outside the capital cities comes mainly through NGOs, although usually ultimately from metropolitan governments.

Nevertheless, Pacific national governments can and do make policies that set a framework for action in their countries, including priorities for funding, e.g. through their National Adaptation Plans.

International Donors

As member states of the United Nations, national governments are the main players for SIDS in the international negotiations aimed at getting action from the large industrialised countries to reduce their greenhouse gas emissions to a levels that would ‘prevent dangerous anthropogenic interference with the climate system’, as the objective of the UNFCCC (1992, Article 2) puts it. A major outcome of these negotiations is the Paris Agreement (2015).

An important issue in the negotiations is funding for adaptation to climate change, since the UNFCCC obliges ‘developed’ countries to provide assistance (including financial assistance) to ‘developing’ countries to help them adapt to climate change [articles 4.4 and 4.8]. Such assistance can be provided either bilaterally or multilaterally.

Several multilateral funds have been set up, either directly by decisions of the UNFCCC Conference of Parties or by groups of concerned countries. These include the Green Climate Fund, the Special Climate Change Fund, the Least Developed Countries Fund and the Adaptation Fund. All of these have project pipelines waiting for money but are critically under-resourced and/or slow to disburse the funds they do have.

In the follow-up negotiations to the Paris Agreement, discussed in a later section, SIDS and other least developed countries continue to push for firmer action (commitments) on finance, and on the related issue of compensation for ‘loss and damage’ these countries have incurred due to climate change caused by the emissions from richer countries.

Regional Organisations

In both the Caribbean and the Pacific, regional (intergovernmental) organisations play a major role as channels for technical assistance (and often also for associated funding programs) in many common areas of need which the small individual governments could not cover all of, as each has only a limited pool of skilled people. Robinson and Gilfillan (2017) explore the extent to which regional organisations such as the South Pacific Regional Environment Program (SPREP) and the Pacific Community (SPC) in the Pacific have been effectively executing their mandates in one such area, namely climate change adaptation, and their future potential to support the design and implementation of national adaptation policies. The authors recommend that, in addition to differentiating organisational mandates, regional organisations should focus on resolving the major climate-related information deficit issues, helping countries to develop ready-to-finance investment projects, building national level capacities to adapt, and supporting the creation of an enabling environment for climate change adaptation.

Community's Own Contributions

In all community-based adaptation projects, including relocations, the expectation of donors, island governments, and fieldworkers is that the community itself will contribute their labour to the project, e.g. (under the direction of an engineer) to clear land, lay water pipes or construct stone walls. Only a few communities have enough cash on hand to make a significant cash contribution towards building materials etc. This might come from income from leasing land to a resort, or (as at Vunidogoloa) the sale of some of their natural resources (Rika 2017), or (particularly in Tonga, Samoa and Fiji) from remittances from former community members now overseas.

Islander Perceptions and Information Sources

Betzold (2015), on the basis of several case studies in SIDS by other researchers, argues that islander lack of awareness of climate change (especially as distinct from climate variability, with which they are all too familiar) is a significant barrier to their implementing adaptation measures (McNaught et al. 2014). Why bother, when more short-term problems like poverty and food security are seemingly more pressing, which is understandable in a context of subsistence-based cultures?

That is why many earlier donor-led programs in the Pacific focussed on spreading information about climate change. This sometimes had the unfortunate result that the project funding ran out with the communities concerned having received lots of background information and a vulnerability assessment, but without much consideration of actions needed or the resources to implement them, as Limalevu et al. (2010) found in Fiji.

Academics generally expect and hope that university students will be better informed than most of the general population, and (as every psychology researcher knows) students are a convenient population for exploring perceptions of any issue of interest. The University of the South Pacific (USP) draws students from 12 Pacific SIDS. Scott-Parker et al. (2017) report that that most of the 1200 USP students they surveyed rank global environmental change as the highest future risk to their countries. The level of awareness of these students stands in remarkable contrast to that found by Lata and Nunn (2011) in some rural villages of Fiji less than 100 km from Suva, the capital of Fiji, which is the location of the main USP campus.

Perceptions shape actions (Adger et al. 2013; see Fig. 3). So the attitudes of these students—who represent the future elite of their region—present unique opportunities for long-range planning of intervention and support strategies to address global environmental change. Of particular note for effective intervention and support is the breadth and trustworthiness of various information sources including scientists and church leaders. In particular, 98% of their sample are certain that climate change is happening, and 87% feel not only a responsibility to do something about it but also that they can personally make a difference. These proportions are strikingly different from those reported by polls in North America (Funk and Rainie 2015) and Australia (CSIRO 2014).

Smaller surveys reported by Beyerl et al. (2018) find that residents of Tuvalu, Tonga and Samoa perceive drought, cyclones and other flood-related problems as a far more imminent danger than rising sea level. But consistent with the survey of USP students, the sample in Samoa which was much younger on average, was much more concerned than that in the other islands about future climate change.

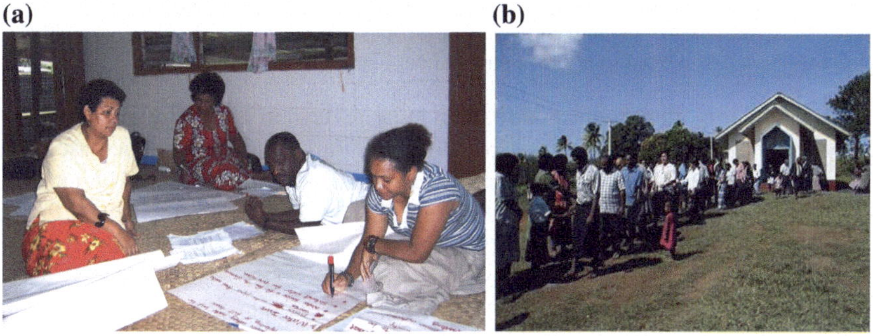

Fig. 3 Perceptions shape actions. **a** Community members in a Fijian village discuss their perception of past and future climate-related hazards—one of the many community discussions that are required for successful adaptation responses [author photo]. **b** Churches (and other religious centres) are important in Pacific communities, so educating their leaders about climatic hazards and responses to them can greatly enhance community resilience [Photo: P. Nunn]

Churches are ubiquitous and well attended in the Pacific islands (Fig. 3b) and thus important in shaping attitudes, including about climate change and how to respond to it (Nunn et al. 2016). Sometimes, strongly held biblical beliefs have led to maladaptive perceptions, e.g. that 'we will be saved from flooding by sea level rise because we are godly people, unlike the sinners that were destroyed when Noah and his family escaped, as told in the book of Genesis.' (Mortreux and Barnett 2009), Related to this is a view that disasters such as TC Winston are a punishment for 'original sin' (Cox et al. 2018). Pacific theologians have attempted to counter such interpretations by reminding their congregations that although God promised Noah that He would not cause such a flood again, this did not rule out such a flood being caused by human action (WCC 2004; Rubow and Bird 2016) and thus that community relocation is not inherently sacrilegious (Rika 2017).

The Paris Agreement and Its Follow-up

While the Paris Agreement (2015) is viewed by Pacific leaders as a positive step forward it will not be enough to limit damage from climate change to acceptable levels, especially for those countries whose very survival is vulnerable to sea level rise. This is because the Agreement leaves each country to set its own emission reduction target and these targets are not legally binding, but only subject to review by the Conference of Parties. Comparison with the latest IPCC Synthesis Report (2014a) shows that the individual commitments that have been tabled to date have no chance of limiting the temperature increase to 2 °C, let alone to the 1.5 °C goal championed by SIDS (UNEP 2018).

As the Director of Oxfam (as reported by Deen 2015) put it:

> This deal offers a frayed life-line to the world's poorest and most vulnerable people. Only the vague promise of a new future climate funding target has been made, while the deal does not force countries to cut emissions fast enough to forestall a climate change catastrophe, [which] will only ramp up adaptation costs further in the future.

Hence the importance attached to the regular review process, where pressure by SIDS and their peers can be put on countries with large emissions to increase their mitigation efforts, and on richer countries for funding for adaptation. Although all these matters remain active in the rules agreed at the Conference of Parties (COP-24) held at Katowice in December 2018, very few industrialised countries have yet increased their commitments on either emissions reduction or on finance.

In an encouraging sign for SIDS, there are strong indications from the Paris conference that many big businesses in the industrialised countries see the needed shift from fossil fuels as both inevitable and a business opportunity which they neglect at their peril (Hamilton 2015). Thus, significant pressure on the governments of large emitting countries to commit to further reduce emissions is likely to come from their own business and financial communities. Some encouragement can also be drawn from the number of subnational governments (cities and States) which have made their own pledges to reduce emissions—often making commitments beyond those made their national governments, especially in the USA where President Trump has announced his intention to exit the Paris Agreement (Mathiesen 2017; Darby 2016).

SIDS negotiated hard in Paris to ensure that the Paris Agreement includes a clause about *loss and damage*, by which the SIDS mean compensation for loss and damage beyond what can readily be 'fixed' by adaptation measures. However this 'win' was undermined, at least in the short term, by an associated COP decision (drafted by the USA) which denies that the rich countries accept any such obligation. This reflects not only a reluctance to make unlimited financial commitments but also very different views about the meaning of the term 'loss and damage' (Boyd et al. 2017). Although there have been many academic papers about loss and damage (McNamara and Jackson 2019; Pill 2020) yielding plenty of ideas, what counts now for the PICs is what commitments can be negotiated.

Fiji was the chair of COP-23, the first island nation to hold that position,[5] and hopes were high that the agenda there would highlight island concerns, in particular loss and damage. But unfortunately the Fiji delegation took too much direction from a bunch of Australian consultants, to the extent that the nominated Fijian chairperson was dismissed by her own government (Darby 2018) and loss and damage barely made it onto the agenda and even then without linking it to funding (Benjamin et al. 2018). Consequently, decisions about loss and damage at COP-24 in December 2018 were largely confined to process for further discussions.

Conclusions: Implications for Research and Policy

The articles referred to in this paper and the negotiations around the Paris Agreement both show that all Small Island Developing States (SIDS) face similar physical and socio-economic threats from climate change and that consequently their responses are also very similar. In this context, the differences between the various regions of SIDS are generally less significant than the differences between SIDS and other developing countries.

One common need is to increase the number of people in SIDS with skills and knowledge needed to adapt to climate change. Given the small population base that SIDS have (by definition) it is fortunate this does not have to imply a large cadre of specialists in climate change, but rather to foster an appreciation of the implications of climate change in the much wider range of professionals and ordinary citizens who are working towards sustainable development in their countries and in their lives. Sharing scarce but often needed specialist skills through regional organisations can be effective in offsetting the pressure on numerically small national agencies. Although engineering and other technical skills are an essential input to many adaptation measures, there has been a tendency to wrongly assume that 'hard' adaptation measures are all that is required. On the contrary, the perceptions and

[5] Although the meeting was actually held in Bonn (Germany) for logistical reasons.

solidarity of the community involved can determine the effectiveness or otherwise of adaptation measures. Fortunately there is a long tradition in the Pacific Islands of resilience in the face of environmental change, drawing on traditional knowledge and social capital; a challenge for today is to maintain and build on those strengths.

Universities and academics and other researchers *in the region* have an important role in recognising the issues and in developing the skills base needed to deal with them. Without indigenous capacity, it is very easy to lose institutional memory of what works and what does not.

At present, too few case studies of adaptation have been documented to assist this institutional memory and to guide future projects away from known causes of failure. It would be particularly useful for researchers to document more cases of successes and failures in community-based adaptation, community relocation (especially in-country relocation),[6] and adaptation (or lack of it) in looser peri-urban communities.

International negotiations and the reduction in greenhouse gas emissions that they can (in principle) produce are key to reducing the threat of climate change to SIDS, and indeed to the rest of the world. The Paris Agreement of 2015 is a positive step in this direction, but SIDS will need to maintain the international pressure for action and not just words. The follow-up mechanisms and further negotiations mandated by the Paris Agreement and the UNFCCC, which underlies it, are also key to securing the international (donor) finance that will be essential for substantial adaptation in the Pacific Islands. Researchers can greatly assist the positions of the islands' negotiators by documenting the economic and social costs incurred by current and projected climate change, the effectiveness (or otherwise) of the financing mechanisms, and the extent to which 'loss and damage' is being incurred.

Acknowledgements This paper is an updated version of an earlier review by Weir and Pittock (2017). I thank Melanie Pill of ANU for her helpful comments on an earlier draft.

References

Adger W, Barnett J, Brown K, Marshall N, O'Brien K (2013) Cultural dimensions of climate change impacts and adaptation. Nat Clim Change 3: 12–117. https://doi.org/10.1038/nclimate1666

Albert S, Bronen R, Tooler N, Leon J, Yee D, Ash J, Boseto D, Grinham A (2018) Heading for the hills: climate-driven community relocations in the Solomon Islands and Alaska provide insight for a 1.5 °C future. Reg Environ Change 18:2261–2272. https://doi.org/10.1007/s10113-017-1256-8

Barnett J (2017) The dilemmas of normalising losses from climate change: towards hope for Pacific atoll countries. Asia Pac Viewpoint 58:3–13. https://doi.org/10.1111/apv.12153

Barnett J, Campbell J (2010) Climate change and small island states: power, knowledge and the South Pacific. Earthscan, London

[6]Lopez-Carr and Marter-Kenyon (2015) have been prominent in urging planned relocation as a priority both for research and action.

Batibasaqa K, Overton J, Horsley P (1999) *Vanua*: land, people and culture in Fiji. In: Overton J, Scheyvens R (eds) Strategies for sustainable development: experiences from the Pacific. UNSW Press, Sydney

Benjamin L, Thomas A, Haynes R (2018) An 'Islands' COP'? Loss and damage at COP23. Rev Eur Compar Int Law 27:332–340. https://doi.org/10.1111/reel.12255

Betzold C (2015) Adapting to climate change in small island developing states. Climatic Change 133:481–489. https://doi.org/10.1007/s10584-015-1408-0

Betzold C, Mohamed I (2017) Seawalls as a response to coastal erosion and flooding: a case study from Grande Comore, Comoros (West Indian Ocean). Reg Environ Change 17:1077–1087. https://doi.org/10.1007/s10113-016-1044-x

Beyerl K, Mieg H, Weber E (2018) Comparing perceived effects of climate-related environmental change and adaptation strategies for the Pacific small island states of Tuvalu, Samoa and Tonga. Island Stud J 13:25–44. https://doi.org/10.24043/isj.53

Bonie JM (2010) Climate change project—Ontong Java: food security—an assessment and plan of action. Episcopal Relief and Development (Anglican Church of Melanesia), Honiara

Boyd E, James RA, Jones RG, Young HR, Otto FEL (2017) A typology of loss and damage perspectives. Nat Clim Change 7:723–729. https://doi.org/10.1038/nclimate3389

Burson B (ed) (2010) Climate change and migration: South Pacific perspectives. Institute of Policy Studies (Victoria University), Wellington, New Zealand

Cambers G, Carruthers P, Rabuatoka T, Tubuna S, Ungaro J (2017) Implementing climate change adaptation interventions in remote outer islands of the Pacific island region. In: Filho WL (ed) Climate change adaptation in Pacific countries. Springer, Cham, Switzerland

Campbell J (2015) Development, global change and traditional food security in Pacific Island countries. Reg Environ Change 15:1313–1324. https://doi.org/10.1007/s10113-014-0697-6

Campbell J, Goldsmith M, Koshy K (2005) Community relocation as an adaptation to the effects of climate change and climate variability in Pacific Island Countries (PICs) final report for APN project 2005-14-NSY-Campbell. University of the South Pacific, Suva. ISBN 978-982-91410-3-3

Campbell J, Warrick O (2014) Climate change and migration issues in the Pacific. United Nations (ESCAP), Apia. http://www.unescap.org/resources/climate-change-and-migration-issues-pacific. Accessed on 20 Mar 2019

Campbell JR (2014) Climate-change migration in the Pacific. Contemp Pac 26:1–28. https://doi.org/10.1353/cp.2014.0023

Cass P (2018) A plan nobody hopes they will need: New Zealand and climate change migration. Pac J Rev 24:138–154. https://doi.org/10.24135/pjr.v24i1.426

Charan D, Kaur M, Singh P (2017) Customary land and climate change relocation—a case study of Vunidogoloa village, Vanua Levu, Fiji. In: Leal Filho W (ed) Climate change adaptation in pacific countries. Springer, Cham, Switzerland

Connell J (2013) Islands at risk: environments, economies and contemporary change. Edward Elgar, Cheltenham, UK

Constable A (2017) Climate change and migration in the Pacific: options for Tuvalu and the Marshall Islands. Reg Environ Change 17:1029–1038. https://doi.org/10.1007/s10113-016-1004-5

Cox J, Finau G, Kant R, Tarai J, Titifanue J (2018) Disaster, divine judgment, and original sin: christian interpretations of tropical cyclone winston and climate change in Fiji. Contemp Pac 30:380–410. https://doi.org/10.1353/cp.2018.0032

Crocombe R (2008) The South Pacific. University of the South Pacific, Suva, Fiji

CSIRO (2014) Annual surveys of Australian attitudes to climate change. CSIRO, Melbourne. http://www.csiro.au/en/Research/LWF/Areas/Social-economic/Climate-change/Climate-attitudes-survey. Accessed on 20 Mar 2019

Curry GN, Koczberski G, Connell J (eds) (2015) Migration, land and livelihoods: creating alternative modernities in the Pacific. Routledge, Abingdon, UK

Curtain R, Dornan M (2019) A pressure release valve? Migration and climate chagne in Kiribati, Nauru and Tuvalu. Australian National University, Canberra. http://devpolicy.org/publications/reports/Migration-climate%20change-Kiribati-Nauru-Tuvalu.pdf. Accessed on 20 Mar 2019

Darby M (2016) Cities rush to join climate drive after Paris Agreement Climate Change News. https://www.climatechangenews.com/2016/08/04/cities-rush-to-join-climate-drive-after-paris-agreement/ Accessed on 20 Mar 2019
Darby M (2018) Fiji climate lead challenged consultants' influence before losing job. Climate Change News. https://www.climatechangenews.com/2018/03/06/fiji-climate-lead-lost-job-challenging-consultants-influence/ Accessed on 20 Mar 2019
Deen T (2015) Paris agreement leaves climate funding in limbo. Caribbean 360, 15 December 2015. http://www.caribbean360.com/news/1128801#ixzz3w8YKARBn. Accessed 20 on Mar 2019
Donner S, Webber S (2014) Obstacles to climate change adaptation decisions: a case study of sea-level rise and coastal protection measures in Kiribati. Sustain Sci 9: 331–345. https://doi.org/10.1007/s11625-014-0242-z
Edwards JB (2013) Climate justice and the Pacific conference of churches: moving the relocation agenda in the Pacific. In: Sustainable alternatives for poverty reduction and ecological justice conference, 27–30 Sept 2012. Orthodox Academy of Crete, Greece
Espinoza MV, Hadj-Abdou L, Brumatt L (2018) Global compact for migration: what is it and why are countries opposing it? The conversation. https://theconversation.com/global-compact-for-migration-what-is-it-and-why-are-countries-opposing-it-106654. Accessed on 5 Apr 2019
Farbotko C, McMichael C, Dun O, Ransan-Cooper H, McNamara KE, Thornton F (2018) Transformative mobilities in the Pacific: promoting adaptation and development in a changing climate. Asia Pac Policy Stud 5:393–407. https://doi.org/10.1002/app5.254
Filia L (2018) Talks for country's first resettlement policy underway. Island Sun, Honiara, 18 October 2018. http://theislandsun.com.sb/talks-for-countrys-first-resettlement-policy-underway/. Accessed 20 on Mar 2019
Funk C, Rainie L (2015) Americans, politics and science issues. Pew Research Center, Washington DC. http://www.pewinternet.org/2015/07/01/chapter-2-climate-change-and-energy-issues/. Accessed on 20 Mar 2019
Gard A, Veitayaki J (2017) In the wake of Winston: climate change, mobility and resiliency in Fiji. Int J Safety Secur Eng 7:157–168. https://doi.org/10.2495/safe-v7-n2-157-168
Gharbaoui D, Blocher J (2017) The reason land matters: relocation as adaptation to climate change in Fiji Islands. In: Milan A (ed) Migration, risk management and climate change: evidence and policy responses. Springer, Switzerland
GOF (2017) Climate vulnerability assessment: making Fiji climate resilient. Government of Fiji, Suva. http://documents.worldbank.org/curated/en/163081509454340771/Climate-vulnerability-assessment-making-Fiji-climate-resilient. Accessed on 20 Mar 2019
GOF (2018) Planned relocation guidelines: a framework to undertake climate change related relocation. Government of Fiji: Suva, Fiji. https://cop23.com.fj/wp-content/uploads/2018/12/CC-PRG-BOOKLET-22-1.pdf. Accessed on 5 Apr 2019
Hagedoorn LC, Brander LM, van Beukering PJH, Dijkstra HM, Franco C, Hughes L, Gilders I, Segal B (2019) Community-based adaptation to climate change in small island developing states: an analysis of the role of social capital. Clim Dev. https://doi.org/10.1080/17565529.2018.1562869
Hamilton C (2015) The Earth has moved: big business's climate shift is now unstoppable. The Conversation. 10 December 2015. (https://theconversation.com/the-earth-has-moved-big-businesss-radical-climate-shift-is-now-unstoppable-52119). Accessed on 20 Mar 2019
Holding S, Allen DM (2016) Risk to water security for small islands: an assessment framework and application. Reg Environ Change 16:827–839. https://doi.org/10.1007/s10113-015-0794-1
Hughes T et al (2018) Spatial and temporal patterns of mass bleaching of corals in the Anthropocene. Science 359:80–83. https://doi.org/10.1126/science.aan8048
IFRC (2018) Disasters and displacement in a changing climate: the role of Asia Pacific National Societies. International Federation of Red Cross and Red Crescent Societies, Geneva. https://media.ifrc.org/ifrc/document/2018-ifrc-climate-change-disasters-displacement-report/. Accessed on 20 Mar 2019
IPCC (2013) Summary for policymakers. In: Stocker TF, Qin D, Plattner G-K, Tignor M, Allen SK, Boschung J, Nauels A, Xia Y, Bex V, Midgley PM (eds) Climate change 2013: the physical science

basis: contribution of working group I to the fifth assessment report of the Intergovernmental Panel on Climate Change. Cambridge University Press, Cambridge, UK

IPCC (2014a) Climate change 2014: synthesis report. contribution of working groups I, II and III to the fifth assessment report of the Intergovernmental Panel on Climate Change. Geneva. http://ar5-syr.ipcc.ch/. Accessed on 20 Mar 2019

IPCC (2014b) Summary for policymakers. In: Field CB, Barros VR, Dokken DJ, Mach KJ, Mastrandrea MD, Bilir TE, Chatterjee M, Ebi KL, Estrada YO, Genova RC, Girma B, Kissel ES, Levy AN, MacCracken S, Mastrandrea PR, White LL (ed) Climate change 2014: Impacts, adaptation, and vulnerability. contribution of working group II to the fifth assessment report of the Intergovernmental Panel on Climate Change. Cambridge University Press, Cambridge, UK

IPCC (2018) Global warming of 1.5 °C: an IPCC special report on the impacts of global warming of global warming of 1.5 °C above pre-industrial levels and related global greenhouse gas emission pathways… Intergovernmental Panel on Climate Change, Geneva. https://www.ipcc.ch/sr15/. Accessed 20 Mar 2019

IPCC (2019) Special Report on Oceans and Cryosphere: summary for policymakers. Intergovernmental Panel on Climate Change: Geneva. https://www.ipcc.ch/srocc/download-report/

Klöck C, Nunn PD (2019) Adaptation to climate change in small island developing states: a systematic literature review of academic research. J Environ Develop 28(2):196–218

Lata S, Nunn P (2011) Misperceptions of climate-change risk as barriers to climate-change adaptation: a case study from the Rewa Delta, Fiji. Clim Change 110:169–186. https://doi.org/10.1007/s10584-011-0062-4

Leal Filho W (ed) (2017) Climate change adaptation in Pacific countries: fostering resilience and improving the quality of life. Springer, Cham, Switzerland

Leckie S, Simperingham E, Bakker J (eds) (2012) Climate change and displacement reader. Routledge, Oxford, UK

Limalevu L, Aalbersberg W, Dumaru P, Weir T (2010) Community adaptation in a Small Island developing country. Tiempo 77:16–19

Long M (2018) *Vanua* in the anthropocene: relationality and sea level rise in Fiji. Symploke 26:51–70

Lopez-Carr D, Marter-Kenyon J (2015) Human adaptation: manage climate-induced resettlement. Nature 517:265–267. https://doi.org/10.1038/517265a

Mathiesen K (2017) US cities and states back Paris deal but ignore climate finance. Climate Change News. https://www.climatechangenews.com/2017/06/15/us-cities-states-back-paris-deal-ignore-climate-finance/ Accessed on 20 Mar 2019

McAdam J (ed) (2010). Climate change and displacement: multi-disciplinary perspectives. Hart Publishing Oxford

McLeman R, Gemenne F (eds) (2018) Routledge handbook of environmental displacement and migration. Routledge, Abingdon, UK

McNamara K, Jacot des Combes H (2015) Planning for community relocations due to climate change in Fiji. Int J Disaster Risk Sci 6:315–319. https://doi.org/10.1007/s13753-015-0065-2

McNamara KE, Jackson G (2019) Loss and damage: a review of the literature and directions for future research. WIRES-Clim Change 10:e654. https://doi.org/10.1002/wcc.564

McNaught R, Warrick O, Cooper A (2014) Communicating climate change for adaptation in rural communities: a Pacific study. Reg Environ Change 14:1491–1503. https://doi.org/10.1007/s10113-014-0592-1

Mimura N, Nurse L, McLean RF, Agard J, Briguglio L, Lefale P, Payet R, Sem G (2007). Small islands. In: Parry ML et al (eds) Climate change 2007: impacts, adaptation and vulnerability. contribution of working group II to the fourth assessment report of the Intergovernmental Panel on Climate Change. Cambridge University Press, Cambridge, UK, pp 687–716

Monson R, Fitzpatrick D (2016) Negotiating relocation in a weak state: land tenure and adaptation to sea level rise in Solomon Islands. In: Price S, Singer J (eds) Global implications of development, disasters and climate change: responses to displacement from Asia-Pacific. Earthscan, Abingdon, UK, pp 240–255

Mortreux C, Barnett J (2009) Climate change, migration and adaptation in Funafuti, Tuvalu. Global Environ Change 19:105–112. https://doi.org/10.1016/j.gloenvcha.2008.09.006

Nabobo-Baba U (2010). The epistemological and philosophical basis of sustainable development practices among indigenous Pacific peoples: A Fiji case study. In: Nabobo-Baba U, Koya CF, Teaero T (eds) Education for sustainable development: continuity and survival in the Pacific. Institute of Pacific Studies, University of the South Pacific, Suva

Nichols A (2019) Climate change, natural hazards and relocation: insights from Nabukadra and Navuniivi villages in Fiji. Clim Change. https://doi.org/10.1007/s10584-019-02531-5

Nunn P (2013) The end of the Pacific? Effects of sea level rise on Pacific island livelihoods. Singapore J Tropical Geogr 34:143–171. https://doi.org/10.1111/sjtg.12021

Nunn PD, Mulgrew K, Scott-Parker B, Hine DW, Marks ADG, Mahar D, Maebuta J (2016) Spirituality and attitudes towards nature in the Pacific Islands: insights for enabling climate-change adaptation. Clim Change 136:477–493. https://doi.org/10.1007/s10584-016-1646-9

Nunn P, Runman J, Falanruw M, Kumar R (2017) Culturally grounded responses to coastal change on islands in the Federated States of Micronesia, northwest Pacific Ocean. Reg Environ Change 17:959–971. https://doi.org/10.1007/s10113-016-0950-2

Nunn P D (2009) Responding to the challenges of climate change in the Pacific Islands: management and technological imperatives. Clim Res 40:211–231. https://doi.org/https://doi.org/10.3354/cr00806

Nurse LA, McLean RF, Agard J, Briguglio LP, Duvat-Magnan V, Pelesikoti N, Tompkins E, Webb A (2014) Small islands. In: Barros VR, Field CB, Dokken DJ, Mastrandrea MD, Mach KJ, Bilir TE, Chatterjee M, Ebi KL, Estrada YO, Genova RC, Girma B, Kissel ES, Levy AN, MacCracken S, Mastrandrea PR, White LL (eds) Climate change 2014: impacts, adaptation, and vulnerability. Part B: regional aspects. Contribution of working group II to the fifth assessment report of the Intergovernmental Panel on Climate Change. Cambridge University Press, Cambridge, UK, pp 1613–1654

Obokata R, Veronis L, McLeman R (2014) Empirical research on international environmental migration: a systematic review. Population Environ 36:111–135. https://doi.org/10.1007/s11111-014-0210-7

Orcherton D, Mitchell D, McEvoy D (2017) Perceptions of climate vulnerability, tenure security and resettlement priorities: insights from Lami Town, Fiji Islands. Aust Geogr 48:235–254

Paris Agreement (2015) Text. UNFCCC, Bonn. http://unfccc.int/files/essential_background/convention/application/pdf/english_paris_agreement.pdf. Accessed on 20 Mar 2019

Piggott-McKellar AE, McNamara KE, Nunn PD, Watson JEM (2019) What are the barriers to successful community-based climate change adaptation? A review of grey literature. Local Environ 24:374–390. https://doi.org/10.1080/13549839.2019.1580688

Piggott-McKellar AE, Nunn PD, McNamara KE, Sekinini ST (2020 in press) Dam(n) seawalls: a case of maladaptation in Fiji. In: Leal W (ed) Climate change adaptation in the Pacific. Springer, Cham, Switzerland

Pill M (2020 in press) Planned relocation in small island developing states: adaptation, loss and damage or something in-between? In: Leal Filho W (ed) Climate change adaptation in the Pacific. Springer, Cham, Switzerland

Pratt S, Harrison D (eds) (2015) Tourism in the Pacific Islands: current issues and future challenges. Routledge, Abingdon, UK

Rasmussen K, May W, Birk T, Mataki M, Mertz O (2011) Prospects for climate change on three Polynesian outliers in Solomon Islands: exposure, sensitivity and adaptive capacity. Geografisk Tidsskrift 111:43–57

Ravuvu AD (1983) *Vaka i Taukei*: the Fijian way of life. University of the South Pacific, Suva, Institute of Pacific Studies

Remling E, Veitayaki J (2016) Community-based action in Fiji's Gau Island: a model for the Pacific? Int J Clim Change Strateg Manag 8:375–398. https://doi.org/10.1108/IJCCSM-07-2015-0101

Rika N (2017) Who pays for climate change? Islands Business (November 2017 issue), pp 4–5:21–22. Suva, Fiji

Robinson S, Gilfillan D (2017) Regional organisations and climate change adaptation in small island developing states. Reg Environ Change 17:989–1004. https://doi.org/10.1007/s10113-016-0991-6
Rubow C, Bird C (2016) Eco-theological responses to climate change in Oceania. Worldviews: Glob Religions Cult Ecol 20:150–168
Schipper L, Ayers J, Reid H, Huq S, Rahman A (eds) (2014) Community-based adaptation to climate change: scaling it up. Earthscan (Routledge), London and New York
Schleussner C-F et al (2018) 1.5 °C hotspots: climate hazards, vulnerabilities, and impacts. Annu Rev Environ Resour 43:135–163. https://doi.org/10.1146/annurev-environ-102017-025835
Scott-Parker B, Nunn P, Mulgrew K, Hine D, Marks A, Mahar D Tiko L (2017) Pacific Islanders' understanding of climate change: where do they source information, and to what extent do they trust it? Reg Environ Change 17:1005–1015. https://doi.org/10.1007/s10113-016-1001-8
Smith R, McNamara KE (2015) Future migrations from Tuvalu and Kiribati: exploring Government, civil society and donor perceptions. Clim Dev 7: 47–59. https://doi.org/10.1080/17565529.2014.900603
Storlazzi CD, Elias E, Berkowitz P (2015) Many atolls may be uninhabitable within decades due to climate change. Sci Rep 5. https://doi.org/10.1038/srep14546
Taylor S, Kumar L (2016) Global climate change impacts on Pacific Islands terrestrial biodiversity: a review. Trop Conserv Sc 9:203–223
Thomas, Benjamin L (2018) Management of loss and damage in small island developing states: implications for a 1.5 °C or warmer world. Reg Environ Change 18:2369–2378. https://doi.org/10.1007/s10113-017-1184-7
Tong, President Anote (2015) Speech at leaders event of COP-21, Paris 30 November 2015. UNFCCC Secretariat. Online at unfccc.int
Tong A, Rytz M (2018) Let Kiribati be a lesson: its fate will be the world's fate if we continue burying our heads in the sand, Islands Business, (November 2018 issue) p6, Suva, Fiji
UNEP (2018) Emissions gap report 2018. United Nations Environment Program, Nairobi. https://www.unenvironment.org/resources/emissions-gap-report-2018
UNFCCC (1992) United nations framework convention on climate change (text). http://unfccc.int/essential_background/convention/items/2627.php
Valmonte-Santos R, Rosegrant M, Dey MM (2016) Fisheries sector under climate change in the coral triangle countries of Pacific Islands: current status and policy issues. Marine Policy 67:148–155. https://doi.org/10.1016/j.marpol.2015.12.022
Veitayaki J (2006) Caring for the environment and the mitigation of natural extreme events in Gau, Fiji Islands: a self-help community initiative. Island Stud J 1:239–252
Wairiu M (2017) Land degradation and sustainable land management practices in Pacific Island Countries. Reg Environ Change 17:1053–1064. https://doi.org/10.1007/s10113-016-1041-0
Warrick O, Aalbersberg W, Dumaru P, McNaught R, Teperman K (2017) The 'Pacific Adaptive Capacity Analysis Framework': guiding the assessment of adaptive capacity in Pacific island communities. Reg Environ Change 17:1039–1051. https://doi.org/10.1007/s10113-016-1036-x
WCC (2004) Otin Tai declaration: a statement and recommendations from the Pacific Churches' consultation on climate change, Tarawa, Kiribati, Geneva, March 2004. https://www.oikoumene.org/en/resources/documents/wcc-programmes/justice-diakonia-and-responsibility-for-creation/climate-change-water/otin-tai-declaration
Weber E (2017) Trade agreements, labour mobility and climate change in the Pacific Islands. Reg Environ Change 17:1089–1101. https://doi.org/10.1007/s10113-016-1047-7
Weir T, Dovey L, Orcherton D (2017) Social and cultural issues raised by climate change in Pacific Island countries: an overview. Regional Environ Change 17:1017–1028. https://doi.org/10.1007/s10113-016-1012-5
Weir T, Dovey L, Orcherton D (2018) Climate change challenges to society and culture in Pacific Island countries. In: Renwick J, Latu K (eds) Pacific Ocean, Pacific climate. SPREP, Te Papa Tongarewa, Wellington, New Zealand

Weir T, Pittock J (2017) Human dimensions of environmental change in small island developing states: some common themes. Reg Environ Change 17:949–958. https://doi.org/https://doi.org/10.1007/s10113-017-1135

Weir T, Virani Z (2011) Three linked risks for development in the Pacific Islands: climate change, disasters and conflict. Clim Dev 3:193–208. https://doi.org/10.1080/17565529.2011.60319

Wilson C (2013) Climate-threatened Solomon Islanders prepare for evacuation. Reuters. http://news.trust.org//item/20131125112530-1wkla/. Accessed on 20 Mar 2019

Dam(n) Seawalls: A Case of Climate Change Maladaptation in Fiji

Annah E. Piggott-McKellar, Patrick D. Nunn, Karen E. McNamara and Seci T. Sekinini

Abstract Coastal communities in Pacific Island Countries are particularly vulnerable to climate change impacts including sea-level rise, coastal erosion, tidal inundation, and the intensification of storm surge activity. In response, adaptation projects across the region have attempted to reduce exposure and overall vulnerability to these coastal pressures. This paper explores what happens once these projects reach communities: are effective and sustainable outcomes achieved, or can the implementation of adaptation projects lead to unintended negative outcomes and result in maladaptation? This paper investigates this issue in relation to two seawall projects implemented in communities on Vanua Levu Island, Fiji. We found that the seawalls have not been successful in achieving their primary aim of safeguarding communities against coastal pressures and have instead resulted in unanticipated negative outcomes for land and livelihood security. Of primary concern is the way that seawalls trap water along their landward sides—acting more like a dam—because of the ineffective design and construction of the seawalls and associated infrastructure. This paper concludes with a call to think more long-term about site-specific adaptation measures that actively involve and are driven by local perspectives in the planning, implementation, and maintenance process.

Keywords Climate change · Maladaptation · Vulnerability · Adaptation · Pacific Island Countries

A. E. Piggott-McKellar (✉) · P. D. Nunn · K. E. McNamara
School of Earth and Environmental Sciences, The University of Queensland, Brisbane, QLD 4072, Australia
e-mail: piggottmckellar@uq.edu.au

P. D. Nunn
e-mail: pnunn@usc.edu.au

K. E. McNamara
e-mail: karen.mcnamara@uq.edu.au

S. T. Sekinini
School of Social Sciences, University of the Sunshine Coast, Sippy Downs, QLD 4556, Australia
e-mail: sts009@student.usc.edu.au

W. Leal Filho (ed.), *Managing Climate Change Adaptation in the Pacific Region*, Climate Change Management, https://doi.org/10.1007/978-3-030-40552-6_4

Vulnerability to Climate Change in Pacific Island Countries

Pacific Island Countries (PICs) are often depicted as one of the most vulnerable regions to climate change (Beyerl et al. 2018; Warrick et al. 2017). Despite such assertions, there is no universally used and recognised definition for vulnerability to climate change (Füssel 2007). The most recent definition by the IPCC exemplifies this ambiguity, defining vulnerability as "The propensity or predisposition to be adversely affected. Vulnerability encompasses a variety of concepts and elements including sensitivity or susceptibility to harm and lack of capacity to cope and adapt" (IPCC 2014: 1775). Despite such definitional ambiguities, there persists a general informal acceptance that vulnerability refers to the exposure of a system coupled with the potential to be harmed (Jackson et al. 2017; Tu'akoi et al. 2018). As such, vulnerability to climate change can refer to both the severity of climate related impacts (i.e. sea-level rise, increased storm surge activity, tidal inundation), coupled with the propensity of people/places to be harmed by such climatic impacts, which can be broadly understood by susceptibility (i.e. institutional, social, historical, economic) and adaptive capacity (i.e. local knowledge, access to information and assets, social capital).

Climate change impacts are already being experienced and projected to increase into the future, with evidence that state-of-the-art climate models are underestimating the severity of future impacts (Schewe et al. 2019). Climate change impacts experienced most severely in PICs include the increased intensification of storm surge activity; sea-level rise and subsequent increased tidal inundation; shoreline erosion and groundwater salinization; changing fish stocks; and less predictable rainfall patterns (Barnett 2011; Chand et al. 2016; IPCC 2014; Kumar and Taylor 2015). These impacts are likely to increase in severity into the future more than necessary owing to the lack of international action on reducing atmospheric GHG quantities through reducing emissions or increasing carbon sinks. Despite PICs being heterogeneous in culture, society, governance, histories, and economy, they are often depicted as a homogenous entity at the forefront of these impacts owing to the similar characteristics that PICs share which enhance their propensity for harm. Such shared characteristics include generally high coastline to land mass ratios and mainly coastal settlements, direct dependence on climate sensitive sectors (such as fishing and agriculture), comparatively low incomes coupled with limited economic opportunity, distance from resource bases and population centres, and inefficient infrastructure. It is this confluence of factors that make PICs recognised as some of those places in the world that are most vulnerable to a changing climate.

While PICs are depicted as highly vulnerable to climate change, such a depiction presents a simplistic notion of vulnerability. Primarily, it disregards the internal adaptive capacity and resilience of island communities and their traditional knowledge systems and coping strategies that have long guided people through extreme environmental changes such as resource constraints and changes in sea level (Barnett and McMichael 2018; Bridges and McClatchey 2009; Granderson 2017: Lefale 2010).

Further, the notion of Pacific Island vulnerability can serve to sideline more intricate aspects that play into vulnerability. Owing to access to resources and developed infrastructure and economies, people in core centres, as opposed to peripheral communities, have differing levels of vulnerability (Nunn et al. 2014; Nunn and Kumar 2018; McNamara et al. 2017). This can also play out in terms of the class, ethnicity, gender, status, or age of a person which can impact their access to information, availability of resources, employment opportunities, and education, all of which can amplify individual and community vulnerability.

Climate Change (Mal)Adaptation

Adaptation to climate change refers to reducing the vulnerability of a system to experienced or anticipated climate change impacts (Barnett and Campbell 2010). As climate change impacts are progressively felt, and projected to become increasingly visible and impactful, it is clear that adaptation policies and projects have been implemented with haste. The depiction of PICs as vulnerable has resulted in significant funding being directed to the region. For example, between 2008 and 2012 adaptation finance to the Pacific Region was just under USD $80 million per annum (Donner et al. 2016). Adaptation implementation comes largely through bilateral funds, and large multilateral funding schemes funded by developed countries and implemented by local governments and NGOs (Spires et al. 2014). A recent review of adaptation in Small Island Developing States (SIDS), of which all PICs identify, shows that documented planned adaptation fit into three categories being structural or physical (i.e. engineering and infrastructural); social (such as educational, behavioural, or informational); or institutional (governance and policies, and law and regulation), with infrastructural and behavioural dominating (Klöck and Nunn 2019).

While there has been increased activity in implementing planned adaptation, whether adaptation is actually reducing the vulnerability of communities has often been questioned (Adger et al. 2005; Barnett and Campbell 2010; Nunn et al. 2014). As almost all of the funding for adaptation comes from developed countries (Nunn 2013), often project objectives are developed by development and donor organisations and agencies. This has resulted is questions being raised about the relevance of such projects to local social, economic, and cultural contexts (Barnett and Campbell 2010; Kumar 2015). Yet, evaluating the success of adaptation in PICs is challenging owing to differing perceptions of what success means, a lack of longitudinal data, and the oft-limited prioritisation of evaluation (Buggy and McNamara 2015; Dumaru 2010; Klöck and Nunn 2019; Remling and Veitayaki 2016). Evaluations of adaptation projects that have been undertaken across the region have shown at times failed outcomes. For example, an analysis of the World Bank Kiribati Adaptation Program (KAP) showed that it was deeply flawed in its design and relevance to the nation's institutional framework and local capacity (Dean et al. 2016). A study in Vanuatu found that adaptation projects have often been unsuccessful resulting from the community viewed as a homogenous entity and not accounting for the nuances and variation in local hierarchies, decision-making processes, and social dynamics (Buggy and McNamara 2015). Aside from the ambiguity of success of project outcomes, there have been broader questions raised about whether a reliance on external aid for adaptation by PIC governments has reduced capacity to independently adapt and diverted attention away from more pressing development needs (Barnett 2008; Webber 2013).

Such concerns about the potential for perverse outcomes from adaptation policies and projects have resulted in a growing scholarship on maladaptation. As Work et al. (2018: 13) state, "We face today new dangers from climate change projects and policies as much as we do from the effects of climate change itself". Challenges in defining exactly what maladaptation is have been raised (Juhola et al. 2016). First, it is important to make the distinction between 'null adaptation' where adaptation has no direct benefit in reducing the vulnerability of the targeted population, as opposed to maladaptation which inadvertently increases vulnerability. Barnett and O'Neill (2010: 211) define maladaptation as an "action taken ostensibly to avoid or reduce vulnerability to climate change that impacts adversely on, or increases the vulnerability of other systems, sectors or social groups". A further definition proposed by Juhola et al. (2016: 139) building off this previous definition, states that "maladaptation could be defined as a result of an intentional adaptation policy or measure directly increasing vulnerability for the targeted and/or external actor(s), and/or eroding preconditions for sustainable development by indirectly increasing society's vulnerability".

Due to the relative early scholarship on maladaptation there have been some ambiguities as to what maladaptation actually means in practice and thus how to identify and measure it. In a manner to develop understanding, Barnett and O'Neil (2010) identify five pathways through which an adaptation can be defined as maladaptive; if the adaptation action: (1) increases GHG emissions, (2) disproportionately burdens the most vulnerable, (3) has high opportunity cost, (4) reduces the incentive to adapt, and (5) creates path dependency. Juhola et al. (2016) advance the scholarship in maladaptation through providing a typology that allows the assessment of actions as maladaptive as either rebounding vulnerability, shifting vulnerability, or eroding conditions for sustainable development. Taking an ex ante approach, in trying to answer under what conditions maladaptation can occur, has become prevalent in the literature. Identifying the potential risk that maladaptation may occur by decision-makers early in the planning process has been documented as an important step in working to avoid the potential of maladaptation occurring (Magnan et al. 2016). While this is undoubtedly an important step, Work et al. (2018) have shown that development partners will often ignore potential detrimental social and environmental impacts, thus not implementing practices to reduce these from occurring, and leading to maladaptation. Further, Owusu-Daaku (2018) argues that adaptation, can be overtaken and perpetuated by actors with underlying or adverse economic interests, and as such maladaptation is a product of a myopic focus dominated by a few actors, positing the term 'maladaptive opportunism'.

Coastal Protection Measures

To adapt to coastal pressures resultant from sea-level rise and associated impacts there are often three categories of initiatives employed: accommodation measures that allow people to adapt in situ, for example, by raising infrastructure; protection measures that employ some form of coastal defence; and retreat which involves the movement of people away from the shoreline (Williams et al. 2018). Coastal protection measures (which are the focus of this research) can be broadly categorised as soft or hard. Soft measures refer to the use of natural materials such as mangroves or other natural vegetation that act as a natural defence buffer while hard measures refer to built infrastructure (Pilkey and Cooper 2012). Hard measures range from locally built ad hoc defences using available materials such riprap, machinery, or boulders, to the construction of planned seawalls or revetments (Shand et al. 2017), in many instances funded through donor funded projects.

Concerns about the suitability of seawalls and revetments implemented in rural settings in developing countries have been expressed. For example, conventional seawalls implemented in a PIC context have been identified as only 'moderately' resilient to climate change impacts with societal outcomes typically average to poor (Shand et al. 2017). Further, the changing dynamics of small islands through natural geomorphic processes coupled with oft-limited scientific knowledge of these locational process can render seawalls and other hard infrastructure inappropriate (Kench 2012). The longevity of seawalls for coastal protection is another challenge as recipient country governments or beneficiary communities have neither the commitment nor the resources to maintain the infrastructure beyond the period of project funding (Nunn 2013). Concerns with seawalls go deeper than an inability to protect rural coastal areas as they represent a more archaic perspective of adaptation through reducing vulnerability to hazards while not acknowledging underlying socio-economic and other contextual drivers of vulnerability.

Despite the ineffectiveness of seawalls there remains a trend within rural areas in particular of PICs, where seawalls have been uncritically implemented as a long-term coastal protection measure (Nunn 2009, 2013). One reason for this trend is that most coastal protection measures in PICs come from donor countries, causing replication of coastal protection measures from wealthier, often continental countries. As Dean et al. (2016: 87) state "there is a naïve and misguided belief of many donors that hard shoreline structures along Pacific Island coasts are the most effective and enduring long-term solutions to shoreline erosion". There appears an allure for donors in building a seawall as it can on the one hand provide initial (yet short-term) comfort to communities, as well as provide a visible outcome suitable for reporting purposes. Softer adaptation measures, such as long-term skills training or meaningful community education building, while requiring more commitment and less observable reward, have been shown to have an increased positive impact for growing adaptive capacity in such situations (Williams et al. 2018).

Research Aim and Questions

Seawalls and revetments have been, both implicitly and explicitly, noted as ineffective and unsuccessful coastal protection measures in PICs (Karlsson and Hovelsrud 2015; Klöck and Nunn 2019; Nunn 2009). While they are acknowledged as unsuccessful long-term coastal protections measures, further questions about the maladaptive potential of seawalls have been raised (Barnett et al. 2013; Dean et al. 2016), in that not only are they unsuccessful but that they can serve to inadvertently increase vulnerability.

With this context in mind, the aim of this research is to examine, through empirical evidence, the outcomes and implications of seawalls on lives and livelihood security. The experiences from two communities—Karoko and Korotasere—on Vanua Levu Island, Fiji who were recipients of seawalls through a donor funded adaptation project, guide this research. This research aims to answer the following questions:

(1) How, if at all, have seawalls successfully reduced exposure to coastal pressures for communities?
(2) How, if at all, has community land and livelihood security been affected by the construction of seawalls in their communities?
(3) Should seawalls be seen as a case of maladaptation to climate change?

This paper first provides an overview of the two case study sites and project background, followed by the methods undertaken. Next, an exploration of the outcomes and implications of the two seawalls in both reducing exposure to coastal pressures, as well as exploring the implications on community livelihood security is provided, showcasing that both the seawalls have not been successful in providing coastal protection and have inadvertently increased the vulnerability of the community due to poor seawall design and construction. A final discussion surrounding the maladaptive nature of seawalls in rural PIC communities, and discussion of more appropriate options to move toward appropriate and sustainable adaptation is presented.

Case Study Sites and Project Background

Fiji is an archipelago situated in the South West Pacific Ocean and home to 330 islands, with Vanua Levu the second largest (Viti Levu being the largest). Approximately 15% of the total 900 000 population live on Vanua Levu. In total, roughly 45% of inhabitants of Fiji live in rural areas, which has declined from 70% in 1960 (World Bank 2018). The case study sites for this research are two villages, Karoko and Korotasere, located on Vanua Levu Island (see Fig. 1). Like most rural areas in Fiji, both villages are coastal and largely subsistence-based relying heavily on marine and terrestrial resources for their livelihoods. Karoko is a significant distance from the nearest town center, Savusavu while Korotasere is situated much closer to Savusavu. A bus services both villages along the only road serving that area.

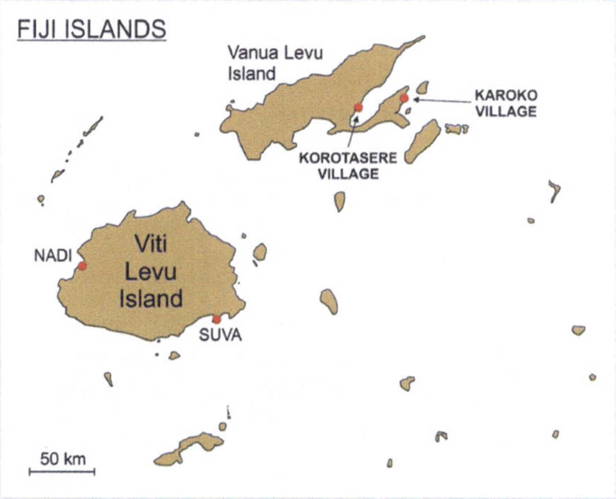

Fig. 1 Map of the two case study sites—Karoko and Korotasere—on Vanua Levu Island, Fiji

Research into climate impacts in the Pacific Region over the past few decades show that increases in sea level have been occurring over the last 150 years and are projected to increase up to 80 cm above the 1985–2005 mean by 2100 (Church et al. 2013). Alongside this, increased surface temperatures and strength of El Niño and La Niña cycles mean that there is likely to be an increase in cyclone intensity creating more impactful storm surges (Walsh et al. 2016). For coastal communities, this means an increased threat of flooding events that can impact livelihoods. Both case studies in the research have a history of flooding in the village. In Karoko this has become exacerbated in recent years as an elderly member of the village explained: "*When I was small we didn't have as much high tide and the village never flooded. But now we see a lot of change*" (Karoko FG).

As both villages have been facing impacts from flooding, there was a desire within the villages for some form of coastal protection measure. In Korotasere a seawall was specifically requested by the village. In Karoko, while there was a desire for some type of coastal protection, a seawall was one option that had been discussed in the village, alongside relocation. Despite this lack of consensus, the implementation of the seawall was welcomed by the village: "*We don't know how to start at a new place so a lot of people lean toward the seawall rather than relocate… we felt good that the seawall was going to be built*" (Karoko FG). This desire of a seawall is representative of the feelings of safety that can be garnered by having a physical structure in place (Lincke and Hinkel 2018). In Karoko the seawall was along the coastal side of the village, while in Korotasere the seawall was constructed along the river.

Both seawalls were funded through The United States Agency for International Development (USAID) under the Coastal Community Adaptation Project (C-CAP) and implemented in both communities at the end of 2015. In its totality, the project was implemented in numerous communities across nine countries throughout the Pacific region, in consultation with respective governments and local partners. The actual construction was undertaken by a local contractor. One of the project's primary objectives is to implement infrastructure adaptations designed to structurally withstand the impacts of climate change and functionally increase community resilience to climate change. In both Karoko and Korotasere, this resulted in the construction of seawalls in the communities.

Methods

The research team consisted of the lead author, who led the discussions with the community members, and the fourth author who helped with translation and independently led discussions, playing a pivotal role in the research process. A local provincial representative acted as gatekeeper for the research team, contacting and arranging site visits with both villages ahead of time. This was done to arrange an appropriate time to visit and ensure that community members were willing, available, and prepared for FG discussions. Ensuring this initial contact is done in an appropriate and respectful manner that is in line with the cultural and societal norms is an essential step in the research procedure when working in PICs. While people who have worked in this region are well aware of the importance of this, the following is presented as a cautionary tale for others. In one of the villages participants noted that another research team had recently arrived unannounced at the village to talk to people without any prior arrangement. They arrived on a Sunday, this being a day of prayer for island communities. As such the community members were not impressed with this visit and stated they refused to talk to them.

Field visits were undertaken during November and December 2017. In Karoko, the village was visited over a two-day period with two formal focus groups (FG) discussions occurring, one women's FG and one men's FG. This was also hoped to occur in Korotasere although on the first day the research team found a funeral was planned for the following day and, as such, only one day was spent in the village and as such one FG was undertaken with a combination of men and women. Site visits to inspect the seawalls as well as walk around the village were also undertaken. Ethical procedures were followed in which participants were given details of the research and asked to sign a consent form. Each FG was recorded with permission from the participants. The qualitative data collected from the field visits was later transcribed and a content analysis undertaken where information was coded according to FG questions to elucidate common responses and experiences.

There were some limitations experienced in this research. Of note was the limited time spent in Korotasere, with only one day spent in the village. Yet, this was unavoidable owing to the unanticipated funeral proceedings planned for the following day, and the time constraints of the researchers. Second, as this research uses qualitative data, there is an assumption that information presented by village participants is true and honest to their individual experiences. Yet, there is potential for positive response bias when implementing such experiential qualitative methods. Finally, there can be challenges associated with the translation of words from the local language into English and vice versa (Rudiak-Gould 2012).

Implications and Outcomes of Seawalls for Communities

With the aim of understanding and evaluating the success of adaptation, this must be done in terms of both reducing exposure from climate related hazards, as well as the outcomes on livelihoods, as this plays an important role in adaptive capacity, resilience, and susceptibility. As such, the implications of the seawalls for affected communities will be explored below in terms of how well the seawalls have reduced exposure to coastal pressures, as well as how the seawalls have impacted community land and livelihoods particularly livelihood security.

Reducing Exposure to Coastal Pressures?

The implementation of the seawalls in both communities was done with the aim of alleviating climate related coastal exposure from increased flooding, tidal inundation, and coastal erosion. Yet, both seawalls were not successful in alleviating these pressures for all members of the community, namely those situated at the front of the village, closest to the seawall. Community members from both Karoko and Korotasere stated that the construction of the seawalls was not adequate as the length did not go far enough and as such water can still enter the village around both ends: "*The end of the seawall, the water comes around it. And it's not done properly. It's not just one end, it's both ends*" (Karoko FG). There were also concerns raised in both villages that the construction of the drainage systems associated with the seawalls was ineffective. These concerns with the infrastructural design are, unfortunately, not unique to these two case studies. Such concerns with incomplete and inadequate design of infrastructure have been noted across the adaptation literature. For example, in a recent review of barriers to effective community level adaptation, challenges associated with infrastructure were noted in 32% of cases examined (Piggott-McKellar et al. 2019).

Infrastructural projects have been noted to be inadequate in many rural areas that lack access to resources to maintain projects, especially when mechanisms to do so are not accounted for and in-built into project design through maintenance training or improving access to resources. In this instance community members have noted that the seawalls have been breaking down in some sections, predominately through rocks falling out of the structure. While these concerns were noted in both villages, such issues have not been rectified. This is owing to communities both having insufficient funds or expertise to solve the problem themselves, but also because they perceive the seawalls as an outside project, the responsibility of the government and other external stakeholders to maintain for the communities benefit.

As a result of the poor design and incomplete infrastructure in both villages, some deleterious outcomes for members of the community have occurred. Water still enters the village as it goes around the seawall, thus not substantially reducing flooding events, one of the primary tasks it was implemented to do. A further associated issue occurred due to the drainage system being built too high, resulting in water having no way to escape once it has entered the village, from both tidal flooding events and heavy rainfall. As such, water is trapped on the landward side of the villages, thus acting like a dam. This issue was brought up in both villages: "*Since they have built the seawall the outcome is that the water is not flowing out, it is getting blocked by the sea wall… the idea of a seawall is for water to stay outside, and that doesn't work here*" (Korotasere FG), and "*So our house on the first row, we will be sitting inside the house on top of the water, flooded, and we will be staring at the pipe nice and dry. We will be watching the water collecting inside the seawall and not going out*" (Karoko FG).

While benefits from the seawalls have been experienced by some community members in that they have reduced the extent of tidal wave surges, owing to poor design and construction of infrastructure, the seawalls have overall been unsuccessful in alleviating coastal pressures in either village. Rather, inadvertent outcomes stemming from poor infrastructural design and implementation have meant water is now trapped inside the village, acting more like a dam.

Implications for Land and Livelihood Security

As described above, the seawalls have not been successful in achieving their goals in reducing exposure to coastal flooding and tidal inundation experienced in both communities. Yet, in addition there have been further inadvertent outcomes on community lives and livelihood security. It was expressed by a participant whose home was close to the front of the village that as the spill over and rain water is retained inside the village, challenges in maintaining their gardens has occurred, resulting in a loss of income: "*the vegetables we grow sometimes we will take to the market and sell, or sell around the village. Now we don't have this source of income and are buying vegetables that is taking a lot of our own money*" (Karoko FG). This is especially pertinent because the limited economic opportunities in both villages due to their distance from core centers, and thus constrained market access.

In a feasibility analysis of coastal protection options across Pacific Islands, Shand et al. (2017) note that one common downfall of seawalls and revetments is that they can reduce beach access for local communities, especially if infrastructure is built without this consideration in mind. This challenge arose in Karoko (where the seawall was built on the coastal side) with beach access reduced for community members: "*Also there was another issue, because they put the steps in the middle of the sea wall and when its flooded sometimes we have to swim to the stairs to get down to the beach instead of just going down like before*" (Karoko FG). Further people expressed that fishing nets are getting caught on the seawall and that it has overall resulted in a negative impact on livelihoods: "*this seawall has become a liability to our children's safety, it damages our nets, our livelihoods are being affected*" (Karoko FG).

Seawalls are often seen as a desirable coastal protection measure for local communities as they present a physical level of comfort, at least initially, that makes people feel safe from increasingly experienced coastal pressures (Morris et al. 2018; Jamero et al. 2018). Korotasere has been asking for a seawall for roughly eight years, prior to this one being implemented, and both communities experienced an initial desire and happiness with the projects in their villages. However, while this is the case, the reality is that while seawalls might present an initial level of comfort, they are invariably an inadequate long-term option for rural communities. This can be shown as over time, due to the negative impacts associated, people have not garnered as sense of safety: "*yes, initially we were happy but it didn't serve its purpose*" (Korotasere FG), and "*So we always have the same feeling like before the seawall was built. Worries and anxiety*" (Karoko FG).

Acknowledging and not undermining the negative outcomes experienced from the seawalls, there have been some more positive outcomes experienced by community members. For example, some people feel the seawall has protected them experiencing the full impacts from large scale tidal surges: "*yes we will be grateful still because at least it stops most of the water and wave surges… but there is still some that gets in around the seawall*" (Karoko FG). Some practical outcomes in terms of the seawall infrastructure allowing easier access to fishing at high tides as people can stand on the seawall, as well as using the sea wall as a place to socialise, as one participant noted: "*And we use it a lot for site seeing because we walk along the sea wall. We take our guests out there and walk around*" (Karoko FG).

A Case of Maladaptation

Across the literature, the implementation of large infrastructural projects to reduce the vulnerability of populations to adverse climate impacts has been questioned (Adger et al. 2005; Dean et al. 2016; Girot et al. 2012; IPCC 2011). After analysing the outcomes from the implementation of seawalls in the two case study sites presented in this research, it is clear that perverse outcomes can arise from infrastructural adaptations. Further, this research showed that the seawall projects had unintended negative impacts on both communities. As such it is argued here that seawalls implemented in PIC communities should be recognised as, in the least, having the potential to be maladaptive.

The maladaptive potential of seawalls will be shown, drawing on empirical data from the two case study sites in this research, as well as other literature. We will use the conceptual underpinnings of maladaptation developed by Juhola et al. (2016), which has previously been used to evaluate the maladaptive potential of other adaptation initiatives (Antwi-Agyei et al. 2018). Juhola et al.'s (2016) framework for assessing maladaptation refers to maladaptation occurring under the conditions of: (1) rebounding vulnerability, whereby the vulnerability is increased for the targeted actor; (2) shifting vulnerability, whereby vulnerability is increased for external actors; and (3) eroding conditions for sustainable development, whereby maladaptation occurs if an adaptation negatively impacts environmental, social, economic, or cultural conditions necessary for sustainable development. The criteria across each of these can be met by increasing vulnerability through either increasing exposure or susceptibility, or decreasing adaptive capacity (Juhola et al. 2016).

Rebounding vulnerability refers to a situation in which a planned adaptation increases the vulnerability of those that the adaptation intervention was implemented to protect. In the case of seawalls, and as this study showcases, perverse outcomes resulted for the communities as a direct result of the seawall projects. In this instance this occurred as the seawalls have not reduced the exposure to coastal pressures, and have inadvertently increased for some village members, specifically those situated closest to the seawall: "*the seawall has flooded the village, not climate change… they said we should relocate from climate change, but we said no, we will relocate because of the sea wall*" (Karoko FG). The implementation of seawalls have also shown to provide a short term, and misleading sense of security. Seawalls are often seen as providing a structurally sound and long-term defence by local communities, when they should be seen as an intermediary activity at most. The expectation of seawalls being a long term solution can divert attention away from adaptation options that can actually reduce the vulnerability of local communities. For example, the time and resources put into building a seawall, which has not had a positive benefit in reducing vulnerability to the community, and has instead added to it, could have been put into a more long-term solution.

The second example of maladaptation is that of shifting vulnerability whereby vulnerability is 'shifted' to another actor. In this instance, drawing on other literature, it can be seen that seawalls can be described as maladaptive under this measure as they can cause downstream unintended negative outcomes through altering natural coastal processes (Shand et al. 2017). Seawalls disrupt natural erosion process which usually provide sediment to adjacent coastal areas through longshore drift, thus resulting in the starvation of sediment for locations at these sites (Linham and Nicholls 2010). Dean et al. (2016) describe the impacts of a seawall constructed alongside an existing seawall that was already causing erosion, thus perpetuating the problem even further. This can lead to impacts on nearby communities who rely extensively on the coast for their livelihoods, and thus shifting the vulnerability onto them.

The third form of maladaptation is that maladaptation occurs through the erosion of the preconditions for sustainable development. These are seen as common pool outcomes that impact society, rather than specific actors themselves. The construction of seawalls as opposed to other more feasible options has been noted to significantly contribute to greenhouse gas (GHG) emissions, through their construction, thus contributing to and exacerbating climate change (Barnett et al. 2013). Further, the construction of seawalls as opposed to natural coastal defences, such as mangrove rehabilitation, have direct negative environmental consequences. As Shand et al. (2017: 972) notes referring to seawalls, "Environmental impacts are likewise average to poor as the natural system is being interrupted by a fixed structure generally with a large occupation area". While these impacts are more localised, they can also contribute to broader environmental concerns as coastal areas are imperative for the health of local marine wildlife, and alternatives such as mangrove restoration can serve to maintain long-term ecosystem health (Morris et al. 2018; Calliari et al. 2019).

Viewing seawalls as maladaptive, or in the least having maladaptive potential is an important realisation going forward. As shown here, seawalls can be maladaptive across all forms through rebounding vulnerability on those the adaptation has been implemented to assist, shifting vulnerability onto other actors, and eroding the conditions for sustainable development.

Conclusions and Future Directions

PICs are described as some of the most vulnerable to climate change which has resulted in adaptation projects implemented across the region with the aim of reducing this vulnerability. Whether such planned adaptations are effective, in that they successfully reduce local vulnerability, has been appropriately and suitably questioned throughout the literature. Additionally, whether such adaptations are instead increasing vulnerability, leading to maladaptation, has been further raised. This research contributes much-needed knowledge about the potential for planned adaptation to be maladaptive using the case of seawalls implemented in two rural communities in Fiji. The construction of the seawalls was part of a wider adaptation project, C-CAP, which was funded by USAID and implemented across the Pacific region in numerous communities across nine countries. These seawall projects were implemented to assist communities in dealing with high exposure to changing coastal conditions as experienced from climate change. This research showed that seawalls have been largely ineffective in protecting recipient communities from associated coastal pressures, especially for people closest to the seawall as water is trapped inside the landward side of the village, acting more like a dam. Poor infrastructural design was identified as the primary cause of this as water can enter around the seawall as its length in insufficient and drainage infrastructure has been ineffectively built. As a result, further inadvertent impacts on land and livelihood security have been experienced by some community members. By exploring this case study and other literature, it is clear how seawalls meet all three categories for maladaptation as proposed by Juhola et al. (2016), as they: (1) can increase vulnerability for those the seawalls were implemented to protect; (2) can increase the vulnerability of down drift communities; and (3) can erode conditions for sustainable development. Due to these deleterious outcomes, the maladaptive potential of seawalls must be recognised going forward, especially in the context of rural communities in small island nations.

Adaptation is an essential component when dealing with climate change, considering climate change impacts are already being experienced and will continue into the future. This research reveals some important insights into how adaptation is currently undertaken and implemented and raises some important considerations and questions related to who is defining 'good' or 'successful' adaptation. In the case studies presented in this research, seawalls could be seen as successful in that they were physically constructed in both communities, yet the outcomes of these have been deleterious to communities whom the projects were purportedly implemented to assist. If adaptation is to be truly meaningful in reducing the vulnerability of communities most affected, a shift is required pertaining to the processes and decision-making that permeate the current adaptation model which prioritises short-term adaptations that are driven by the perspectives, objectives, and criteria of donors and implementing agencies. For example, hard infrastructural adaptations, especially when implemented in island nations, have been documented extensively to be ineffective, while communities at the frontline of climate change impacts are continuously left out of decision-making processes. As such, a move away from the current focus on tangible short-term responses to long-term adaptations that account for the breadth of factors that influence and drive vulnerability and include and account for the diverse perspectives of those affected is needed. If such a paradigm shift is not made, there is a genuine risk that planned adaptations will instead increase people's vulnerability above that already experienced by climate change. As such, the authors reiterate and join the call amongst others to plead "for the anticipation of the risk of maladaptation to become a priority for decision-makers and stakeholders at large, from the international to the local levels" (Magnan et al. 2016: 661).

References

Adger N, Arnell N, Tompkins A (2005) Successful adaptation to climate change across scales. Glob Environ Change 15(2):77–86

Antwi-Agyei P, Dougill AJ, Stringer LC, Codjoe S (2018) Adaptation opportunities and maladaptive outcomes in climate vulnerability hotspots of northern Ghana. Clim. Risk Manag 19(C):83–93

Barnett J (2008) The effect of aid on capacity to adapt to climate change: insights from Niue. Polit Sci 60(1):31–45

Barnett J (2011) Dangerous climate change in the Pacific Islands: food production and food security. Reg Environ Change 11(Supplement 1):229–237

Barnett J, Campbell J (2010) Climate change and small island states: power, knowledge, and the South Pacific. Earthscan, London

Barnett J, McMichael C (2018) The effects of climate change on the geography and timing of human mobility. Popul Environ 39(4):339–356

Barnett J, O'Neill S (2010) Maladaptation. Glob Environ Change 20(2):211–213

Barnett J, Waller S, Rogers S, O'Neill S (2013) Reducing the risk of maladaptation in response to sea-level rise and urban water scarcity. In: Moser SC, Boykoff MT (eds) Successful adaptation to climate change: linking science and policy in a rapidly changing world. Routledge, Abingdon, Oxon, pp 37–49

Beyerl K, Mieg HA, Weber E (2018) Comparing perceived effects of climate-related environmental change and adaptation strategies for the Pacific small island states of Tuvalu, Samoa and Tonga. Island Stud J 13(1):25–44

Bridges KW, McClatchey WC (2009) Living on the margin: ethnoecological insights from Marshall Islanders at Rongelap atoll. Glob Environ Change 19(2):140–146

Buggy L, Mcnamara KE (2015) The need to reinterpret "community" for climate change adaptation: a case study of Pele Island, Vanuatu. Clim Dev 8(3):1–11

Calliari E, Staccione A, Mysiak J (2019) An assessment framework for climate-proof nature-based solutions. Sci Total Environ 656:691–700

Chand SS, Tory KJ, Ye H, Walsh KJE (2016) Projected increase in El Niño-driven tropical cyclone frequency in the Pacific. Nat Clim Change 7:123–127

Church JA, Clark PU, Cazenave A, Gregory JM, Jevrejeva S, Levermann A, Merrifield MA, Milne GA, Nerem RS, Nunn PD, Payne AJ, Pfeffer WT, Stammer D, Unnikrishnan AS (2013) Sea level change. In: Stocker TF, Qin D, Plattner G-K, Tignor M, Allen SK, Boschung J, Nauels A, Xia Y, Bex V, Midgley PM (eds) Climate change 2013: the physical science basis (Working Group I contribution to the fifth assessment report of the intergovernmental panel on climate change. Cambridge University Press, Cambridge, pp 1137–1308

Dean A, Green D, Nunn P (2016) Too much sail for a small craft? Donor requirements, scale, and capacity discourses in Kiribati. In: Stratford E (ed) Island geographies: essays and conversations. Routledge, London, pp 67–88
Donner SD, Kandlikar M, Webber S (2016) Measuring and tracking the flow of climate change adaptation aid to the developing world. Environ Res Lett 11(5):054006/1–054006/9
Dumaru P (2010) Community-based adaptation: enhancing community adaptive capacity in Druadrua Island, Fiji. Wiley Interdisc Rev Clim Change 1(5):751–763
Füssel H-M (2007) Vulnerability: a generally applicable conceptual framework for climate change research. Glob Environ Change 17:155–167
Girot P, Ehrhart C, Oglethorpe J (2012) Integrating community and ecosystem-based approaches in climate change adaptation responses. Viewed 10 February 2019 at, http://careclimatechange.org/files/adaptation/ELAN_IntegratedApproach_150412.pdf
Granderson AA (2017) The role of traditional knowledge in building adaptive capacity for climate change: perspectives from Vanuatu. Weather Clim Soc 9(3):545–561
IPCC (2011) Special report on managing the risks of extreme events and disasters to advance climate change adaptation (SREX), Summary for Policy Makers. IPCC, Kampala
IPCC (2014) Climate change 2014: synthesis report. Contribution of Working Groups I, II and III to the fifth assessment report of the intergovernmental panel on climate change. In: Core Writing Team, Pachauri RK, Meyer LA (eds) IPCC, Geneva, Switzerland
Jackson G, McNamara K, Witt B (2017) A framework for disaster vulnerability in a small island in the Southwest Pacific: a case study of Emae Island, Vanuatu. Int J Disaster Risk Sci 8(4):358–373
Jamero ML et al (2018) Community-based adaptation in low-lying islands in the Philippines: challenges and lessons learned. Reg Environ Change 18(8):2249–2260
Juhola S, Glaas E, Linnér BO, Neset TS (2016) Redefining maladaptation. Environ Sci Policy 55(1):135–140
Karlsson M, Hovelsrud GK (2015) Local collective action: adaptation to coastal erosion in the Monkey River Village, Belize. Glob Environ Change 32:96–107
Kench P (2012) Compromising reef island shoreline dynamics: legacies of the engineering paradigm in the Maldives. In: Cooper JAG, Pilkey OH (eds) Pitfalls of shoreline stabilization: selected case studies. Springer, Dordrecht, pp 165–186
Klöck C, Nunn PD (2019) Adaptation to climate change in small island developing states: a systematic literature review of academic research. J Environ Dev. https://doi.org/10.1177/1070496519835895
Kumar S (2015) Green climate fund faces slew of criticism. Nature 527(7579):419–420
Kumar L, Taylor S (2015) Exposure of coastal built assets in the South Pacific to climate risks. Nat Clim Change 5(11):992–996
Lefale P (2010) Ua 'afa le Aso Stormy weather today: traditional ecological knowledge of weather and climate. The Samoa experience. Clim Change 100(2):317–335
Lincke D, Hinkel J (2018) Economically robust protection against 21st century sea-level rise. Glob Environ Change Hum Policy Dimensions 51:67–73
Linham MM, Nicholls RJ (2010) Technologies for climate change adaptation: coastal erosion and flooding. In: Zhu X (ed) TNA guidebook series. UNEP Riso Centre on Energy, Climate and Sustainable Development, Roskilds, Denmark
Magnan AK et al (2016) Addressing the risk of maladaptation to climate change. Wiley Interdisc Rev Clim Change 7(5):646–665
McNamara KE, Clissold R, Piggott-McKellar AE, Buggy L, Azfa A (2017) What is shaping vulnerability to climate change? The case of Laamu Atoll, Maldives. Island Stud J. https://doi.org/10.24043/isj.67
Morris RL, Konlechner TM, Ghisalberti M, Swearer SE (2018) From grey to green: efficacy of eco-engineering solutions for nature-based coastal defence. Glob Change Biol 24(5):1827–1842
Nunn PD (2009) Responding to the challenges of climate change in the Pacific Islands: management and technological imperatives. Clim Res 40(2/3):211–231

Nunn PD (2013) The end of the Pacific? Effects of sea level rise on Pacific Island livelihoods. Singap J Trop Geogr 34(2):143–171

Nunn PD, Kumar R (2018) Understanding climate-human interactions in Small Island Developing States (SIDS): implications for future livelihood sustainability. Int J Clim Change Strat Manag 10(2):245–271

Nunn PD, Aalbersberg W, Lata S, Gwilliam M (2014) Beyond the core: community governance for climate-change adaptation in peripheral parts of Pacific Island Countries. Reg Environ Change 14(1):221–235

Owusu-Daaku KN (2018) (Mal)Adaptation opportunism: when other interests take over stated or intended climate change adaptation objectives (and their unintended effects. Local Environ 23(9):934–951

Piggott-McKellar AE, McNamara KE, Nunn PD, Watson JEM (2019) What are the barriers to successful community-based climate change adaptation? A review of grey literature. Local Environ 24(4):374–390

Pilkey OH, Cooper JAG (2012) "Alternative" shoreline erosion control devices: a review. In: Cooper JAG, Pilkey OH (eds) Pitfalls of Shoreline stabilization: selected case studies. Springer, Dordrecht, pp 187–214

Remling E, Veitayaki J (2016) Community-based action in Fiji's Gau Island: a model for the Pacific? Int J Clim Change Strat Manag 8(3):375–398

Rudiak-Gould P (2012) Promiscuous corroboration and climate change translation: a case study from the Marshall Islands. Glob Environ Change Hum Policy Dimensions 22:46–54

Schewe J, Gosling SN, Reyer C, Zhao F, Ciais P, Elliott J, Francois L, Huber V, Lotze HK, Seneviratne SI, van Vliet MTH, Vautard R, Wada Y, Breuer L, Büchner M, Carozza DA, Chang J, Coll M, Deryng D, de Wit A, Eddy TD, Folberth C, Frieler K, Friend AD, Gerten D, Gudmundsson L, Hanasaki N, Ito A, Khabarov N, Kim H, Lawrence P, Morfopoulos C, Müller C, Schmied HM, Orth R, Ostberg S, Pokhrel Y, Pugh TAM, Sakurai G, Satoh Y, Schmid E, Stacke T, Steenbeek J, Steinkamp J, Tang Q, Tian H, Tittensor DP, Volkholz J, Wang X, Warszawski L (2019) State-of-the-art global models underestimate impacts from climate extremes. Nat Commun 10(1):1005

Shand T, Carley J, Whalley O, Estigarribia L, Blacka M (2017) Affordable coastal protection in the Pacific: impacts of local resource availability and transport costs [online]. In: Australasian Coasts and Ports 2017: working with nature. Barton, ACT: engineers Australia, PIANC Australia and Institute of Professional Engineers New Zealand, pp 971–977. Viewed 26 November 2018 at https://search.informit.com.au/documentSummary;dn=932578655232845;res=IELENG. ISBN:9781922107916

Spires M, Shackleton S, Cundill G (2014) Barriers to implementing planned community-based adaptation in developing countries: a systematic literature review. Clim Devel 6(3):277–287

Tu'akoi S, Vickers MH, Tairea K, Aung YYM, Tamarua-Herman N, Ofanoa M, Bay JL (2018) The significance of DOHaD for small island developing states. J Dev Origins Health Dis 9(5):487–491

Walsh KJE, McBride JL, Klotzbach PJ, Balachandran S, Camargo SJ, Holland G, Knutson TR, Kossin JP, Lee TC, Sobel A, Sugi M (2016) Tropical cyclones and climate change. Wiley Interdisc Rev Clim Change 7(1):65–89

Warrick O, Aalbersberg W, Dumaru P, McNaught R, Teperman K (2017) The 'Pacific adaptive capacity analysis framework': guiding the assessment of adaptive capacity in Pacific island communities. Reg Environ Change 17(4):1039–1051

Webber S (2013) Performative vulnerability: climate change adaptation policies and financing in Kiribati. Environ Plan A 45(11):2717–2733

Williams AT, Rangel-Buitrago N, Pranzini E, Anfuso G (2018) The management of coastal erosion. Ocean Coast Manag 156:4–20

Work C, Rong V, Song D, Scheidel A (2018) Maladaptation and development as usual? Investigating climate change mitigation and adaptation projects in Cambodia. Clim Policy 1–16

World Bank (2018) Rural population (% of total population). Viewed 5 April 2019 at https://data.worldbank.org/indicator/sp.rur.totl.zs

An Overview of the Information Presented in Nationally Determined Contributions (NDCs) for Fiji, Papua New Guinea, Solomon Islands and Vanuatu

Linda Flora Vaike, Diana Hinge Salili and Morgan Wairiu

Abstract The Paris Agreement requires that signatories develop and communicate their post 2020 climate actions in the form of Nationally Determined Contributions. This paper encompasses an overview of the nationally determined contributions communicated by the Pacific Small Islands Developing States with a focus on the four (4) Melanesian countries; Fiji, Papua New Guinea, Solomon Islands and Vanuatu. It identifies the gaps and challenges the nationally determined contributions potentially entail and is a contribution to the development of discourse on nationally determined contributions for the four (4) Melanesian countries. The methodology used is a situational analysis of the NDCs communicated by Fiji, Papua New Guinea, Solomon Islands and Vanuatu. The paper explores the questions 'are the current nationally determined contributions communicated by the four (4) Melanesian countries consistent, comprehensive and feasible?' and 'what can be done to improve the information provided in these nationally determined contributions?' The paper proposes key mitigation and adaptation interventions for NDC enhancement and is intended to be used by the four (4) countries to enhance the implementation of their nationally determined contributions.

Keywords Nationally determined contributions · Fiji · Papua New Guinea · Solomon Islands · Vanuatu

L. F. Vaike (✉) · D. H. Salili · M. Wairiu
Pacific Center for Environment and Sustainable Development (PaCE-SD), University of the South Pacific, Suva, Fiji
e-mail: linda.vaike@usp.ac.fj

D. H. Salili
e-mail: diana.salili@usp.ac.fj

W. Leal Filho (ed.), *Managing Climate Change Adaptation in the Pacific Region*, Climate Change Management, https://doi.org/10.1007/978-3-030-40552-6_5

Introduction

Intended Nationally Determined Contributions (INDCs) were introduced at the 2013 19th session of the conference of the Parties (COP 19) in Warsaw. The term "contributions" was developed as a compromise between Quantified Emissions Limitation and Reduction Objectives (QUELROs) and Nationally Appropriate Mitigation Actions (NAMAs) (Höhne et al. 2014). When introduced, INDCs were developed for the purpose of identifying actions at the national level that different governments intend to implement. For the purpose of facilitating action and discussions at COP21 in Paris, parties[1] were requested to submit their INDCs in the first 4 months of 2015. INDCs can therefore be seen as a basis for post 2020 global emissions reduction commitments under the Paris Agreement (PA). After the PA came into force on the 4th of November 2016, the INDCs were converted to Nationally Determined Contributions (NDCs) and are specifically presented in Article 4 of the PA (Pouffary et al. 2016). According to Article 4 paragraph 2 of the PA, parties are required to successively communicate NDCs and as such, will implement the objectives of their NDCs through national mitigation and adaptation actions (United National Framework Convention on Climate Change 2016a, b). In adherence to paragraph 12 of Article 4 of the PA all NDCs that are submitted will be recorded in a public registry[2] that is maintained by the secretariat (United Nations Framework Convention on Climate Change 2016b). It is a requirement under the PA that successive NDCs reflect an increase in national efforts beyond previous NDCs and that developing countries such as the Pacific Small Island Developing States (PSIDS) continue to enhance mitigation efforts.

Even more important for PSIDS and particularly the region of Melanesia, are the provisions under paragraph 5 of Article 4 stating that developing country parties shall be provided support to implement Article 4 (in accordance with Articles 9, 10 and 11)[3] Parties are to communicate NDCs every 5 years as stipulated in paragraph 9 of Article 4 of the PA. With regards to accounting, provisions under the PA outline all parties shall account for their NDCs, and that in the process, they should promote environmental integrity, transparency, accuracy, completeness, comparability and consistency (United Nations Framework Convention on Climate Change 2016a, b). The inclusion of adaptation actions in the NDCs is based on the negotiation outcomes from COP20 in Lima, that confirm that adaptation actions can be included in country NDCs (Paw et al. 2017; Pouffary et al. 2015).

[1] *Parties* refers to the countries who have signed up to the UNFCCC.

[2] Public Registry as described in Article 4 paragraph 12 of the Paris Agreement. The COP, by its decision 1/CP.21 paragraph 29, requested the Subsidiary Body for Implementation (SBI) to develop modalities and procedures for the operation and use of the public registry referred to in Article 4, paragraph 12, of the Paris Agreement, for consideration and adoption by the Conference of the Parties serving as the meeting of the Parties to the Paris Agreement (CMA) at its first session.

[3] Articles 9, 10 and 11 focus on support and financial resources, technology transfer and capacity building respectively.

Fiji, Papua New Guinea, Solomon Islands and Vanuatu have communicated NDCs that reflect energy and the forestry sector as priority areas for action. However, recent studies have shown that the strong shift by the international community to focus on mitigation has not sufficiently translated renewable energy into local contexts. This has resulted in PSIDS not fully attaining a self-defined energy identity (Michalena and Hills 2018). The highly dependent nature of conditional targets is also a major barrier for the implementation of NDCs (Senshaw and Kim 2018).

The information provided in their NDCs contains varying degrees of consistency, comprehensiveness and feasibility. This paper focuses on the four countries in the Melanesian region of the Pacific owing to its diversity in terms of culture, language, environment, emissions, landmass and the limited capacity to develop, communicate and report on successive NDCs (United Nations Development Programme (UNDP) 2016). The Melanesian group also boasts the highest population in the PSIDs region (Fiji Bureau of Statistics 2015; Solomon Islands National Statistics Office 2009; Vanuatu National Statistics Office 2016) and as such, the implications of the lack of clarity, transparency and understanding (CTU) and communicating NDCs that are not feasible will significantly affect a huge proportion of the population of PSIDS.

This study involved desktop research that was carried out between October 2018 and March 2019. The scope of this paper encompasses an overview of the NDCs communicated by the PSIDS with a focus on 4 Melanesian countries namely Fiji, Papua New Guinea, Solomon Islands and Vanuatu. The objective is to identify key gaps and recommend key mitigation and adaptation intervention to include in NDC enhancement. The paper raises the questions 'are the current NDCs communicated by the four Melanesian countries consistent, comprehensive and feasible?' and 'what can be done to improve the information provided in these NDCs?' In the context of this paper, 'consistent' refers to a comparison of the information provided in the NDCs submitted by the four countries (in the absence of agreed[4] guidance on 'reference'). 'Comprehensive' draws on the 'information' and considers mitigation and adaptation options presented, while considering initial communications. 'Feasible' refers to the feasibility of implementing proposed targets and adaptation and mitigation actions against 'conditional' funds.

A Description of the (Intended) Nationally Determined Contributions for Fiji, Papua New Guinea, Solomon Islands and Vanuatu

The process taken by each country to develop and submit initial NDCs included different levels of consultations at the national level and data collection and analysis. Although guidelines for developing NDCs exist (South Pacific Environment Programme 2015), national circumstances dictated the processes involved including the

[4]Commonly agreed under UNFCCC guidelines.

level of information sharing and associated policies and guidelines. The initial availability of data was an issue as there were no established and stringent processes in place to guide the collection and collation of data. All countries included components of mitigation that included renewable energy options, all requiring high initial capital, the feasibility of which is heavily reliant on 'conditional' funds. Based on these national circumstances, different levels of reporting and data sharing are evident in the different NDCs. Mitigation and adaptation priorities identified at the national level (*Refer to* Table 1) through national planning and budgetary processes provided scaffolding for the development of NDCs. The section below is a descriptive summary highlighting national circumstances (and processes) as well as linkages with proposed actions, inevitably including gaps arising from the NDCs.

Fiji

The different sections in the NDC communicated by Fiji include national circumstances, a specific section labelled NDC, Mitigation (electricity, energy efficiency and transport), Adaptation, key challenges and proposed way forward. Information provided under national circumstances include physical characteristics, population and economy. This includes references to its national climate change policy of 2012, the Mauritius Strategy 2005–2015, The Barbados Plan of Action 1994, the Samoa Pathway and the UNFCCC's Sustainable Energy for All (SE4All) global report. It highlights the significant sectoral and institutional reforms focusing on improving community livelihoods. Further references to national policies relating to agriculture, land use, forestry, fisheries and water including the sustainable use of natural resources are made. It identifies climate change as one of the largest threats to sustainable development and makes references to the damaging impacts of natural hazards.

Information provided under the NDC section includes the mention of both conditional (only with the availability of external funding) and unconditional (available in country) means of emissions reductions and presents the energy sector as the focus of the NDC. A target of 30% emissions reduction is presented, 10% of which is proposed to be achieved by the implementation of the Green Growth Framework (GGF) and 20% with external funding (should this be made available) amounting to US$500 million. The period of defining actions is 10 years beginning in 2020 and ends in 2030 with 2013 as the reference year. A table containing information on the type and level of commitment, estimated quantified emissions impact, baseline and coverage is provided highlighting the target for renewable energy share in electricity generation to approach 100% by 2030 from approximately 60% during the baseline year (2013). This will reduce the BAU emissions by 20%. A reduction in the cost of imported fuels by the year 2030 is proposed simultaneously with a reduction in dependence on imported fuel as a source of energy. The major assumption is that the conditional funding will be obtained to finance mitigation options under the energy

Table 1 Mitigation and adaptation priorities presented in national mitigation and adaptation policies and frameworks (Fiji Government 2012, 2017; Government of Papua New Guinea 2014; Solomon Islands Government 2012; Vanuatu Government 2012)

Country	National mitigation priorities	National adaptation priorities
Fiji	Energy contributes 24.6% of global greenhouse gas emissions National Mitigation priorities for Fiji include: • Efficient electricity generation and distribution • Losses to reduce fossil fuel consumption • Promotion of renewable energy • Minimisation of deforestation related to hydroelectricity and • Dam construction	National adaptation priorities for Fiji are listed in the Fiji Climate Change Policy 2017 and include: • Agriculture • Human health/welfare • Marine and fisheries • Forestry • Communications • Water resources and water infrastructure • Tourism • Urban development and housing
Papua New Guinea	**Energy** Promote low-carbon growth and investments in low carbon infrastructure and technology development, renewable energies, energy efficiency, transport, waste management, manufacturing and construction, and industrial processing sectors **Forestry** Promote the reform process for SABLs so that they become climate compatible, including assessing former SABLs on their potential for payment for ecosystem services (e.g., REDD+ initiatives) **Sustainable Land Use Planning** Promote establishment of nation-wide sustainable land use planning, in considering climate change resilience and maximising payment for ecosystem services under REDD+ management as national land use priorities	**National Climate Compatible Development Management Policy** **Adaptation responses**: resilience to climate change impacts in the natural and built environment is significantly enhanced **Adaptive capacity**: ability of government and partners to prevent and respond to climate change impacts is maximized **Risk management** • Quantifying & prioritizing hazards • Identifying & selecting Interventions • Monitoring & evaluation **Adaptive governance** • Sectoral coordination • Institutional strengthening • Data management

(continued)

Table 1 (continued)

Country	National mitigation priorities	National adaptation priorities
Solomon Islands	**The Solomon Islands national development strategy (2011–2020)**: • Shift from non-renewable to renewable energy sources • Improving energy efficiency (domestic shipping, land transport, aviation, communications) • Promotion of energy conservation and efficiency through school formal (school curricula) and information capacity building initiatives	**Solomon Islands national development strategy (2011–2020)** Undertake vulnerability and adaptation assessments in identified provinces and communities and establish procedures for identifying climate change projects that meet national needs and for submitting them to potential donors to support establishment of management regimes to cater for effects of climate change, such as disaster preparedness and resettlement of small island communities **NAPA priorities** • Agriculture and food security • Water supply and sanitation • Human settlement • Human health • Education, awareness and information on climate change
Vanuatu	**The national energy road map (2013–2020)** The national energy road map, which sets out a clear strategy and action plan for the development and use of alternative and sustainable energy sources, has an ambitious goal of reducing the country's high reliance on imported fossil fuel by meeting 65% of its electricity needs from renewable energy sources	**NAPA priorities include** • Agriculture and food security • Sustainable tourism development (Sustainable transport) • Community based marine resource management. • Sustainable forest management • Integrated water resources management

sector. Electricity and energy efficiency have been identified as low cost and achievable considering the successes and major strides taken in the renewable energy sector (with the exception of solar PV).

Reducing vulnerability and increasing resilience is the focus of adaptation and rehabilitation plans that are being developed focus on building back better. The final section of the NDC identifies key challenges that include the need for greater understanding of climate change impacts to improve development planning, strengthening the role of local governments to build resilience, the need to build cyclone resistant infrastructure, the need to ensure more robust development planning and budgetary processes and the need to develop an integrated approach at the policy and operational levels. The importance of forging strong overarching partnerships is also stressed along with proposed ways forward that have been allocated time frames ranging from short to the medium and long term (Fiji Government 2015).

Papua New Guinea

The NDC communicated by Papua New Guinea is introduced with information on national circumstances in the context of its national development goals. PNGs estimated seven (7) million people are mostly employed in the informal sector with an almost entirely sustainable, and fossil fuel free existence. The country's Vision 2050 is a forty (40) year development plan strategy that is intended to shift socio economic growth towards a sustainable development pathway, focusing on community inclusivity in development planning, and capitalising on strengths like significant human and natural resource capital. Other sections in the NDC include PNG's mitigation contribution (including existing national GHG emissions and BAU projections of emissions), gases considered, expected trajectory, assumptions and methods for establishing BAU emissions, Mitigation opportunities, electricity supply, energy efficiency, transport and forestry. This is followed by a section on methodology and assumptions and options for mitigation contributions for INDC. A section on adaptation is presented at the end concluding with Measuring, Reporting and Verification (MRV) approach for mitigation actions and equity and ambition.

National emissions are estimated from previously submitted national communications including national development plans and third party reports. Emissions are estimated to be 1.4Tonnes per person, per year (noting considerable[5] uncertainty) which are low by global standards CO_2 and CH_4 (for indigenous oil and gas production) are the gases considered in this NDC. Changes in the mining, oil and gas sectors make it difficult to project emissions but long term projections suggest that as more opportunities for Liquefied Petroleum Gas (LPG) exports emerge, emissions from this local resource will be significantly reduced (although uncertainties owing to the unknown extent of economically viable gas and oil reserves still exist). The limitations of mitigation options save for those in the energy sector is acknowledged

[5]Considerable' is not clearly defined in the NDC document.

as REDD+ initiatives are largely dependent of the success of implementation actions. Shifting from non-renewable energy sources to renewable energy sources in the electricity sector has been identified as beneficial in the long term. It is implied that this will be implemented simultaneously with energy efficiency initiatives as the move to renewable energy will require such initiatives. The increasing preference for individual transportable has been identified as a limiting factor mitigation options. The NDC has considered the vast rainforest resources in PNG and presents options to reduce emissions from deforestation and land degradation through increasing REDD+ initiatives. Forest inventories are ongoing and it is anticipated that date collected will inform more accurate documentation on the potential emissions reductions that PNG can achieve through REDD+ initiatives. The methodology that was used to calculate emissions excluded emissions from shipping, aviation and burning of fossil fuels as per IPCC guidelines. Drivers of domestic carbon emissions were identified and an approximate was calculated for each emission activity. PNG has proposed to use a net approach to calculate emissions from land use, land use change and forestry (LULUCF) as per the IPCC's 2006 guidelines. Emissions to obtain a BAU scenario were based on current population and economic growth projections. PNG has proposed a 2020–2030 timeframe to implement its aim of totally shifting to renewable energy in the electricity sector conditional to availability and access to external funds. There is also the acknowledgement that while energy efficiency initiatives will prove beneficial in the long term, high capital requirements have prevented the shift in the past. The need to improve human resource capabilities and data collection is mentioned as a gap that needs to be addressed for proposed NDC targets to be met. The challenge for the oil and gas sector in PNG is that it generates jobs that provide income for many families and communities as well as a major income generator for the government. Changes in emissions under this sector will only be possible if the move is global and results in a no markets for hydrocarbons post 2050. It is proposed that the introduction of trains and trams alongside higher energy efficient vehicles may be the response needed to reduce emissions in this sector. PNG is yet to quantify how much external funding is required for the electricity sector. However, it is anticipated that the May 2015 enacted climate change act will contribute to facilitating data collection and initiatives to achieve this. Tracking and monitoring progress on contributions has been identified as another activity that will be implemented once proper data collection processes have been set up. The NDC clearly outlines that PNG cannot be expected to engage in mitigation initiatives to significantly reduce emissions, in the absence of international assistance. The adaptation section of the NDC presents coastal flooding, inland flooding and droughts as the hazards that are most prominent, with climate change exacerbating these and potentially introducing new threats associated with climatic conditions. It is suggested that this is likely to result in increased exposure to malaria in the highlands, changes in agricultural production and negative impacts on coral reefs. Risk management has been prioritised in PNGs adaptation strategies, hence the inclusion of adaptation in the NDC was deemed vital. Similar to mitigation, it is anticipated that external aid will be required to effectively implement adaptation actions (Government of Papua New Guinea 2016).

Solomon Islands

The Solomon Islands' NDC is introduced with geographical features, a brief on the climate and demographic characteristics as well as information on the economy and major income generating avenues. A mitigation section containing information on mitigation status and context as well as actions is presented followed by a section on adaptation containing information on current and near term adaptation planning and action, adaptation gaps, barriers and needs, financing needs for priority adaptation interventions addressing gaps in climate resilient programs and innovative financing approaches. A means of implementation and equity section complete the NDC. Information on Mitigation defining actions will be measured over five year periods beginning 2020 and ending in 2030. The year of reference in 2015 with a proposed conditional 27% reduction in GHG emissions by 2025 and 45% by 2030. It is anticipated that access to international assistance will ensure a reduction in emission of more than 50% by the year 20,150. The NDC covers fossil fuels (90% of reported national inventory) and forest sequestration. The sector of focus is the energy sector including power (for electricity) and transportation. Renewable energy and energy efficiency options including hydro power and solar farms are the focus of unconditional contributions (currently implemented by government). The successful implementation of conditional mitigation actions require capacity building, technology transfer and financial support in the form of grants and from various bilateral programs, accessed in a timely manner.

Adaptation gaps, barriers and needs include institutional challenges, adaptation knowledge sharing between stakeholders and capacity building. The translation of climate science to local languages, cultural barriers to effectively implement awareness and community capacity to carry out local level vulnerability assessments have also been identified. The Solomon Islands is looking at a community based 'whole of island' vulnerability, adaptation and management mapping approach that will attract direct access to finance for community based resilience projects. There are plans in place to strengthen institutional structures and build community capacity to achieve this. There is also the proposal to establish a national climate change trust fund supported by market based mechanisms. Solomon Islands is a small contributor to GHG emissions and while the government has engaged in a number of initiatives to progress the shift from non-renewable to renewable forms of energy, long term sustainable development that is simultaneous with a reduction in GHG emissions is dependent on the availability and access to external funding (Solomon Islands Government 2016).

Vanuatu

The NDC communicated by Vanuatu includes national circumstances, mitigation contributions, adaptation priorities and goals, fairness, equity and ambition and an annex. This NDC is specific to the energy sector with references to Vanuatu's Energy Roadmap and national communications. The transition from non-renewable to renewable energy sources will require significant conditional external support to meet targets proposed in Vanuatu's Energy Roadmap as well as those presented by Vanuatu's renewable readiness Assessment by the International Renewable Energy Agency (IRENA). Alignments with Vanuatu's Priority Action Agenda and Vanuatu's Rural Electrification Nationally Appropriate Mitigation Actions (NAMA) design document are also presented. Mitigation initiatives associated with the generation of power for electricity are targeted at 100% below BAU emissions and 30% for the entire energy sector. Planned mitigation initiatives include doubling wind installed capacity by 2025, installing grid connected solar PV by 2025, commissioning a first stage geothermal plant by 2025 and substituting fossil fuels with coconut oil based electricity. These proposals require substantial external funding, technology transfer, institutional support and training. Additional planned interventions include REDD+ initiatives under the forestry sector. Data from utilities, customs department and other development partners and CSOs was used to analyse sector emissions (with acknowledgements of possible uncertainties).

Adaptation priorities (as identified by the NAPA process) include 11 thematic areas which were refined to 5 thematic areas for the NDC (Refer to Table 1). Further findings from the NAPA that are presented in the NDC include cross cutting themes like awareness raising at all levels, capacity building (including institutional capacity), technology transfer, research and development, promotion of appropriate traditional knowledge, education and training, consideration of marine and terrestrial biodiversity and mainstreaming climate change and disaster risk reduction. References to the National Climate Change and Disaster Risk Reduction policy (Vanuatu Government 2016b) are also made in this regard. Information on fairness, equity and ambition includes the anticipation of the bleak climate projections for Vanuatu, and shift towards focusing on reducing vulnerabilities and risks and increasing resilience and the acknowledgement that while Vanuatu's emissions are miniscule, efforts are being made to improve data collections methods for NDCs and invest in both mitigation and adaptation initiatives. Progress on the implementation of the NDC however, is heavily reliant on external funding (Vanuatu Government 2016a).

Table 1 provides an overview of the information provided in national mitigation and adaptation policy frameworks and documents for Fiji, Papua New Guinea, Solomon Islands and Vanuatu.

Table 2 contains a brief summary of mitigation and adaptation priorities provided in the NDCs for Fiji, Papua New Guinea, Solomon Islands, and Vanuatu.

Clarity, Transparency and Understanding (CTU)

Article 3, paragraph 2, of the Paris Agreement calls on Parties submitting (I) NDCs to do so "...in a manner that facilitates CTU...". While this call can be traced back to Article 2 of the Convention, the Cancun 2010 and Lima 2014 conferences, it is

Table 2 National mitigation priorities and priorities listed in NDCs (Fiji Government 2012, 2017; Government of Papua New Guinea 2014; Solomon Islands Government 2012; Vanuatu Government 2016a)

Country	Mitigation priorities in NDCs	Adaptation priorities in NDCs
Fiji	**Energy** Measures to reduce emissions from the energy sector Mitigation targets for the period of 2020–2030 **Target 1**: to reduce 30% of BAU CO_2 emissions from the energy sector by 2030 **Target 2**: as a contribution to Target 1, to reach close to 100% renewable energy power generation (grid-connected) by 2030, thus reducing an expected 20% of energy sector CO_2 emissions under a BAU scenario **Target 3**: as a contribution to Target 1, to reduce energy sector CO_2 emissions by 10% through energy efficiency improvements economy wide, implicitly in the transport, industry, and electricity demand-side sub-sectors	
Papua New Guinea	**Energy** Renewable energy for electricity generation **Forestry sector** The implementation of REDD+ activities	

(continued)

Table 2 (continued)

Country	Mitigation priorities in NDCs	Adaptation priorities in NDCs
Solomon Islands	**Energy** Hydro and Solar power for electricity generation	• Agriculture and food security • Water and sanitation • Human settlements and human health • Education awareness and human health • Loy-lying and artificially built-up islands • Waste management • Coastal protection • Fisheries and marine resources • Infrastructure development • Tourism
Vanuatu	**Energy** – Shift towards renewable energy for electricity generation **Forestry** – Measures to reduce deforestation and promote good land care and accepted mitigation practices according to REDD+ principles	• Agriculture and food security • Sustainable tourism development • Community based marine resource management • Sustainable forest management • Integrated water resource management • Awareness raising at all levels • Capacity building including institutional capacity • Research and development • Promotion of appropriate traditional knowledge and practices • Technology transfer • Education and training • Mainstreaming of CC and DRR • Consideration of marine and terrestrial biodiversity issues

at the discretion of Parties to come up with the guidance on information that will facilitate CTU.

A synthesis report of written submissions made after COP22 noted Parties' views on important linkages of the guidance for CTU with guidance on features and accounting (1/CP.21) and other articles of the agreement (United Nations Framework Convention on Climate Change 2015). Although it is not clear how specific elements of the guidance will look, there is a consistent view among Party submissions that CTU guidance will enable understanding of NDCs and its features (Moarif 2017). There are however earlier attempts at the sort of information that would facilitate CTU including target(s), timeframes, coverage, and details of baselines and accounting assumptions for mitigation contributions (Hood et al. 2015). The four countries studied and the Alliance of Small Island States (AOSIS) did not submit views on this issue making it difficult to conclude individual country as well as the regional negotiating block's views.

While debates on guidance for CTU continues, it is perhaps important to relook at the Decision 1/CP.21 and the successive decisions of the Ad Hoc Working Group on the Paris Agreement (APA) (APA 1.7 Agenda Item 3 2018).

Decision 1/CP.21, paragraph 27, states:

> ……Parties communicating their nationally determined contributions, in order to facilitate clarity, transparency and understanding, may include, as appropriate, inter alia, quantifiable information on the reference point (including, as appropriate, a base year), time frames and/or periods for implementation, scope and coverage, planning processes, assumptions and methodological approaches including those for estimating and accounting for anthropogenic greenhouse gas emissions and, as appropriate, removals, and how the Party considers that its nationally determined contribution is fair and ambitious, in the light of its national circumstances, and how it contributes towards achieving the objective of the Convention as set out in its Article 2.

Drawing on the synthesis report and the various decisions of the COP after Paris, it is still not clear how the final guidance on information for CTU will look. It is however becoming clearer from the synthesis conducted by Moarif (2017) that there is general agreement that features "refer to the characteristics of NDCs including (but not limited to) nationally determined, representing a progression, reflecting highest possible ambition and communicated every five years. The synthesis report is important as it reflected Party views post Paris.

Taking all the above into consideration, and as a preliminary overview of the NDCs of the four countries studied, all were nationally determined although a lot of the NDCs for the PSIDS were rushed and therefore excluded significant sources of emissions and emission reduction opportunities: see https://pacificclimatechange.net/project/regional-pacific-ndc-hub (Regional Pacific NDC Hub 2018). Despite the missing emission sources, the PSIDS are already showing great leadership in having ambitious targets (see Table 2), especially when they contribute very little to total global emissions. Countries may want to consider adding the missing emission sources to their next NDCs as a means to reflect higher ambitions; important sectors for consideration include the transport, land use and forestry. In terms of the time frame for communication (every five years), all four countries have time frames of 2020–2030.

Are Current NDCs Consistent, Comprehensive and Feasible?

A more comprehensive analysis of Information to facilitate CTU can only be made when the guidance to facilitate CTU is agreed upon. Countries have, however, communicated their first NDCs and are therefore required to achieve their targets. Notwithstanding the guidance or the lack thereof, and in an attempt to inform the development of successive communication of NDCs, consistency, comprehensiveness and feasibility are important considerations that will help CTU. This is particularly true for comparison between successive NDCs as well as global monitoring and reporting on collective progress towards achieving the temperature goal of the PA.

The NDCs of the four countries show both consistencies and inconsistencies in the type of information provided, the methodology used and the general structure. The following sections present an overview of the NDCs for the four countries studied. It is important to note that, without an agreed reference guidance, information contained in the NDCs are compared with each other.

Mitigation

The mitigation sections show inconsistencies in both structure and content. Fiji, Solomon Island and Vanuatu presented the information in tabular form while PNG presented the information in plain text. In terms of content, the four countries mitigation targets were mostly in the energy sector (Refer to Table 2).

The methods used to calculate emissions are presented in varying formats and degrees of clarity although the IPCC Guidelines is primarily used. PNG and Vanuatu specifically mentioned using the 2006 and 1996 guidelines respectively. It is also clear that countries mostly used secondary data, in many cases information contained in reports prepared by development partners. While secondary data provides a good source of information already at the disposal of countries, national capacities will have to be further built and institutional arrangements strengthened to enable countries to continuously monitor the progress of implementation of NDCs. Continuous monitoring will also ensure consistencies in data generation, tracking, analysis and reporting for successive NDCs.

There is also a difference in the way countries report emissions and the units of measurement. Vanuatu reports CO_2 in kilowatt hour (kWh). Fiji communicated its target in Gigagrams (Gg) of CO_2. Solomon Islands and PNG used tonnes of CO_2 equivalent (tCO_2e) and Metric tons of CO_2 (Mt).

In light of the need to communicate successive NDCs that reflect an increase in national efforts, all the current NDCs contain areas that countries consider and/or intend to include to further increase ambitions. These include market-based mechanisms, agriculture and land-use.

The synergies between the mitigation components of the NDCs and national policies, plans and frameworks are clear. Although the synergies were not presented as separate sections, they are embedded in different sections in the NDC documents.

Adaptation

All four countries have adaptation as a standalone component of their NDCs although presented in different forms: Fiji, PNG and SI presented the information in paragraphs while Vanuatu in tabular form.

Without any common guidance on how information on adaptation should be presented (United Nations Framework Convention on Climate Change 2014), the four countries have all taken different approaches to presenting their adaptation priorities. Similar to the mitigation components, it is clear that there are synergies between the adaptation components of the NDCs and national climate change and development instruments including National Adaptation Plan of Action (NAPA), National Communications and National Climate Change Policies (or its equivalent) (see Tables 1 and 2).

In terms of adaptation priorities, Solomon Islands and Vanuatu provided information on specific sectors (see Table 2) and costing estimates required to meet 'some' of the adaptation priorities identified in the NDCs. PNG, provided information on the different types of hazards affecting the country and the need for financial, capacity building and technical support for adaptation. Fiji placed more emphasis on the required governing and institutional arrangements and identified the sectors for which adaptation measures were already implemented.

Finance

All four countries have both conditional and unconditional components in their NDCs. Given the special circumstances facing the countries, timely and full implementation of NDCs will not be possible without external support. Conditional requirements were either presented as stand-alone sections or embedded in different parts of the NDCs and differ in type including capacity building, technology transfer, finance and technical support.

In terms of the unconditional support, Fiji, Solomon Islands and Vanuatu have provided information on the national financial contributions that will help meet their

Table 3 Table showing conditional finance for Fiji, Papua New Guinea (PNG), Solomon Islands and Vanuatu

Country	Cost estimate for mitigation (conditional) in USD million	Cost estimate for adaptation (conditional) in USD million
Fiji	\$500	Not Stated
Solomon Islands	\$170,700,000	\$126,650,000
Vanuatu	\$180	\$9.5 per year
PNG	Not stated	Not stated

Source NDC for Fiji, Solomon Islands (SI) & Vanuatu
Note For SI, the cost estimate was calculated from the total of specific projects planned

mitigation and adaptation targets. These mostly comprise of ongoing mitigation projects. Although the unconditional contributions are labelled as "national contributions", a lot of the projects that were tagged under this category still rely on external funding and technical support for their full implementation.

In terms of finances Fiji, Solomon Islands and Vanuatu presented the total values in USD that will be needed to fully implement their NDCs (see Table 3). PNG highlighted the need for external assistance without any mention of the monetary value. Costing adaptation needs, although will be challenging considering future uncertainties and long-term impacts, is crucial for national planning and budgetary processes. For the four countries studied and any other Developing Country Party, proper accounting of external support is also important to track the \$100 billion by 2020 commitment and the principle of Common but Differentiated Responsibilities and Respective Capabilities as set out in the Convention.

Drawing on the information provided in the NDCs, and considering these were initial communications from the four countries, it is certain that current NDCs are comprehensive when referenced against the commonly agreed elements echoed through Party submissions as well as those already defined by various articles of the Convention and the PA. Although there are varying degrees of inconsistencies in the information communicated, it is assumed that communication of successive NDCs will be further improved. The degree of consistency of NDCs can be clearly defined once clear guidelines are agreed upon.

The feasibility of the NDCs are highly dependent on external support. It is apparent that activities that are currently ongoing are from funds committed as part of climate change programmes/projects. Capacity building, technology transfer, finance and technical support is therefore needed for full implementation of NDCs.

Recommendations to Improve Information Provided in NDCs

The following are some recommendations, for the four NDCs, that countries *may* want to consider as they communicate and enhance successive NDCs:

- Review and reflect on the process and time involved to develop and communicate initial NDCs. Important lessons from the process will assist countries improve successive NDCs. This is particularly important for communicating of NDCs every 5 years.
- Include important emission sectors that are missing in current NDCs, in successive NDCs as a way to increase ambitions.
- Use the same units for reporting emission inventories and stocks for ease of monitoring and accounting. Using the same units will also enable individual countries to compare future NDCs with initial or baseline emission targets and levels of resilience in terms of adaptation implementation.
- Identify capacity, technical, technological and funding support for full implementation of NDCs. For financial support, available sources of funding should be identified and sourced to allow for timely implementation of NDCs.
- Develop national NDC implementation plans and strategies to guide NDC implementation. Such a plan should include all support (capacity, technical, technological and funding support) needed for full implementation and achievement of NDC targets.
- Consider other national alternative models of support that can assist with implementation of both conditional and unconditional components of NDCs. Green bonds and trust funds are options worth considering.[6]

Although the above recommendations are listed as separate points, all of them are interlinked and have far reaching implications on planning processes in individual countries as well as global commitments.

Study Limitations

It is important to note that the lack of commonly agreed guidelines poses significant limitations to conduct a thorough analysis of the type of information that should be provided in the NDCs. Comparing the four NDCs against each other and some of the requirements already spelt out in various Decisions under the UNFCCC may not fully cover what is expected for CTU.

Issues such as consistency, comprehensiveness and feasibility are not terms used in the Paris Agreement and Decision texts stemming from current negotiations surrounding the guidelines. To minimise potential mixed interpretations of the terms,

[6]Fiji has set up a Green bond. See: https://www.worldbank.org/en/news/press-release/2017/10/17/fiji-issues-first-developing-country-green-bond-raising-50-million-for-climate-resilience.

this paper maintains the definitions as given in the introduction and further acknowledges that the definitions used may not fully capture important considerations in the wider discourse surrounding NDCs.

This study should nonetheless be seen as an initial critical attempt at looking at the information already communicated in (I)NDCs of the four countries.

Conclusion

The PSIDS have shown great commitment and leadership in the fight against climate change. The timely communication of INDCs and subsequent conversion into NDCs is yet another testament to the commitment towards emission reductions and global efforts to limit global temperatures to below 2 °C, and particularly 1.5 °C.

Fiji, PNG, Solomon Islands and Vanuatu are among the PSIDS who communicated their NDCs. Article 4 of the PA calls for guidance to facilitate CTU. Parties to the convention are, however, still negotiating specific elements of the guidance. Without the agreed guidance, it is difficult to provide input on the sort of information Parties are required to include in their NDCs. One thing that is certain is that successive NDCs will have to be consistent, comprehensive and feasible in order to ease other monitoring and reporting obligations at both the national and international levels.

Drawing on the information in the initial (I)NDCs communicated by Fiji, PNG, Solomon Islands and Vanuatu, it is certain that the NDCs are comprehensive in terms of coverage of mitigation targets and adaptation needs, considering these were first attempts at developing them.

This study noted varying degrees of inconsistencies across the NDCs. These inconsistencies range from the forms (tabular and non-tabular) and types of information used to more technical issues such as the method used to calculate emissions and inventories, conversion units for emission targets as well as support needed. Information that are common across the four NDCs include separate sections for mitigation targets and adaptation needs. There are also clear synergies between the NDCs and other national policies and frameworks.

The feasibility of the four NDCs is dependent on conditional support. Without external support, the countries will not be able to fully implement the NDCs. It is however strongly recommended that countries also consider alternative options of financing such as green bonds and trust funds to support NDC implementation.

Recommendations to further guide the implementation of current NDCs and development of successive NDCs, although given, are only outlined for consideration by the four countries. Although based on the information provided in the four NDCs studied, the recommendation are relevant to all small island developing states, including other PSIDS.

This paper maintains the "Nationally Determined" nature of NDCs and emphasizes the discretion on Parties to own the process while keeping in mind the importance of having comprehensive, consistent and feasible NDCs that will help achieve CTUs.

References

Draft Text on APA 1.7 Agenda Item 3 (2018, December 5) Retrieved from United Nations Framework Convention on Climate Change: https://unfccc.int/sites/default/files/resource/APA11_7_DT_i3.pdf

Fiji Bureau of Statistics (2015) Household income expenditure survey—preliminary findings—Release 1. FBoS Release No: 98. Accessed from https://www.statsfiji.gov.fj/index.php/statistics/population-censuses-and-surveys

Fiji Government (2012) Republic of Fiji National Climate Change Policy, Secretariat of the Pacific Community, Suva, Fiji Islands

Fiji Government (2015) Fiji nationally determined contributions. UNFCCC NDC registry. Accessed from https://www4.unfccc.int/sites/ndcstaging/PublishedDocuments/Fiji%20First/FIJI_iNDC_Final_051115.pdf

Fiji Government (2017) Fiji NDC implementation roadmap 2017–2030 setting a pathway for emissions reduction target under the Paris Agreement, Suva, Fiji Islands

Government of Papua New Guinea (2014) National Climate Compatible Development Management Policy), Office of Climate Change and Development, Boroko, Papua New Guinea. Accessed from https://www.pacificclimatechange.net/sites/default/files/documents/National_Climate_Change_Policy1.pdf

Government of Papua New Guinea (2016) Intended Nationally Determined Contribution (INDC) Under the United Nations Framework Convention on Climate Change. UNFCCC NDC Registry (Interim). Accessed from https://www4.unfccc.int/sites/ndcstaging/PublishedDocuments/Papua%20New%20Guinea%20First/PNG_INDC%20to%20the%20UNFCCC.pdf

Höhne et al (2014) Intended nationally determined contributions under the UNFCCC discussion paper, ECOFYS Sustainable Energy for Everyone, Cologne, Germany. Accessed from https://www.ecofys.com/files/files/ecofys-giz-2014-intended-nationally-determined-contributions-under-unfccc.pdf

Hood C, Adkins L, Levina E (2015) Overview of INDCs submitted by 31st August 2015. OECD/IEA Climate Change Expert Group Paper, No. 2015/4, OECD Publishing, Paris. Accessed from https://www.oecd-ilibrary.org/docserver/5jrqm0bw3mvl-en.pdf?expires=1553939342&id=id&accname=guest&checksum=919AD1459EFA6FFF8BC10BC94AB36190

Michalena E, Hills J (2018) Paths of renewable energy development in small island developing states of the South Pacific. Renew Sustain Energy Rev 82(1). https://doi.org/10.1016/j.rser.2017.09.017 Accessed from https://www.sciencedirect.com/science/article/pii/S1364032117312467

Moarif S (2017) "Information needed to faciliate the clarity, transparency and understanding of mitigation contributions", OECD/IEA Climate Change Expert Group Papers, No.2017/1, OECD Publishing, Paris. Accessed from https://www.oecd.org/environment/cc/Information-needed-to-facilitate-the-CTU-of-mitigation-contributions.pdf

Paw et al (2017) Beyond headline mitigation numbers: we need more transparent and comparable NDCs to achieve the Paris Agreement on climate change, J Clim Change. Accessed from https://link.springer.com/article/10.1007/s10584-017-2122-x

Pouffary et al (2015) Pacific guidelines for INDCs: a guide to the development of intended nationally determined contributions and their further implementation, SPREP, Apia, Samoa. Accessed from http://prdrse4all.spc.int/system/files/indc-guidelines.pdf

Pouffary et al (2016) From Paris to Marrakesh or the challenge of implementation: twenty-second conference of the Partiesto the United Nations Framework Convention on Climate Change (COP22, CMP12, CMA1), Institut de la Francophonie pour le développement durable (IFDD) 2016 56 rue Saint-Pierre, 3e étage, Quebec City (Quebec) G1K 4A1, Canada. Accessed from http://www.zora.uzh.ch/id/eprint/130661/1/AM6_695_Guide_COP22_OIF_eng_web.pdf

Regional Pacific NDC Hub—a regional arm of the NDC Partnership. (2018, March 3). Retrieved 10 Apr 2010, from Pacific Climate Change Portal: https://pacificclimatechange.net/project/regional-pacific-ndc-hub

Senshaw DA, Kim AW (2018) Meeting conditional targets in nationally determined contributions of developing countries: renewable energy targets and required investments of GGGI Member and partner countries. Energy Policy 116. https://doi.org/10.1016/j.enpol.2018.02.017 Accessed from https://www.sciencedirect.com/science/article/pii/S0301421518300946

Solomon Islands Government (2012) Solomon islands national climate change adaptation policy, Honiara, Solomon Islands

Solomon Islands Government (2016) Solomon Islands intended nationally determined contributions. UNFCCC NDC Registry (interim). Accessed from https://www4.unfccc.int/sites/ndcstaging/PublishedDocuments/Solomon%20Islands%20First/SOLOMON%20ISLANDS%20INDC.pdf

Solomon Islands National Statistics Office (2009) Population and housing census national report, vol 2. Accessed from https://www.statistics.gov.sb/statistics/demographic-statistics/census

South Pacific Environment Programme (2015) Pacific guidelines for INDCs: a guide to the development of Intended nationally determined contributions and their further implementation. Accessed from https://www.sprep.org/publications/pacific-guidelines-for-indcs-a-guide-to-the-development-of-intended-nationally-determined-contributions-and-their-further-implementation

United Nations Development Programme (UNDP) (2016) Regional dialogue on (intended) nationally determined contributions for Pacific Islands, meeting report. Accessed from https://www.undp.org/content/dam/LECB/events/2016/20161206-regional-dialogue-indcs-pacific-islands-nadi-fiji/undp-lecb-regional-dialogue-indcs-pacific-islands-nadi-fiji-meeting-report-20161206.pdf

United Nations Framework Convention on Climate Change (2014) Report on sources of support for the preparation of Intended Nationally Determined Contributions (INDCs). Accessed from https://unfccc.int/files/focus/mitigation/application/pdf/support_for_indcs.pdf

United Nations Framework Convention on Climate Change (2015) Synthesis report on the aggregate effect of the intended nationally determined contributions. Accessed from https://unfccc.int/resource/docs/2015/cop21/eng/07.pdf

United Nations Framework Convention on Climate Change (2016a) The Paris agreement. Accessed from https://unfccc.int/sites/default/files/english_paris_agreement.pdf

United Nations Framework Convention on Climate Change (2016b) Aggregate effect of the intended nationally determined contributions: an update. Accessed from http://unfccc.int/resource/docs/2016/cop22/eng/02.pdf

Vanuatu Government (2012) Vanuatu national energy road map (2013–2020). Accessed from https://www.iea.org/media/pams/vanuatu/Vanuatu_National_Energy_Roadmap20132020.pdf

Vanuatu Government (2016a) Intended nationally determined contributions. UNFCCC NDC Registry (Interim). Accessed from https://www4.unfccc.int/sites/ndcstaging/PublishedDocuments/Vanuatu%20First/VANUATU%20%20INDC%20UNFCCC%20Submission.pdf

Vanuatu Government (2016b) Vanuatu climate change and disaster risk reduction policy 2016–2030, Suva, Fiji Islands

Vanuatu National Statistics Office (2016) Vanuatu Post TC Pam Mini Census Report, vol 1. Accessed from https://vnso.gov.vu/images/PublicDocuments/Census/2016/2016_Mini_Census_Main_Report_Vol_1.pdf

iTaukei Ways of Knowing and Managing Mangroves for Ecosystem-Based Adaptation

Jasmine Pearson, Karen E. McNamara and Patrick D. Nunn

Abstract Global concerns for Pacific Island Countries under a new climate regime and increasing development challenges has prompted many external agencies to intervene with climate change adaptation programs. Despite extensive funding and efforts, many external interventions tend to overlook the importance of Indigenous and local knowledge, and working in partnership with local people to co-produce sustainable and effective adaptation strategies. In many Pacific countries, mangroves deliver ecosystem goods and services that are essential to the livelihoods of local people and can enhance resilience to climate change. This paper explores how iTaukei (Indigenous Fijian) communities have sustainably managed mangrove ecosystems over time, and how this knowledge and experiences can enable future ecosystem-based adaptation options that are more sustainable and effective. Across six rural villages in western Vanua Levu, a series of semi-structured household interviews ($n = 41$) were undertaken, coupled with participant observation. The findings demonstrate the importance of understanding, respecting and utilising Indigenous knowledge for managing and protecting local ecosystems as part of communities' response to climate change adaptation.

Keywords Climate change · Fiji · Mangroves · Ecosystem-based adaptation · Indigenous knowledge

---

J. Pearson (✉) · K. E. McNamara
Faculty of Science, School of Earth and Environmental Sciences, The University of Queensland, St Lucia Campus, Brisbane 4072, Australia
e-mail: jasmine.pearson@uq.edu.au

K. E. McNamara
e-mail: karen.mcnamara@uq.edu.au

P. D. Nunn
School of Social Sciences, University of the Sunshine Coast, Sippy Downs Campus, Sunshine Coast 4556, Australia
e-mail: pnunn@usc.edu.au

W. Leal Filho (ed.), *Managing Climate Change Adaptation in the Pacific Region*, Climate Change Management, https://doi.org/10.1007/978-3-030-40552-6_6

Introduction

Pacific Island Countries (PICs) have experienced significant environmental changes throughout history from gradual variations over time to extreme weather events (Nunn 2007; Salick and Ross 2009). Despite this, Pacific Islander people have long demonstrated resilience when it comes to coping with these changes through their vast local knowledge systems, and their context-specific management of local ecosystems (Bridges and McClatchey 2009; Lefale 2010; Chand et al. 2014; McNamara and Prasad 2014). There is an expanding body of literature that draws on the value of local and Indigenous knowledge in climate change adaptation, and its role in the broader discipline of environmental management (Veitayaki 2002; Percival 2008; Williams and Hardison 2013). Studies predict that the effects of climate change will increase and even accelerate over time (IPCC 2014), adding pressure to the availability of natural resources and the sustainability of Pacific livelihoods (Campbell and Barnett 2010). Given this challenge, innovative solutions that allow coastal communities in PICs to respond and live with changed local conditions and environments are needed.

Current approaches to adaptation in the Pacific have been largely dominated by top-down decision-making from a number of non-government organisations (NGOs), international donors, and government agencies (Remling and Veitayaki 2016). Despite their efforts and extensive funding, many external interventions have failed to address the long-term needs of Pacific Island communities and embed sustainable livelihood solutions (Dean et al. 2017). Studies indicate that the limited effectiveness and overall success of many projects is linked to a lack of appropriate engagement, communication and cooperation between stakeholders and local communities throughout both the development and implementation phases of projects (Ireland 2012; Spires et al. 2014; McNaught et al. 2014). Many organisations tend to underestimate the importance of not only appreciating and understanding local knowledge towards environmental management, but also the value of building such knowledge into local adaptation projects (Nunn 2009). These parties are often under pressure to produce quick and visible short-term outcomes, hindering any desirable long-term benefits for local communities (Remling and Veitayaki 2016). As a result, Pacific Islanders voices are rarely heard when discussing adaptation options, meaning that projects are usually shaped based on the interests and concerns of more developed countries that fund them (Dreher and Voyer 2015).

In light of this, there has been a broader global movement to involve Indigenous knowledge into climate change discourse and adaptation strategies. At the Indigenous Peoples' Global Summit on Climate Change 2009, Indigenous representatives from around the world expressed their deep concerns about the climate crisis, calling for urgent collective action and offering their traditional knowledge, innovations and practices relevant to climate change (Anchorage Declaration 2009). At the Conference of Parties (COP17) 2011, the Indigenous peoples of Abya Yala demanded that Indigenous knowledge be respected, protected and promoted during adaptation and mitigation. In 2014, artist and activist Kathy Jetñil-Kijiner captured worldwide attention with her moving opening statement at the UN Climate Summit. Through the power of poetry and storytelling, she called out world leaders for ignoring the threat of climate change and urged them to take action so that people living in island nations could not only survive, but to thrive (Jetñil-Kijiner 2017). Climate change has brought the importance of Indigenous knowledge to the forefront due to its roots in achieving sustainability and balance (Koya et al. 2018).

Ecosystem-based Adaptation (EbA) has recently emerged as an innovative approach to climate change adaptation (Reid and Swiderska 2008; Roberts et al. 2012; Nobel et al. 2014), mainly due to its role in reducing coastal vulnerability and enhancing ecosystem resilience (Sierra-Correa and Cantera Kintz 2015). EbA is often referred to as "the use of biodiversity and ecosystem services as part of an overall adaptation strategy to help people adapt to the adverse effects of climate change" (Sierra-Correa and Cantera Kintz 2015: 387). Building on this definition, Seddon et al. (2016) defines 'effective' EbA as "an intervention which has restored, maintained or enhanced the capacity of ecosystems to produce services on which local human communities depend for their wellbeing, adaptive capacity or resilience, and which reduces vulnerability, and allows the ecosystem to withstand climate change impacts and other stressors." One of the key principles of EbA is the inclusion of local, traditional and Indigenous knowledges. Yet, recent literature suggests that past EbA projects often fail to adequately involve local communities and their knowledge systems throughout key decision-making processes (Nalau et al. 2018b). This calls for more research that explores the role of local and Indigenous knowledge systems in enabling sustainable and effective EbA solutions, while recognising and respecting the needs, views and rights of Indigenous peoples.

Study Aims

The aim of this study is to demonstrate how Indigenous ways of knowing and managing ecosystems can contribute to future ecosystem-based adaptation. This will be achieved through the following key objectives:

1. Explore how iTaukei communities use and value mangrove ecosystems;
2. Examine how these values inform the traditional management techniques used by iTaukei communities to protect, conserve and restore mangrove ecosystems; and
3. Discuss the implications of these findings for enabling sustainable and effective ecosystem-based adaptation options in the Pacific.

Literature Review

Mangroves in Fiji

In Fiji, mangroves are a vital ecosystem for coastal communities, providing an array of ecological services that help sustain the lives of people (Veitayaki et al. 2017). Fijian mangroves provide a habitat for a diverse range of marine life including unique and rare flora and fauna species, and a critical breeding and nursing ground for crustaceans, molluscs, and fish (Spalding et al. 2010). Around 80% of Fiji's population are located in coastal communities and are at least partly reliant on marine resources for subsistence (Techera and Troniak 2009). In addition to this, mangroves have also been utilised for timber and non-timber products, medicines, dyes and other resources used directly by Fijian people (Lal 2003). The natural barrier created by mangroves is also crucial for protecting the coastline of Fiji from projected sea-level rise, storm surges, coral damage and other impacts from climate change (Agrawala et al. 2003; Veitayaki et al. 2017; Carranza Ortiz et al. 2018).

Even though mangroves have long been directly utilised and are unarguably important to the people of Fiji, they continue to be cleared for large-scale coastal development and disturbed by associated human-induced impacts. These include the intertwining stressors of logging, overfishing, industrialisation, agriculture and pollution (Gilman et al. 2016; Spalding et al. 2010; Bhattarai and Giri 2011). Climate change also threatens the future status of Fijian mangroves communities, especially with rising sea levels and the increase of extreme weather events (Blankespoor et al. 2017). Mangroves have evolved to cope with changes in salinity, soil, sediments and the dynamics of the coastline and there may be no other group of plants that is able to adapt to such extreme conditions (Kathiresan and Bingham 2001). Despite this, local development and surrounding human activities may limit their adaptation abilities (McLeod and Salm 2006; DasGupta and Shaw 2017).

The Role of Indigenous Knowledge in Environmental Management

Despite the rich cultural diversity across the Pacific Islands region, there is a shared holistic worldview of deep connectedness and inseparability between people and nature (Voi 2000; Friedlander 2018; Koya et al. 2018). Teresia Teaiwa encapsulated this with the following quote: "We sweat and cry salt water, so we know that the Ocean is really in our blood" (Hau'ofa 1998: 392). Due to their intrinsic connections with land and sea, many Pacific Islanders hold a wealth of knowledge on their local ecosystems and how to manage them. Although they are beginning to struggle to cope with the impacts of climate change, these coastal communities are not inherently vulnerable (Hay 2013). In fact, they are traditionally known to demonstrate strong adaptation skills when it comes to harsh environmental changes over time

(McNamara and Prasad 2014; Taylor et al. 2016). It is the colonisation, globalisation and development initiated by Western contact that has instigated significant social and economic changes (Campbell 2009), marginalising Indigenous knowledges and ways of being in the process (Veitayaki 2004; Koya et al. 2018). It wasn't until the early 1990s that non-scientific knowledge systems began to gain worldwide recognition in international agreements and policy (UN 1992, 2012, 2017; IPCC 2014). Since then, there has been a multitude of studies that have explored the role of local, traditional and Indigenous knowledges as a critical resource tool in climate change adaptation, and the broader field of environmental management.

In Tarawa, Kiribati, local communities demonstrated the ability to relay valuable information about spawning migrations and identify changes in the behaviour and abundance of important fish species as a result of human activities (Johannes and Yeeting 2001). In a 2010 study, Lauer and Aswani (2010) found that local people in the Solomon Islands were able to accurately detect long-term ecological changes in seagrass meadows. In Fiji, previous studies have demonstrated the ability of local people to identify changes in local ecosystems and the species within them. For example, a 2010 study undertaken on Viti Levu and Vanua Levu islands found that local people were able to provide important information on the presence, behaviour and ecology of shark species (Rasalato et al. 2010). Another study found that communities in both Fiji and Vanuatu were able to develop effective coping strategies due to the passing down of knowledge and experiences with cyclones (McNamara and Prasad 2014). Owing to their local knowledge and ongoing interaction with mangroves, Fijians also hold valuable insights on how these environments and their resources have changed over time (Johnston 2014, 2015). Research that explores local Fijian knowledge of mangrove ecosystems could therefore provide significant contributions to improving current mangrove management and their long-term sustainability under a new climate regime.

Ecosystem-Based Adaptation to Climate Change

EbA has gained growing support as a more sustainable and cost-effective solution to responding to climate change in contrast to 'hard infrastructure' adaptation options such as sea walls, breakwaters and other forms of engineered coastal protection which are often proposed by policy-makers (Munang et al. 2013; Iacob et al. 2014; Reid and Shafiqul Alam 2016). According to Munang et al (2013), EbA offers a robust and adaptive strategy for reaching climate change adaptation and mitigation targets, as well as the opportunity to achieve local development goals. Although there is uncertainty surrounding the adaptive capacity of ecosystems under future climate conditions, they will be more resilient to climate change if they are kept in a healthy condition in which negative human impacts are reduced (Munang et al. 2013; Reid and Shafiqul Alam 2016). Indigenous knowledge performs a key role in the long-term success of EbA projects, as local and traditional communities can deliver valuable information on the functioning of local ecosystems, the various services they provide and how this has benefited local communities over time (Granek et al. 2010).

Some examples of EbA that have been implemented in the past include mangrove restoration to help buffer against storm surges and sea-level rise, watershed management to protect against droughts and floods, and the sustainable management of fisheries and forests to ensure food security (Chong 2014). In the Mekong Delta of Viet Nam, mangrove planting and rehabilitation involving local communities has taken place in order to increase local resilience to climate change (Schmitt et al. 2013). Ongoing monitoring and assessment of the project has showed promising results by enhancing the protection function of the mangrove forest belt, while also providing livelihoods for local people (Schmitt et al. 2013). Mangrove conservation and restoration has also become a particularly popular form of EbA in the Pacific region as the ability of mangroves to withstand harsh environmental conditions makes them an obvious choice for EbA in vulnerable coastal areas (DasGupta and Shaw 2017).

In Fiji, there have been multiple projects implemented by NGOs which take an ecosystem-based approach to climate change adaptation. In conjunction with the Global Environment Facility (GEF) and United Nations Environment Program (UNEP), WWF established the 'Building Mangrove Resilience Project' in 2010 which included the implementation of multiple projects in Fiji including Verata; Tikina Wai on Viti Levu and Kubulau on Vanua Levu (Heileman and Cabanban 2013). The main aim of the project was to build the capacity of local resource managers to assess the vulnerability of mangroves to climate change, and design adaptation strategies through coastal planning and collaboration with local communities (Ram 2010). This was part of a broader project called "Coastal Resilience to Climate Change" in which a series of pilot activities were implemented in Fiji, Tanzania and Cameroon (Ellison 2012). Although there is a lack of outcome details on the Fiji project specifically, WWF published the 'Climate Change Vulnerability Assessment and Adaptation Planning for Mangrove Systems' which drew on the collective experiences and lessons learnt from the pilot projects (Cook 2012; Ellison 2012).

'Mangroves for Fiji' is a privately owned ecosystem-based project involving the large-scale replantation of mangroves in villages on Viti Levu and Vanua Levu, Fiji. The project was initiated by Beqa Adventure Divers (BAD) in partnership with Projects Abroad as a carbon offsetting program in which the businesses intend to completely offset their emissions. BAD are sponsoring the replanting of 33 ha of mangrove forests per year in cooperation with local communities, NGOs and government partners (BAD 2015). In Vanua Levu, mangrove planting is overseen by local planter, Napolioni Yauraki, in the Naibulu Villages in which the proceeds contribute to the local church hall, and assistance with general family expenses (BAD 2015). Although 'Mangroves for Fiji' demonstrates an example of an inclusive, 'bottom-up' approach to EbA, details on project outcomes have not been provided.

Despite its growing popularity in international climate policy, there is still a lack of understanding of the barriers and limitations involved with EbA (Nalau et al. 2018a). A series of challenges often arise as a result of conflicting interests between stakeholders, poor governance, limited scientific data, and lack of political will (Granek et al.

2010; Aswani et al. 2012; Leslie et al. 2015). Previous ecosystem-based approaches tend to focus on protecting biodiversity and ecosystem functions, with a lack of consideration for the priorities of local communities in Pacific Island countries (Aswani et al. 2012). According to Wiener et al (2011), one of the major challenges with ecosystem-based management is that cultural diversity among stakeholders is often not assessed when integrating science management. To overcome this issue, projects conducted in partnership with local communities are likely to be more effective and sustainable than top-down approaches. This is especially the case in the Pacific Island region where local people tend to hold stewardship over a substantial amount of natural resources (Aswani et al. 2012).

In a recent review on EbA literature, Nalau et al (2018a) outlines a number of financial, institutional, social, cultural, physical, biological and knowledge constraints highlighted from previous studies. One of the key constraints found was a lack of monitoring and evaluation methodologies (Nalau et al. 2018a). Another common problem has been the failure to effectively respect local and Indigenous knowledge systems, and collaborate with local people. According to Leslie et al. (2015), a prominent strategy for ecosystem-based interventions in Fiji has been the initial collection of biological and social data by scientists, followed by the presentation of results to the community, and then finally, local knowledge comes into the equation after implementation. Other studies show that researchers often want to 'validate' traditional knowledge before it is recognised to be as important as Western scientific knowledge (Nalau et al. 2018a). This demonstrates that need for improved adaptation responses that put local, traditional and Indigenous knowledge systems at the forefront rather than considering them as an afterthought only when external parties deem them useful.

Method and Study Site

Sampling Strategy

Purposive sampling was used as the key sampling method for semi-structured interviews during this research. Purposive sampling is a non-random measure used to target certain individuals or groups that may have a unique, different or important perspective on the research topic (Robinson 2014; Suen et al. 2014). Owing to the nature of the study, the primary target was local people who visit mangroves and use their resources on a regular basis as they were considered more likely to have significant local knowledge of and experiences with mangroves. The 'snowballing' technique was also used after each interview by asking participants to recommend acquaintances who might have the knowledge or experience for the study (Robinson 2014). This method has been used by similar studies in the past as an effective tool for selecting local 'experts' on surrounding environments (Lauer and Aswani 2010; de Oliveira Braga and Schiavetti 2013).

Semi-structured Household Interviews

From October 2017 to December 2017, a total of 41 semi-structured household interviews were held with key knowledge holders across all six village sites. Semi-structured interviews are versatile and can produce in-depth insights (Creswell et al. 2003; Chen et al. 2016), offering the opportunity to improvise follow-up questions (Kallio et al. 2016). In contrast to surveys, this method is also more culturally appropriate as storytelling through verbal communication or *talanoa* is a common practice of information dissemination in Fiji. *Talanoa* is a concept prevalent in many Pacific Island societies which involves a formal or informal conversation, or exchange of ideas which is always carried out face-to-face (Vaioleti 2006). This method removes the distance between researcher and participant, and aims to build respect, trust and understanding and sharing (of stories, emotions, experiences) between those involved (Halapua 2000). Interviews were held in either the participant's house or community hall, each taking approximately 30–45 min. With the consent of participants, interviews were recorded so that they could be transcribed later and analysed. The interview guide involved mostly open-ended questions that allowed participants to tell their stories and elaborate on key issues as they saw fit (Bourke 2014).

All participants interviewed in this study were iTaukei (Indigenous Fijian). As shown in Table 1, the majority of participants (83%) were female and over 50 years old. Traditionally, Fijian women are typically the ones to collect *qari* (mudcrabs)

Table 1 Socio-demographic information of interview participants

Village	Bua	Dalomo	Denimanu	Koroinasolo	Navunievu	Tiliva
Total population size	324	140	175	248	223	75
No. of households	53	14	52	30	32	18
No. of participants	9	4	4	7	12	5
Gender						
Female	8	3	4	6	9	4
Male	1	1	0	1	3	1
Age group						
18–29	0	0	1	0	0	0
30–39	0	0	0	0	1	0
40–49	0	2	0	1	2	1
50–59	4	1	3	3	2	2
60–69	1	1	0	2	5	2
70–79	2	0	0	1	2	0
80+	2	0	0	0	0	0

and other resources from mangrove ecosystems, whereas men are usually responsible for travelling further to sea by boat to go fishing in outer coral reef areas (Ram-Bidesi 2015).

Participant Observation

Participant observation was used throughout interviews as this is an important method for understanding participant behaviours and what influences them (Dahlke et al. 2015). While interviews focus on talking and listening to people, participant observation takes this a step further by "watching, sensing, feeling, and being present with people" (Aagaard and Matthiesen 2015, p 41). Key notes were taken during interviews to document important observations and interactions between participants. In addition to this, transect walks were undertaken with participants through mangrove environments to enable a visual understanding of the location, abundance and diversity of local mangrove communities. This helped to contextualise the interviews by experiencing the mangrove environments personally and understanding which parts of the mangroves were being utilised. Informal observations were also made during fieldwork while staying in local villages to better understand local culture and how participants maintain their livelihoods.

Data Analysis

After data collection, the results from the interviews were translated, transcribed and analysed using the qualitative data research software, NVivo 11. NVivo has proven to be an effective and appropriate software for organising and analysing interview data in similar research projects (Gero et al. 2015; Magee et al. 2016). Thematic Content Analysis within NVivo was used to analyse the interview results in a systematic manner. This technique goes beyond just counting words or phrases within the data by identifying recurrent themes (Guest et al. 2011: 3). Using NVivo, initial codes were generated to organise interesting and relevant features of the data in a systematic manner. Transcripts were classified under participant attributes such as age, gender and village location. Queries were used to search for and identify any links between different codes. Field notes were also incorporated into the analysis process to fill in any missing gaps from the transcripts and ensure data triangulation.

Research Limitations

Due to the nature of the study, there was an uneven distribution of female and male participants. The majority of the participants were women as they were considered key knowledge holders by the community due to their traditional role in utilising mangroves. This gender bias means we were unable to effectively compare male and female responses equally. The position of the researcher as a young, Australian female from an outside country would have influenced the data collection and analysis to a certain extent. With this position, comes the recognition that the researcher has their own potential biases, ideas and interpretations of the data, which may impact the study (Cohen et al. 2011). Although the positionality of the researcher cannot be removed from the study, it is important to take a reflexive approach when undertaking ethical research. The language barrier is another minor limitation of this study. Although there was a local translator present for all interviews, there is always the risk that some questions and responses will be misinterpreted.

Study Site

Fiji is situated in the Southwest Pacific Ocean and encompasses a total of 332 islands, stretching over 18,270 km^2. The 2017 census shows that the population of Fiji is around 884,887 (FBoS 2018). Viti Levu (10,388 km^2) and Vanua Levu (5587 km^2) are the two largest and most developed islands. All six study sites were located within the Bua province, one of Fiji's fourteen provinces, on western Vanua Levu Island. This province covers an area of 1378 km^2 and is home to an extensive mature mangrove area (Nair 2004). The 2017 census indicated a population of 15,466 people living in Bua Province (FBoS 2018). While the population of Bua province is mainly Fijian *iTaukei*, there is a small proportion of Fijian-Indians as well; and the majority of the population identify as Christian (Bakker 2014). Figure 1 illustrates the locations of the villages of Bua, Dalomo, Koroinasolo, Navunievu, Tiliva and Denimanu located on Yadua Island in the southwest corner.

Valuing Mangrove Ecosystems

This section outlines the ways in which iTaukei communities perceive the value of mangroves and their ecosystem services. It is important to understand these values as they usually underpin the reasoning for traditional resource management techniques. Participants provided an extensive range of benefits they retrieved from mangroves and associated resources, highlighting their role in sustaining iTaukei livelihoods.

Participants (n = 17) indicated that their community had utilised mangrove wood for building products including house posts, bridges, and canoes. Mangrove wood was more commonly used in past times to build the traditional Fijian '*bure*' (thatched house) which are not so commonly constructed now that alternative building materials are available. When asked if the local mangrove uses had changed over time, the use of the *bure* was the only major change observed by participants. Firewood (n = 31) was also commonly cited as an important use of mangrove wood by local people

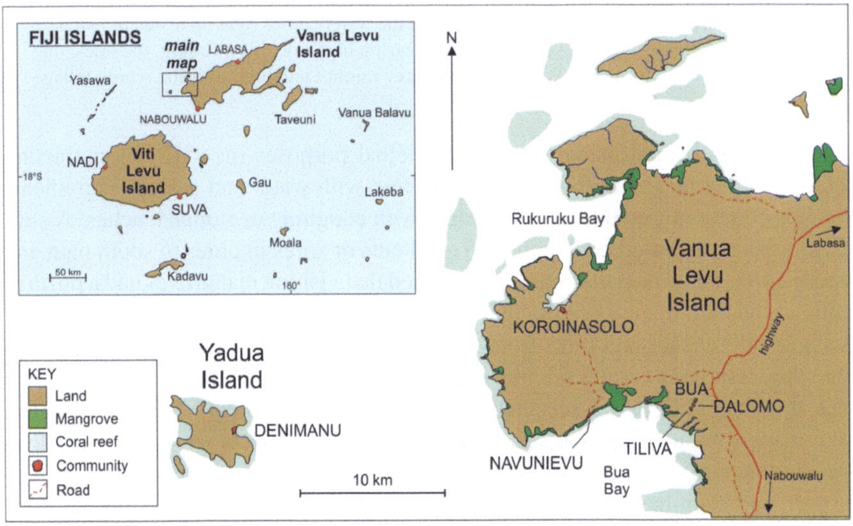

Fig. 1 Map of study site: Bua, Dalomo, Denimanu, Koroinasolo, Navunievu and Tiliva villages, Vanua Levu Island, Fiji

who commonly cook food with woodstoves at home, with some noting they would only use the 'old' and 'dying' mangroves for wood instead of the new ones (n = 3). The majority of participants also mentioned the value of mangrove resources for local subsistence (n = 33). Almost all participants received direct financial benefits from mangrove ecosystems (n = 39), which was mainly attributed to the selling of resources collected from mangroves including mudcrabs and fish.

Participants (n = 20) valued mangroves for their role in providing habitat for marine organisms such as mudcrabs and fish species, with some mentioning their role as a nursery for these species to lay their eggs (n = 7). Participants also recognised the role of mangroves in preventing soil or coastal erosion (n = 12), reducing pollution from entering waterways (n = 1) and even oxygen production (n = 1). In addition to their array of benefits and everyday uses for local people, some participants also highlighted the role of mangroves in protection against the impacts of climate change. For instance, coastal protection was mentioned as a key benefit of mangroves (n = 12), in terms of protecting the coastline from extreme weather events and sea-level rise. Of these participants, most were from Navunievu (n = 6) or Denimanu (n = 2), both villages where houses are extremely close to the shoreline. It was clear that the livelihoods of these two communities were being compromised by the encroaching impacts of climate change. Two participants from Navunievu highlighted the role of mangroves in protecting against sea-level rise in the following quotes:

> We need to replant more mangroves. We are moving to another higher place because of the waves. The waves come straight into the village. It is better for us to plant the mangroves to stop the waves coming into the village.—P27

> We need to keep on replanting. During high tide, the water comes right up inside the village. If you are here this afternoon, you will notice. During high tide this morning, the tide came right up to the breadfruit tree. When there is heavy rainfall and high tide, the whole village is flooded. The seawall is there but it is broken.—P37

Participants valued mangroves for medicinal purposes (n = 15) by extracting liquid from the *titi* (mangrove seedling) to mix with water and various ingredients which is commonly used to assist children with coughing or stomach aches. A similar process is also used and applied to open cuts or sores in order to sooth pain and prevent infection. One participant also claimed that visiting mangroves had a positive influence on their mental health. Some participants (n = 5) indicated that they would use smaller branches and roots from the mangroves, and tie them together to create what they referred to as a 'crab trap' which was used to carry mudcrabs back home once they had collected a sufficient number. Other uses included using branches from mangroves to create baskets, toys, and even home décor (n = 3). 'Mangrove dye' was mentioned as a common resource for painting, dyeing material, furniture varnish, and even hair colouring (n = 17). This process involved boiling the bark of the mangroves until it turns into red, black or brown-coloured dye (see Fig. 2). Materials from mangrove wood are also commonly used for making *salusalu* (traditional garlands), which are worn by local people during ceremonies and other special occasions (n = 6).

When asked to rate the importance of mangroves on a scale of 1 to 3 (1 = not important at all, 2 = somewhat important and 3 = very important), all participants chose '3' indicating that mangroves were highly important to them. When asked to compare this importance with other ecosystems such as coral reefs and forests, most

Fig. 2 Traditional Fijian artwork made with mangrove dye, Bua village

participants responded with 'they are all the same' (n = 31). This is consistent with the idea that Pacific Island peoples mostly feel that they are part of nature rather than separate from it (Nunn et al. 2016). These perspectives on the value and importance of mangrove ecosystems inform the motivation behind iTaukei communities to implement the management strategies discussed in the following section.

iTaukei Mangrove Management Strategies

Throughout Fiji, *iTaukei* people have customary ownership of wetland areas where management needs to abide by traditional customs and meet the social, economic and cultural needs of communities (Ellison 2009). In this traditional system, tribal lands and coastal waters are interpreted by Indigenous Fijians as a single unit called '*vanua*', which can also be divided into multiple sub-clans or 'mataqali' (Veitayaki 2002). Each vanua has a Chief who holds ownership and responsibility over the land, water, resources and people within the vanua (Cooke et al. 2002). For iTaukei people, the concept of vanua symbolises the interconnectivity of community and the environment, in which nature is perceived as part of the self (Long 2017). This section outlines the techniques and strategies used by participants to conserve, protect and manage mangrove ecosystems at a community level. It was found that all villages had some sort of traditional management techniques in place.

Protected *Tabu* Areas

The most common management technique that was mentioned in interviews was the implementation of '*tabu*' areas (n = 14) in Bua (n = 5), Dalomo (n = 1), Denimanu (n = 2), Koroinasolo (n = 3) and Navunievu (n = 2) villages. These are essentially protected areas traditionally used throughout Fiji to temporarily restrict the collection of resources from a given area, usually with permission of the Ratu (village Chief). The tabu can apply to a particular part of land or sea, or a specific species and in some cases, is permanently reserved for chiefs only (Siwatibau 1984). The rules and regulations around this management technique differed between villages but there was a general consensus that this method had been successful in preserving and restoring marine resources. It is common in Fiji for fishing grounds owners to implement this technique for special occasions including weddings, births or death-related ceremonies (Ravuvu 1983). For example, in Bua village, participants indicated that it was tradition for a tabu to be put in place for mangrove or coral reef areas every time someone in the village passed away. One participant from Bua provides insight into this in the following passage:

> It is our culture. It is a tradition. When someone dies in the village, we have a tabu for dogo (mangroves) so we can go there for fishing and catch plenty of fish. For 100 days, we do a tabu for the river and the mangrove – then after 100 days, we can do the fishing.—P8

Another Bua participant corroborates this by saying '*the river and the mangroves cannot be touched during this time—it is a tradition*" (P6). This emphasises the cultural significance of adhering to traditions which involve managing mangroves and associated resources for iTaukei communities. In Koroinasolo, a village meeting was held to discuss a tabu strategy once concerns were raised about depleting marine resources. From this meeting, it was decided that a 5-year tabu was put into place which begins from the coral reefs and goes through to the mangroves (P20). One participant retells the benefit of this management strategy:

> It has been in place for a few years. Before, there was no *bêche-de-mer (sea cucumber)* just around the seawall near the village. From the tabu, we realised the *bêche-de-mer* was coming back. So there is a good reason for having the tabu. Even the fish are growing bigger now after the tabu.—P20

In Denimanu village, local people had put in a type of rotating system around their island for tabu areas. One participant explains this: "*There is a tabu every 5 years for the fish, land, mangroves, everything! We rotate the tabu area (around the island) every 5 years. It comes around.*" This demonstrates the nuances in traditional management techniques between villages, with each community taking their own unique approach to the implementation of tabu areas. The villages did, however, share a common understanding of the importance in collectively preserving resources and allowing ecosystems to replenish over time through cultural traditions. This mirrors previous studies that state Fijian customary livelihoods practice enables the preservation of specific techniques and skills that are unique to a particular village which have been passed down across generations (Derrick 1957; Seruvakula 2000; Movolo et al. 2017).

Community Replanting of Mangroves

It was found that community mangrove replantation was a common form of mangrove management in some villages (n = 12), with participants from Navunievu (n = 7), Bua (n = 4), and Koroinasolo (n = 1) villages all indicating that they had successfully planted mangroves at some point. Four of these participants said that they '*used to*' replant mangroves, implying that replantation was not as common anymore. Two participants from Navunievu village said they took part in the mangrove replanting, but only when they were still at a young age (P26, P37). When asked about who decided to replant the mangroves, these participants said it was a community decision and that only local people were involved. Another participant from Navunievu said '*we used to plant mangroves to protect the village from the high tide*' as they pointed to a healthy nearby mangrove community (P27). Another Navunievu participant said they had recently planted over 40 mangroves but local children had uprooted them (P30). Furthermore, they also said there had been discussion of a community plan to plant more mangroves which was yet to be implemented (P30).

Participants from Bua indicated that they had planted mangroves to prevent soil erosion occurring along the Bua River (n = 3), stating that mangroves '*keep the soil*

Fig. 3 Bua River, Bua village, Vanua Levu Island

from washing away' (P6). When asked about who taught them how to replant the mangroves, one participant from Tiliva village said that their grandparents taught them how to put the mangroves in the soil (P12). A participant from Bua village said they learnt how to replant mangroves by watching people planting small mangroves near the sea (P4). Another participant from Bua (P6) said '*when we see the seeds of the mangroves that bare out from the tree, we get those seeds and replant them*' indicating that this knowledge stems from ongoing interactions and experience with mangroves. The same participant also added that they used to replant mangroves in Tiliva village but when the mangroves died, they began using stones to prevent soil erosion along the river instead (P6). Figure 3 shows the combination of stones and mangroves along the riverbanks of Bua River. Other participants indicated that they had been struggling with replantation in Navunievu and Denimanu villages. Moreover, all participants from Denimanu village expressed a strong desire for external assistance with mangrove replantation (n = 4).

Knowledge Sharing

Another common management strategy mentioned by participants was the process of knowledge transmission. In Fiji, traditional and Indigenous knowledge is typically transferred both orally and through observations of cultural practices and regular day-to-day activities (Ravuvu 1983, 1987; Seruvakula 2000). When asked about where their knowledge on the use, value and importance of mangrove ecosystems had come from, most participants said that this had been passed on by their parents (n = 17), friends (n = 5) or grandparents (n = 2). The remaining participants said they had learnt about mangroves through first-hand experience (n = 17). This process is vital for informing management strategies such as tabu areas. The shared knowledge on the values of mangroves forms the basis of the traditional management techniques designed to ensure they are maintained sustainably over time. When participants were asked if they had passed on their knowledge to anyone else themselves, most responded with 'yes, they had' or 'yes, they will' (n = 39). Moreover, some participants also specified that they had stressed the importance of mangroves through this knowledge sharing, encouraging future generations to protect and take care of these valuable ecosystems (n = 5).

Future Recommendations: A Way Forward for Ecosystem-Based Adaptation

The internal management strategies outlined by iTaukei communities in this study all align with the principles of EbA through the acts of restoring, conserving and enhancing the capacity of mangrove ecosystems to provide services for local people. Like many other Indigenous societies around the world, iTaukei communities have been successfully employing these principles long before the term EbA emerged in Western scientific literature (Veitayaki et al. 2018). This paper argues that future EbA projects should be people-centred, building on local and Indigenous knowledge systems that are relevant and appropriate for climate change adaptation. In contrast to top-down approaches, scientific knowledge should be integrated to fill in any gaps, along with financial assistance and additional resources where required. In light of potential limitations and unsustainable practices, local and Indigenous knowledges should be carefully evaluated, and blended with external knowledge when necessary (Remling and Veitayaki 2016). Veitayaki et al. (2018) suggests that some Fijian communities require training, awareness and support on specific modern challenges including how to convince people to support appropriate marine resource use that complements their customary system.

Initial consultation and ongoing collaboration with local communities throughout EbA is essential for external parties to understand and appreciate the value of these knowledge systems. It is necessary for all stakeholders involved to discuss and develop project outcomes together to ensure that everyone is given the opportunity to effectively communicate their needs, values and interests in the early stages of decision-making. Conflicting perceptions between stakeholders can restrict the collaboration of knowledge sources so it is important to maintain regular interactions with all parties involved (Wiener et al. 2011). When local people see that their values and ideas are being represented in a project, they are more likely to trust the process, to support its implementation, and to sustain it (Espinosa-Romero et al. 2011), the latter being a hugely important issue in community uptake of external interventions (Nunn 2009). This also allows local people to feel a sense of ownership over projects, encouraging them to carry on with adaptation long after external parties have intervened. 'Lack of ownership' has been a significant challenge for climate change adaptation in Pacific Island communities, which is mostly due to the majority of adaptation plans being presented in English, and communicated through inappropriate cultural contexts (Nunn 2009, 2013).

Moreover, an adaptive management approach should be taken to enable ongoing monitoring and assessments of adaptation projects. Too often, external parties do not maintain ongoing partnerships with local people or efficiently evaluate project outcomes (Nalau et al. 2018a). Taking an adaptive management approach to EbA allows for incremental adjustments to be made over time based on careful monitoring of key parameters used to measure the success of a given project (Susskind et al. 2012). This is paramount for EbA that is sustainable and effective in the long-term, ensuring that the needs, interests and values of local communities continue to be addressed over time. Many documents do not offer explicit examples of how EbA has been employed in the past, or detail any of the challenges involved (Nalau et al. 2018a). It is recommended that project outcomes should be documented and made available for policymakers, scientists, academics and other relevant stakeholders to consider during the planning of ecosystem-based adaptation programs. This will provide valuable insight into the success stories and lessons learnt of previous projects, giving decision-makers a guideline on how to improve future approaches to ecosystem-based adaptation.

Conclusion

As the impacts of climate change begin to take their toll on Pacific Island countries, sustainable and effective adaptation options that address the long-term needs of coastal communities are urgently required. Past ecosystem-based approaches to climate change adaptation have failed to adequately respect and appreciate Indigenous peoples and their knowledge, values and rights. It is crucial that Indigenous and local communities are considered key decision makers in future interventions to ensure that their knowledge, interests and rights are respected in the process. Findings from this case study demonstrate that iTaukei communities possess valuable knowledge on the sustainable management of mangrove ecosystems, which has been passed down by generations through oral transmission. This demonstrates that Indigenous and traditional communities have been utilising the principles of EbA to sustainably manage resources for countless generations. With this understanding in mind, external parties should advocate for a balance of both Indigenous and scientific knowledge in future ecosystem-based adaptation options so that people living in island nations can not only survive, but thrive. The findings of this study can be utilised by policymakers, academics, scientists, and relevant stakeholders in the planning and development of EbA solutions, not only in Fiji, but in other island communities that are also beginning to feel the impacts of climate change. If employed correctly, EbA has the power to conserve ecosystems, and improve the livelihoods of local people in a cost-effective and sustainable manner.

Acknowledgements We would like to thank the local communities of Bua, Dalomo, Denimanu, Koroinasolo, Navunievu and Tiliva for warmly welcoming the researchers into their villages for the duration of data collection. We are eternally grateful to the participants who were kind enough to share their knowledge and experiences that form the basis of this article. A special thank you to the Ramasima family from Bua village for being wonderful hosts, and for Titilia and Metuisela Mocevakaca for their assistance with translation and facilitating data collection. I would also like to thank my PhD advisers for their ongoing research guidance, and contributions to this paper. We are also grateful to the School of Earth and Environmental Sciences, University of Queensland for funding this fieldwork research.

References

Aagaard J, Matthiesen N (2015) Methods of materiality: participant observation and qualitative research in psychology. Qual Res Psychol 13(1):1–14. https://doi.org/10.1080/14780887.2015.1090510

Agrawala S, Ota T, Risbey J, Hagenstad M, Smith J, van Aalst M, Koshy K, Prasad B (2003) Development and climate change in Fiji: focus on coastal mangroves. OECD, Paris, France

Anchorage Declaration (2009) The anchorage declaration of indigenous peoples on climate change, Anchorage. Alaska, USA 20–24 April 2009

Aswani S, Christie P, Muthiga NA, Mahon R, Primavera JH, Cramer LA, Hacker S (2012) The way forward with ecosystem-based management in tropical contexts: reconciling with existing management systems. Marine Policy 36(1):1–10. https://doi.org/10.1016/j.marpol.2011.02.014

Bakker ML (2014) Bua: a profile of the demographic and socio-economic characteristics of the population of the province based on the 1996 and 2007 census data. Bureau of Statistics, Suva, Fiji

Beqa Adventure Divers (BAD) (2015) Mangroves for Fiji. From: http://mangrovesforfiji.com/about-us/

Bhattarai B, Giri C (2011) Assessment of mangrove forests in the Pacific region using Landsat imagery. J Appl Remote Sens 5(1):053509-053509-053511. https://doi.org/10.1117/1.3563584

Blankespoor B, Dasgupta S, Lange G-M (2017) Mangroves as a protection from storm surges in a changing climate. J Human Environ 46(4):478–491. https://doi.org/10.1007/s13280-016-0838-x

Bourke B (2014) Positionality: reflecting on the research process. Qual Rep 19(33):1–9

Bridges KW, McClatchey WC (2009) Living on the margin: ethnoecological insights from Marshall Islanders at Rongelap atoll. Glob Environ Change 19(2):140–146

Campbell J (2009) Islandness: vulnerability and resilience in Oceania. Shima: Int J Res Island Culture 3(1):85–97

Campbell J, Barnett J (2010) Climate change and small island states: power, knowledge and the South Pacific. Routledge, London

Carranza Ortiz G, Gómez-Mendoza L, Caetano E, Infante Mata D (2018) Vulnerability of human communities in Mexican mangrove ecosystems: an ecosystem-based adaptation approach. Investigaciones Geograficas 95(95): 18 pp

Chand SS, Chambers LE, Waiwai M, Malsale P, Thompson E (2014) Indigenous knowledge for environmental prediction in the Pacific Island countries. Weather Clim Soc 6(4):445–450. https://doi.org/10.1175/WCAS-D-13-00053.1

Chen Y, Fang W, Li L, Morissey P, Nie C (2016) Social attitudes in contemporary China. Routledge, Abingdon, New York

Chong J (2014) Ecosystem-based approaches to climate change adaptation: progress and challenges. Int Environ Agreements Politics Law Econ 14(4):391–405. https://doi.org/10.1007/s10784-014-9242-9

Cohen L, Manion L, Morrison K (2011) Research methods in education, 7th edn. Routledge, Abingdon, London, New York

Cook J (2012) Adaptation stories from around the globe. Increasing Mangrove resilience to climate change. Retrieved from: http://www.climateprep.org/stories/2016/3/7/increasing-mangrove-resilience-toclimate-change

Cooke AJ, Polunin NVC, Moce K (2002) Comparative assessment of stakeholder management in traditional Fijian fishing-grounds. Environ Conserv 27(3):291–299

Creswell JW, Plano Clark VL, Gutmann ML, Hanson WE (2003) Advanced mixed methods research designs. Handbook of mixed methods in social and behavioral research. Sage Publications, USA, pp 209–240

Dahlke S, Hall W, Phinney A (2015) Maximizing theoretical contributions of participant observation while managing challenges. Qual Health Res 25(8):1117–1122. https://doi.org/10.1177/1049732315578636

DasGupta R, Shaw R (2017) Participatory Mangrove management in a changing climate. Springer, Tokyo

de Oliveira Braga H, Schiavetti A (2013) Attitudes and local ecological knowledge of experts fishermen in relation to conservation and bycatch of sea turtles (reptilia: testudines), Southern Bahia, Brazil. J Ethnobiol Ethnomed 9(1):15. https://doi.org/10.1186/1746-4269-9-15

Dean A, Green D, Nunn PD (2017) Too much sail for a small craft? Donor requirements, scale, and capacity discourses in Kiribati. Island geographies: essays and conversations. Routledge, Abingdon, pp 67–88

Derrick RA (1957) A history of Fiji. Government Press, Suva

Dreher T, Voyer M (2015) Climate refugees or migrants? Contesting media frames on climate justice in the Pacific. Environ Commun 9(1):58–76

Ellison JC (2009) Wetlands of the Pacific Island region. Wetlands Ecol Manage 17(3):169–206. https://doi.org/10.1007/s11273-008-9097-3

Ellison JC (2012) Climate change vulnerability assessment and adaptation planning for Mangrove systems. World Wildlife Fund (WWF), Washington, DC, 130 pp

Espinosa-Romero MJ, Chan KMA, McDaniels T, Dalmer DM (2011) Structuring decision-making for ecosystem-based management. Mar Policy 35(5):575–583. https://doi.org/10.1016/j.marpol.2011.01.019

Fiji Bureau of Statistics (FBoS) (2018) Census 2017 of population & housing. From: https://www.statsfiji.gov.fj/. Last accessed 27th Jan 2019

Friedlander AM (2018) Marine conservation in Oceania: past, present, and future. Mar Pollut Bull 135:139–149

Gero A, Fletcher S, Rumsey M, Thiessen J, Kuruppu N, Buchan J, Willetts J (2015) Disasters and climate change in the Pacific: adaptive capacity of humanitarian response organizations. Clim Dev 7(1):35–46. https://doi.org/10.1080/17565529.2014.899888

Gilman E, Van Lavieren H, Ellison J, Jungblut V, Wilson L, Areki F, Brighouse G, Bungitak J, Dus E, Henry M, Kilman M (2016) Pacific Island Mangroves in a changing climate and rising Sea-UNEP regional seas reports and studies, no. 179

Granek EF, Polasky S, Kappel CV, Reed DJ, Stoms DM, Koch EW, Kennedy CJ, Cramer LA, Hacker SD, Barbier EB, Aswani S, Ruckelshaus M, Perillo GME, Silliman BR, Muthiga N, Bael D, Wolanski E (2010) Ecosystem services as a common language for coastal ecosystem-based management. Conserv Biol 24:207–216

Guest G, MacQueen KM, Namey EE (2011) Introduction to applied thematic analysis. In: Applied thematic analysis. SAGE Publications, Thousand Oaks, 3 pp

Halapua S (2000) Talanoa process: the case of Fiji. East West Centre, Hawaii

Hau'ofa E (1998) The Oceania in Us. Contemp Pac 10(2):391–410

Hay JE (2013) Small island developing states: coastal systems, global change and sustainability. Sustain Sci 8(3):309–326

Heileman S, Cabanban A (2013) Terminal evaluation of the UNEP GEF medium sized project: coastal resilience to climate change: developing a generalizable method for assessing vulnerability and adaptation of Mangroves and associated ecosystems (E. Office, Trans.). In: UNEP (Ed.). WWF US

Iacob O, Rowan JS, Brown I, Ellis C (2014) Evaluating wider benefits of natural flood management strategies: an ecosystem-based adaptation perspective. Hydrol Res 45(6):774–787. https://doi.org/10.2166/nh.2014.184

Intergovernmental Panel on Climate Change (IPCC) (2014) Climate change 2014: synthesis report. Contribution of Working Groups I, II and III to the Fifth Assessment Report of the Intergovernmental Panel on Climate Change [Core Writing Team, R.K. Pachauri and L.A. Meyer (eds.)]. IPCC, Geneva, Switzerland, 151 pp

Ireland P (2012) Nepalganj, the centre of the world: local perceptions of environmental change and the roles of climate-change adaptation actors. Local Environ 17(2):187–201. https://doi.org/10.1080/13549839.2012.660907

Jetñil-Kijiner K (2017) Iep jaltok: poems from a Marshallese daughter. The University of Arizona Press, Tucson

Johannes RE, Yeeting B (2001) I-Kiribati knowledge and management of Tarawa's Lagoon resources. Atoll Res Bull 481–493:20–24

Johnston I (2014) Disaster management and climate change adaptation: a remote island perspective. Disaster Prev Manage 23:123–137. https://doi.org/10.1108/dpm-062013-0096

Johnston I (2015) Traditional warning signs of cyclones on remote islands in Fiji and Tonga. Environ Hazards 14(3):210–223. https://doi.org/10.1080/17477891.2015.1046156

Kallio H, Pietilä A-M, Johnson M, Kangasniemi M (2016) Systematic methodological review: developing a framework for a qualitative semi-structured interview guide. J Adv Nurs 72(12):2954–2965. https://doi.org/10.1111/jan.13031

Kathiresan K, Bingham BL (2001) Biology of mangroves and mangrove ecosystems. Adv Mar Biol 40:81–251

Koya CF, Vaka'uta L, Lagi R (2018) Reflections from Oceania on indigenous epistemology, the ocean and sustainability. Tidalectics: imagining an Oceanic worldview through art and science. MIT Press, Cambridge, pp 127–132

Lal P (2003) Economic valuation of mangroves and decision-making in the Pacific. Ocean Coast Manage 46(9–10):823–844. https://doi.org/10.1016/S09645691(03)00062-0

Lauer M, Aswani S (2010) Indigenous knowledge and long-term ecological change: detection, interpretation, and responses to changing ecological conditions in Pacific Island communities. Environ Manage 45(5):985–997. https://doi.org/10.1007/s00267-0109471-9

Lefale PF (2010) Ua 'afa le Aso Stormy weather today: traditional ecological knowledge of weather and climate. The Samoa experience. Clim Change 100(2):317–335

Leslie H, Sievanen L, Crawford TG, Gruby R, Villanueva-Aznar HC, Campbell LM (2015) Learning from ecosystem-based management in practice. Coast Manage 43(5):471–497. https://doi.org/10.1080/08920753.2015.1051424

Long M (2017) Vanua in the anthropocene: relationality and Sea level rise in Fiji. symplokē 26(1–2):51–70

Magee AD, Verdon-Kidd DC, Kiem AS, Royle SA (2016) Tropical cyclone perceptions, impacts and adaptation in the Southwest Pacific: an urban perspective from Fiji, Vanuatu and Tonga. Nat Hazards Earth Syst Sci 16(5):1091–1105. https://doi.org/10.5194/nhess-16-1091-2016

McLeod E, Salm RV (2006) Managing mangroves for resilience to climate change. IUCN Resilience Science Group Working Paper Series - No 2. World Conservation Union (IUCN). Gland, Switzerland, 64 pp

McNamara KE, Prasad SS (2014) Coping with extreme weather: communities in Fiji and Vanuatu share their experiences and knowledge. Clim Change 123(2):121–132. https://doi.org/10.1007/s10584-013-1047-2

McNaught R, Warrick O, Cooper A (2014) Communicating climate change for adaptation in rural communities: a Pacific study. Reg Environ Change 14(4):1491–1503. https://doi.org/10.1007/s10113-014-0592-1

Movono A, Dahles H, Becken S (2017) Fijian culture and the environment: a focus on the social and ecological connectedness of Fijian culture. J Sustain Tourism 26(3):451–469. https://doi.org/10.1080/09669582.2017.1359280

Munang R, Thiaw I, Alverson K, Mumba M, Liu J, Rivington M (2013) Climate change and Ecosystem-based adaptation: a new pragmatic approach to buffering climate change impacts. Curr Opin Environ Sustain 5(1):67–71. https://doi.org/10.1016/j.cosust.2012.12.001

Nair V (2004) Setting priorities for marine conservation in the Fiji Islands Marine ecoregion. WWF SPPO, 6 pp

Nalau J, Becken S, Mackey B (2018a) Ecosystem-based adaptation: a review of the constraints. Environ Sci Policy 89:357–364

Nalau J, Becken S, Schliephack J, Parsons M, Brown C, Mackey B (2018b) The role of indigenous and traditional knowledge in ecosystem-based adaptation: a review of the literature and case studies from the Pacific Islands. Weather Clim Soc 10(4):851–865

Noble IR, Huq S, Anokhin YA, Carmin J, Goudou D, Lansigan FP, Osman-Elasha B, Villamizar A (2014) Adaptation needs and options. In: Climate change 2014: impacts, adaptation, and vulnerability. Part A: global and sectoral aspects. contribution of working group II to the fifth assessment report of the intergovernmental panel on climate change [Field, C.B., V.R. Barros, D.J. Dokken, K.J. Mach, M.D. Mastrandrea, T.E. Bilir, M. Chatterjee, K.L. Ebi, Y.O. Estrada, R.C. Genova, B. Girma, E.S. Kissel, A.N. Levy, S. MacCracken, P.R. Mastrandrea, and L.L. White (eds.)]. Cambridge University Press, Cambridge and New York, pp 833–868

Nunn PD (2007) Climate, environment, and society in the Pacific during the last Millennium. In: Patrick DN (ed) Developments in Earth and environmental sciences, vol 6. Elsevier, Amsterdam, p v-302

Nunn PD (2009) Responding to the challenges of climate change in the Pacific Islands: management and technological imperatives. Clim Res 40(2–3):211–231. https://doi.org/10.3354/cr00806

Nunn PD (2013) The end of the Pacific? Effects of sea level rise on Pacific Island livelihoods. Singap J Trop Geogr 34(2):143–171. https://doi.org/10.1111/sjtg.12021

Nunn PD, Mulgrew K, Scott-Parker B, Hine DW, Marks ADG, Mahar D, Maebuta J (2016) Spirituality and attitudes towards Nature in the Pacific Islands: insights for enabling climate-change adaptation. Clim Change 136(3–4):477–493. https://doi.org/10.1007/s10584-016-1646-9

Percival GS (2008) An assessment of Indigenous Environmental Knowledge (IEK) in the Pacific region to improve resilience to environmental change. The Climate Change Research Centre. University of New South Wales, Kensington

Ram R (2010) Country paper of Fiji Islands—water resource management for responding to climate change. Land & Water Resource Management Division, Department of Agriculture, Ministry of Primary Industries, Fiji Islands

Ram-Bidesi V (2015) Recognizing the role of women in supporting marine stewardship in the Pacific Islands. Mar Policy 59:1–8

Rasalato E, Maginnity V, Brunnschweiler JM (2010) Using local ecological knowledge to identify shark river habitats in Fiji (South Pacific). Environ Conserv 37(1):90–97. https://doi.org/10.1017/S0376892910000317

Ravuvu A (1983) Vaka i Taukei: The Fijian way of life. Institute of Pacific Studies of the University of the South Pacific, Suva, p 87

Ravuvu AD (1987) The Fijian ethos. Institute for Pacific Studies, University of the South Pacific, Suva

Reid H, Swiderska K (2008) Biodiversity, climate change and poverty: exploring the links. IIED Briefing Paper. International Institute for Environment and Development

Reid H, Shafiqul Alam S (2016) Ecosystem-based approaches to adaptation: evidence from two sites in Bangladesh. Clim Dev 1–19. https://doi.org/10.1080/17565529.2016.1167663

Remling E, Veitayaki J (2016) Community-based action in Fiji's Gau Island: a model for the Pacific? Int J Clim Change Strat Manage 8(3):375–398

Roberts D, Boon R, Diederichs N, Douwes E, Govender N, Mcinnes A, Mclean C, O'Donoghue S, Spires M (2012) Exploring ecosystem-based adaptation in Durban, South Africa: "learning-by-doing" at the local government coal face. Environ Urbanization 24(1):167–195

Robinson O (2014) Sampling in interview-based qualitative research: a theoretical and practical guide. Qual Res Psychol 11(1):25–41. https://doi.org/10.1080/14780887.2013.801543

Salick J, Ross N (2009) Traditional peoples and climate change. Glob Environ Change 19(2):137–190

Schmitt K, Albers T, Pham TT, Dinh SC (2013) Site-specific and integrated adaptation to climate change in the coastal mangrove zone of Soc Trang Province, Viet Nam. J Coast Conserv 17(3):545–558. https://doi.org/10.1007/s11852-013-0253-4

Seddon N, Reid H, Barrow E, Hicks C, Hou-Jones X, Kapos V, Rizvi AR, Roe D (2016) Ecosystem-based approaches to adaptation: strengthening the evidence and informing policy: research overview and overarching questions. IIED, London

Seruvakula SB (2000) Bula vakavanua. Institute for Pacific Studies, University of the South Pacific, Suva

Sierra-Correa PC, Cantera Kintz JR (2015) Ecosystem-based adaptation for improving coastal planning for sea-level rise: a systematic review for mangrove coasts. Mar Policy 51:385–393. https://doi.org/10.1016/j.marpol.2014.09.013

Siwatibau S (1984) Traditional environmental practices in the South Pacific—case study of Fiji. Ambio 13(5–6):365–368

Spalding M, Kainuma M, Collins L (2010) World atlas of Mangroves. Taylor and Francis, USA, p 336

Spires M, Shackleton S, Cundill G (2014) Barriers to implementing planned community based adaptation in developing countries: a systematic literature review. Clim Dev 6(3):277–287. https://doi.org/10.1080/17565529.2014.886995

Suen LJ, Huang HM, Lee HH (2014) A comparison of convenience sampling and purposive sampling. Hu Li Za Zhi 61(3):105–111. https://doi.org/10.6224/JN.61.3.105

Susskind L, Camacho AE, Schenk T (2012) A critical assessment of collaborative adaptive management in practice. J Appl Ecol 49(1):47–51

Taylor M, McGregor A, Dawson B (2016) Vulnerability of Pacific Island agriculture and forestry to climate change. Secretariat of the Pacific Community, Noumea, New Caledonia

Techera EJ, Troniak S (2009) Marine protected areas policy and legislation gap analysis, Fiji Islands

United Nations (UN) (1992) Convention on Biological Diversity. Article 8(j). UN. In: Convention on Biological Diversity

United Nations (UN) (2012) Knowledge for the twenty-first century: indigenous knowledge, traditional knowledge, science and connecting diverse knowledge systems. UNEP/IPBES.MI/2/INF/9. from: https://www.ipbes.net/system/tdf/downloads/UNEP_IPBES_MI_2_INF_9_EN.pdf?file=1&type=node&id=14024

Vaioleti TM (2006) Talanoa research methodology: a developing position on Pacific research. Waikato J Educ 12(1)

Veitayaki J (2002) Taking advantage of indigenous knowledge: the Fiji case. Int Soc Sci J 54(173):395–402

Veitayaki J (2004) Building bridges: the contribution of traditional knowledge to ecosystem management and practices in Fiji. Bridging scales and epistemologies: linking local knowledge and global science in multi-scale assessments', Alexandria, Egypt, 17–20

Veitayaki J, Waqalevu V, Varea R, Rollings N (2017) Mangroves in small Island development states in the Pacific: an overview of a highly important and seriously threatened resource. In: DasGupta R, Shaw R (eds) Participatory Mangrove management in a changing climate: perspectives from the Asia-Pacific. Springer, Tokyo, pp 303–327

Veitayaki J, Ledua E, Nakoro A, Hong HP, Han DP, Moon S, Breckwoldt A (2018) Future use of past practices: policy implications of insights from two community-based marine resources management initiatives in Fiji. In: Chircop A, Coffen-Smout S, McConnell M (eds) Ocean yearbook, vol 32, pp 376–406

Voi M (2000) Vaka Moana—the ocean roads. Culture and sustainable development in the Pacific. ANU E Press and Asia Pacific Press, Canberra, 207 pp

Wiener CS, Rivera MA, Toonen RJ, Leong JAC, Kosaki RK, Karl S, Keller K, Johnson H (2011) Creating effective partnerships in ecosystem-based management: a culture of science and management. J Mar Biol, 8 pp. https://doi.org/10.1155/2011/241610

Williams T, Hardison P (2013) Culture, law, risk and governance: contexts of traditional knowledge in climate change adaptation. Clim Change 120(3):531–544

Planned Relocation from the Impacts of Climate Change in Small Island Developing States: The Intersection Between Adaptation and Loss and Damage

Melanie Pill

Abstract An increasing number of people have to abandon their homes and livelihoods due to the adverse impacts of climate change. Human mobility has always been part of people's lives, however, some movements, especially planned relocation in the context of climate change, have become involuntary. Non-economic losses occur and the question is whether the relocation of entire communities is still and adaptation response or falls under the realm of loss and damage (L&D) from climate change. This chapter explores the intersection between migration as an adaptation response and L&D with a focus on small island developing states. It analyses when human mobility can no longer be described as adaptation as non-economic losses become too high. It shows that existing frameworks are inadequate to assess community relocation in the context of L&D and non-economic losses. The chapter concludes that there is a spectrum leading from human mobility as an adaptation response to forced migration as L&D. It develops a new framework to assess planned relocation projects and provides concrete recommendations to reduce non-economic losses.

Keywords Climate change · Loss and damage · Planned relocation · Adaptation · Small island developing states · Human mobility · Non-economic losses

Introduction

Throughout history, people have always moved for various reasons, mostly in search for better livelihoods and living conditions but also due to armed conflict and persecution. Over the last decades, there has been an increasing movement of people which is believed to be attributable to the adverse effects of climate change (Asian Development Bank 2011; Nansen Initiative 2015; Platform of Disaster Displacement 2019). It is challenging to determine how many people move because of climate change as a single factor as it is more often than not a combination of many (Nansen Initiative

M. Pill (✉)
Fenner School of Environment and Society Canberra, Australian National University, Acton, Australia
e-mail: melanie.pill@anu.edu.au

W. Leal Filho (ed.), *Managing Climate Change Adaptation in the Pacific Region*, Climate Change Management, https://doi.org/10.1007/978-3-030-40552-6_7

2015; Campbell 2014). Often individuals themselves do not identify climate change as their primary motive to leave, even though it is clearly associated with changes in weather patterns. However, McLeman and Hunter (2010), for example, reviewed several existing case studies that show movement of people due to changed environmental circumstances. Although these case studies are not just related to Small Island Developing States, they are equally valid. Examples include the dry-season migration in the West African Sahel, hurricane-related population displacements in the Caribbean basin, winter migration of 'snowbirds' to the US Sun-belt and drought migration on the North American Great Plains in the 1930s. They showed that human mobility could indeed be attributed to specific changes in weather patterns. In another paper, McLeman and Smit (2006) developed a model that describes the relation of migration and climatic changes. While many people choose to move on their own volition in search for better living conditions (Campbell 2014), cases have started to emerge where whole communities are at threat having their livelihoods inundated by water; a growing trend in island states, particularly in the Pacific region. Planned relocation is increasingly mentioned as a possible adaptation response to the adverse impacts of climate change in literature, governments' climate change strategies as well as the United Nations Convention Framework on Climate Change (UNFCCC) (Nichols 2019). In a case study on the Solomon Islands, Albert et al. (2018) describe how families move due to rising sea levels without a government-led process. In contrast, the Government of Fiji, for example, launched their planned relocation guidelines in 2018 at the 24th meeting of the Conference of the Parties (COP) in Katowice that intend to outline best practice in relocation projects (Government of Fiji 2018).

Nations have also started adding human mobility into existing policies and frameworks, giving evidence for the importance of relocation of communities (UNFCCC 2018b). In Fiji, the government, in collaboration with the German Gesellschaft für internationale Zusammenarbeit (GIZ), have already relocated or is in the process of relocating entire communities as current settlements become uninhabitable due to sea level rise (GIZ 2017). Existing cases, especially in the Pacific, have shown the long and intense consultation process that is required even though communities themselves approached the government to facilitate relocation. Charan et al. (2017) report that many community members in the Pacific suffer from the move or are still reluctant to relocate. Planned relocation due to the impacts of sea level rise is not isolated to Pacific Islands. In the Caribbean, the Bahamas are considering the relocation of communities away from the shoreline in the future if conditions worsen (Thomas and Benjamin 2018). The relocation process is associated with personal losses to individuals that are not easily replaceable, if at all, or not replicable in the new living conditions (Serdeczny et al. 2016a). These so called non-economic losses fall under the heading of loss and damage (L&D) from climate change impacts under the UNFCCC (UNFCCC 2018a). If non-economic losses fall under L&D and these are caused by planned relocation, then it requires re-thinking whether planned relocation is in fact fully an adaptive response to new living circumstances. This can be particularly questioned when considering that anthropogenic climate change and its impacts is a result of the world's largest emitters, whereas island states have contributed negligible emissions (Campbell and Barnett 2010). Therefore, planned relocation as an adaptation response needs to be revisited if undertaken due to anthropogenic climate change. This book chapter will examine whether planned relocation can be labelled as an adaptation response or should be considered failed adaptation, in which case it falls under L&D. In doing so, the evolution of the terms human mobility and displacement in the reports by the Intergovernmental Panel on Climate Change (IPCC) will be analysed. It further looks at how displacement is addressed under the UNFCCC, in particular within Cancun Adaptation Framework and the work plan of the Warsaw Mechanism of Loss and Damage (WIM). The chapter includes an analysis of current identified non-economic losses that are associated with planned relocation and proposes actions to reduce these losses. Existing frameworks by Hino et al. (2017), Dow et al. (2013) and Boyd et al. (2017) are analysed to evaluate whether they are appropriate to determine whether planned relocation is an adaptive response or L&D. This chapter concludes that they are unsuitable for the assessment and therefore, a new framework is developed which is described as spectrum. Planned relocation cases can be placed on this spectrum and allow for an assessment of the extent of L&D due to non-economic losses. Finally, counter actions to reduce non-economic losses are proposed.

Data was triangulated from a variety of sources:

- Literature review on grey literature especially reports published by the UNFCCC, the IPCC and the WIM to illustrate the evolution of the term "human mobility"
- Literature review to identify existing planned relocation cases that were a result of impacts from climate change in island states
- Literature review on existing frameworks related to human mobility, adaptation and L&D to illustrate their unsuitability for L&D and non-economic losses
- 43 semi-structured stakeholder interviews, conducted in the time between July 2018 and February 2019 with international negotiators for small island developing states, NGOs and SIDS government representatives working on L&D.

Although this chapter will look at planned relocation as an adaptation response in small island developing states with a focus on the Pacific region, findings are likely to be applicable to all islands, island states or other communities being relocated.

Human Mobility

The concept of human mobility was for the first time mentioned and acknowledged under the United Nations Convention Framework on Climate Change (UNFCCC) in the Cancun Adaptation Framework (CAF) at the 16th session of the Conference of the Parties in 2010. It was mentioned in section II in paragraph 14(f) as migration, displacement and planned relocation (UNFCCC 2011). Although the acknowledgement illustrated the importance and noteworthiness of the issue of displacement (Warner 2012), the language around it is weak and only "invites" Parties to undertake measures that "enhance displacement, migration and planned relocation". The inclusion in the CAF also meant that displacement of any kind could be eligible for climate finance for adaptation efforts (Serdeczny 2017). Interestingly, in 2012 at COP18 in Doha, human mobility was then mentioned under L&D in decision 3/CP.18 under paragraph a 7(vi). The paragraph acknowledges that "further work is needed on enhancing the understanding of how "impacts of climate change are affecting patterns of migration, displacement and human mobility" (UNFCCC 2013a). In 2013 the Warsaw International Mechanism was established which in its initial two-year rolling work plan included displacement as action area 6: "Enhance the understanding of and expertise on how the impacts of climate change are affecting patterns of migration, displacement and human mobility; and the application of such understanding and expertise" (UNFCCC 2014). In the subsequent five-year rolling work plan, human mobility is included in strategic work stream D: "Enhanced cooperation and facilitation in relation to human mobility, including migration, displacement and planned relocation". The inclusion of human mobility in both, the provisions for adaptation and the L&D work program shows the divide of where planned relocation belongs. Adger et al. (2007), not without caution, recognised planned relocation as an adaptation response. However, with discussion progressing, planned relocation gradually finds its way under the very discourse of L&D. Projects of planned relocation, however, are currently executed as adaptation efforts.

Literature defines human mobility with different characteristics such as spatial, temporal, scale, planned or ad hoc and forced or voluntary. Table 1 provides a summary of the different types of relocation accompanied with a short description. All of them can be a combination (Warner et al. 2013).

A common definition used for community relocation is by Campbell and Warrick (2014), developed in their review for the 2014 United Nations conference on small island developing states. Community relocation means "the forced permanent or long-term movement of an entire community (or a significant part of it) from one location to another". Relocation can be positive when the structure and function of a community is maintained but also be negative when the relocation to the new place meant significant changes to the old way of living (Campbell and Warrick 2014). It could mean that physically the move was successful, but social and cultural structures were damaged. Relocation can be on a national or international level (Warner et al. 2013). In some cases, such as the Pacific, a third level of community relocation should to be acknowledged due to the deep connection to land as well as customary land ownership rights. Therefore, Gharbaoui and Blocher (2018) and Weir et al. (2017) identify the relocation within a community's own land as another possible scenario.

Human Mobility in the Context of Loss and Damage

Compared to adaptation and mitigation, formally, L&D has only relatively recently entered the international climate change negotiations. The concept itself, however, is not new. It precedes the establishment of the UNFCCC (Roberts and Huq 2015) and was mentioned throughout negotiations (UNFCCC 2013a, 2008, 2011). On behalf of

Table 1 Different types of relocation

Characteristic	Description
Spatial (local, regional internal or external/international)	Meaning the movement within or outside national boundaries
Temporal (permanent or temporary)	Meaning the length of time that individuals stay away from their usual homes. This may be because homes have become inhabitable or because they are unwilling to return
Scale (individuals, households, communities, towns, cities, countries)	Meaning how many people are moving
Forced versus voluntary	Meaning the extent to which the decision was based on personal desire to move or determined by external circumstances were beyond control
Planned versus ad hoc	Meaning whether the decision to move was made and prepared for over a longer period or had to be made in a hurry. Most of the time these types of relocations refer to the movement of a collective of individuals due to the changes in the environment and surroundings that make their current homes uninhabitable, or movement due to extreme weather events such as hurricanes

Author, adapted from Warner et al. (2013) and Martin et al. (2013)

the Alliance of Small Island Developing States (AOSIS), Pacific Islands have taken the lead on fighting, for not only the inclusion and recognition of L&D in international negotiation texts, but also advocating for a funding mechanism. A proposal in 1991 to the Intergovernmental Negotiating Committee (INC)[1] suggested a mechanism that had an insurance component, a disaster risk component and a compensation component for the adverse impacts of climate change, sea level rise in particular, to be paid for by developed countries (AOSIS (Vanuatu) 1991). The proposal was not included in the UNFCCC convention text (UNFCCC 1992). L&D did not emerge until 2007 at COP 13 when L&D was for the first time included in the Bali Action Plan (UNFCCC 2008). AOSIS submitted a proposal for a L&D financial mechanism to the ad hoc working group of long-term co-operative action (AWG-LCA), responsible for the implementation of the Bali Action Plan (Roberts and Zakieldeen 2018). The submission suggested a multi-window mechanism with the inclusion of three components: (a) insurance, (b) rehabilitation and compensation and (c) risk management (AOSIS 2008). In 2013 the Warsaw International Mechanism of Loss and Damage (WIM) was established consisting of an Executive Committee (ExCom) and pursuant to decision 3/CP18, "enhance knowledge and understanding of comprehensive risk management approaches, strengthening dialogue, coordination, coherence and synergies among

[1]The Intergovernmental Negotiating Committee was formed to negotiate the United Nations Framework Convention on Climate Change.

Table 2 Overview of L&D types

Loss and damage associated with the impacts of climate change	Example most relevant to the Pacific
Sudden-onset or extreme events	Tropical cyclones, storm surges, floods
Slow-onset events	Sea level rise, salinisation
Economic losses	Infrastructure, tourism, property agricultural production
Non-economic losses	Life, human mobility, cultural heritage, territory

Adapted from UNFCCC (2018a)

relevant stakeholders and enhancing action and support, including finance, technology and capacity-building" (UNFCCC 2013a). In its five-year rolling work plan of the WIM work stream D is dedicated to human mobility, including migration, displacement and planned relocation (UNFCCC 2017). As part of this work stream, the ExCom formed the Taskforce on Displacement that developed recommendation for integrated approaches to address displacement (UNFCCC 2018b).

In 2015 at COP21 in Paris and after a strong push from AOSIS, L&D received its own article[2] in the Paris Agreement formally recognising it under the Convention, separate to adaptation. Although an official definition is not yet agreed upon, the WIM ExCom has identified impact types of losses, damages and events. Table 2 gives a brief overview.

At COP 24 in Katowice, L&D received its own heading under the transparency framework next to mitigation and adaptation pursuant to decision G. 115. This is significant as from then onwards it is part of the global stocktake and countries are able to report on past, present and potential future L&D (Serdeczny et al. 2018; UNFCCC 2019a; UNFCCC 2019b).[3]

This brief historical overview has illustrated that, over the existence of the UNFCCC, L&D has always been mentioned and recognised, even though not formally as a separate concept to adaptation. The distinction of L&D from adaptation is also confirmed by 43 stakeholder interviews that were conducted with experts on L&D. Stakeholders included representatives of small island states government representatives including the Caribbean and the Pacific, NGOs and multilateral organisations all working on L&D. 19 out of 43 of respondents described L&D broadly as "something that goes beyond adaptation" or "when you can't adapt anymore, this is when L&D occurs". The remaining interviewees referred to L&D as "all impacts

[2]Article 8 of the Paris Agreement: "Parties recognize the importance of averting, minimizing and addressing loss and damage associated with the adverse effects of climate change, including extreme weather events and slow onset events, and the role of sustainable development in reducing the risk of loss and damage".

[3]For a detailed outline of the history of loss and damage see Roberts and Huq (2015).

of climate change", "disaster risk reduction", "the loss of culture" or mentioned the concept of "avert, minimize, address" as used as a term under the UNFCCC.

Human Mobility in the Context of Adaptation

For some, human mobility can in many circumstances be seen as a normal part of life and in fact, Pacific islands in particular have patterns of high human mobility (Birk and Rasmussen 2014; Farbotko and Lazrus 2012; Barnett and McMichael 2018). McAdam (2014) also describes human mobility as an adaptive response and part of a "historical continuum" that is a core part of Pacific islanders' history and a legitimate solution to natural disasters and slow onset events. Human mobility can be beneficial when individuals are economically better off due to better living conditions or employment that can be used to send home remittances (Birk and Rasmussen 2014). Adger et al. (2014) argue that when planned and adequately integrated within adaptation policies, mobility can be effective, but also acknowledges that movements can come at considerable costs to the individual. Collier (2013) supports this and concludes that migrants incur psychological costs that are not easily compensated by the benefits they enjoy from moving. Problems during the relocation process stem from insufficient community consultation, especially with already marginalised groups such as women, limited guidance from the government and missing follow-up support (Connell and Coelho 2018).

Literature and government reports show adaptation as a response to climate change and L&D incurred from climate change impacts these concpets start to blur in the context of human mobility. This can be seen through the evolution of the wording used in the IPCC reports. Planned relocation has never entirely been seen as an acceptable adaptive response and is in the IPCC's fourth assessment report already described to have "enormous economic, cultural and human costs" as well as "unacceptable impacts in terms of human rights and sustainability" (Adger et al. 2007).

Adaptation is meant to improve or at the very least, maintain current living conditions and in doing so, change the environment around the individual. Planned relocation, however, forces the individual to adapt to the changing environment. This is also supported in earlier works by Adger et al. (2005) who mention that managed retreat is not entirely chosen out of free will, by neither the government, the individual nor the community. They also argue that an adaptation response that does not have direct benefits to the community has to be considered as unsuccessful.

The IPCC (2014) defines adaptation also as "the adjustment in natural or human systems in response to actual or expected climatic stimuli or their effects". However, the part on *how* to adjust "human systems" is by far less clear than how to modify the natural environment *around* the human system. The IPCC (2014) further mentions planned relocation as an adaptation strategy in various places but with caution and reservation. In Chap. 12 under "Human Security" it is described as an adaptation strategy that has the potential to reduce risk to climate change impacts and that planned relocation can be "effective if undertaken in a sensitive manner". At the same time, however, they admit that scientific literature sees it as a failure to adapt. Another indirect reference to human mobility can be found in Chap. 5.5.6 under the heading "Constraints and Limits". Chap. 14 lists possible adaptation options for social systems and mentions labour migration as an important strategy, however, stresses that migration and relocation do have implications for family relations, health, and human security. The IPCC 1.5 Special Report SPM outlines that there are "limits to adaptation and adaptive capacity for some human and natural systems". It acknowledges that constraints to adaptation exist and these are exacerbated by higher levels of global warming. However, the report also states that migration and relocation might be an adaptation option (Hoegh-Guldberg et al. 2018).

Research has also shown that migration is perceived by people as the least favourable adaptation option. It is only considered as a last resort and in situ adaptation might be preferred (Nansen Initiative 2015). Barnett et al. (2015) draw an important distinction that illustrates the divide of planned relocation as adaptation response. They describe limits and barriers to adaptation in which barriers can be overcome with "concerted effort". Limits, however, occur when the response fails "to protect what stakeholders value". Further, barriers and limits to adaptation are context-dependent and differ in spatial and temporal aspects. Dow et al. (2013) also describe adaptation as an attempt "to keep risks to valued objectives at a tolerable level". To add to the controversy, Connell and Coelho (2018) describe planned relocation as a complex process associated with social, environmental, legal and political issues.

Human mobility and displacement are increasingly included in countries' national policies and frameworks and for some small island states also in their National Determined Contributions (NDC) (Thomas and Benjamin 2018; UNFCCC 2018b). The Task Force on Displacement of the WIM, found that out of 37 countries and territories that have national adaptation policies, plans or strategies in place, 81% refer to human mobility. Similarly, out of 193 countries and territories having submitted NDCs, 20% refer to human mobility.[4] The planned relocation guidelines of the Fijian Government, for example, define community relocation as "the voluntary, planned and coordinated movement of climate-displaced persons within (a State) to suitable locations, away from risk-prone areas, where they can enjoy the full spectrum of rights including housing, land and property rights and all other livelihood and related rights" (Government of Fiji 2018).

Despite the inclusion of planned relocation in governmental adaptation policies, planned relocation still has distinct characteristics from adaptation. This is demonstrated by the fact that relocation, especially planned relocation, can be found in adaptation and L&D discussions in international negotiations. The inclusion of human mobility as a separate workstream in the WIM work plan for L&D acknowledges that planned relocation goes beyond the adaptation discourse and recognises that there are irrecoverable losses to individuals.

---

[4]These figures do not make a distinction between small island developing states or other signatories to the Convention.

Non-economic Losses in the Context of Planned Relocation

Attributes of Non-economic Losses

Planned relocation is highly likely associated with non-economic losses. They have certain characteristics and manifest themselves in different ways. Non-economic losses are context-dependent, incommensurable and not easily, if at all, determined or traded in a market. Serdeczny et al. (2016a, b) and Kirsch et al. (2001) explain that these losses go unnoticed to an outsider who is not immersed in them. This is a particular problem in the Pacific where the people are defined by their land and vice versa (Weir et al. 2017). Non-economic losses also lack a common unit of measurement and do not have a monetary value attached to them. Losses with sentimental value are experienced and felt differently from individual to individual. A single definition does not seem appropriate but rather a descriptive list of what they are and could be. Therefore, in the context of planned relocation they would have to be assessed on a case by case and possibly even on an individual-by-individual level. This might sound impractical but critical to protect the well-being of the individual. It is also recognised that women suffer disproportionally from non-economic losses and climate change because of their already more vulnerable status in society which is due to unequal distribution of tasks and carer responsibilities (Nansen Initiative 2015; Connell and Coelho 2018; Campbell et al. 2005).

Addressing Non-economic Losses in the Context of Planned Relocation

Minimising non-economic losses is complicated, especially when they can differ for each person. So far, there are no practical solutions explicitly developed that can help address non-economic losses in planned relocation projects. Reports and literature refer to "avert, minimise and address" non-economic losses but do not outline *how* to "avert, minimize and address" even though it is now acknowledged that they will certainly occur and fall under unavoidable damage L&D (Durand and Huq 2015). Non-economic losses are also included in the five-year rolling work plan of the WIM in strategic work stream B. Priority activity four in work stream B specifically asks for collaboration with the Task Force on Displacement to assess non-economic losses in the context of human mobility (UNFCCC 2017).

Scholars recognise the difficulty to address non-economic losses with economic assessment tools or policies due to their very characteristics (Serdeczny et al 2016a; Serdeczny et al 2016b; Serdeczny 2017) but do not outline concrete actions.

Non-economic losses can be categorised into certain themes which are consolidated in the first column in Table 3 based on existing literature. Column two assigns a level of L&D before action to reduce non-economic losses is taken. Column three

Table 3 Types of non-economic losses

Non-economic losses as identified in literature	L&D before action to reduce non-economic losses	Action to avert, minimise or address non-economic loss	Residual L&D from non-economic loss (high, medium, low)
Physical well-being	High for males Extremely high for females	Integrate physicians and doctors in the relocation process; ensure adequate number of female doctors as well recognition of different female medical needs	Low for males Medium for females
Psychological well-being	High for males Extremely high for females	Integrate psychologists in the relocation process; ensure adequate number of female psychologists	Low for males Medium for females
Territory (sovereignty, sense of place)	High	Choose a place close to the original settlement Discuss land rights Provide as many options as possible for relocation and allow residents to provide options Ensure women are included in the decision-making process	Low-medium
Cultural heritage	High	Choose a place close to the original settlement Discuss land rights Provide as many options as possible for relocation and allow residents to provide options Ensure women are included in the decision-making process	Medium-low

(continued)

Table 3 (continued)

Non-economic losses as identified in literature	L&D before action to reduce non-economic losses	Action to avert, minimise or address non-economic loss	Residual L&D from non-economic loss (high, medium, low)
Indigenous knowledge	High	Communicate with major knowledge holders of the community to preserve knowledge and develop a collection	Medium
Destruction of cultural sites and culturally important landscapes	Extremely high	Protect and retain with infrastructure Where possible consider relocating with the new settlement	Low-medium
Loss of identity and ability to solve problems collectively	High for males Extremely high for females	Find a way to make decisions together and decide on a way forward Facilitation of community meetings to discuss changes Ensure women are included in the discussions and have a voice Facilitate community meetings after relocation to ensure longevity and preempt conflict	Medium for males High for females
Loss of knowledge/ways of thinking that are part of lost livelihood systems	High	Find a way to make decisions together and decide on a way forward Facilitation of community meetings to discuss changes Ensure women are included in the discussions and have a voice Facilitate community meetings after relocation to ensure longevity and preempt conflict	Medium

(continued)

Table 3 (continued)

Non-economic losses as identified in literature	L&D before action to reduce non-economic losses	Action to avert, minimise or address non-economic loss	Residual L&D from non-economic loss (high, medium, low)
Social cohesion, peacefully functioning society	High	Find a way to make decisions together and decide on a way forward Facilitation of community meetings to discuss changes Ensure women are included in the discussions and have a voice Facilitate community meetings after relocation to ensure longevity and preempt conflict	Medium
Education	High for males Extremely high for females	Prioritise building of educational facilities in the new location Ensure continued education during relocation by integrating teaching staff or freeing up existing staff in the village Ensure attending school is compulsory for female and male students Consider escorts to educational facilities for female students	Low-medium for males Medium for females
Traditions/religion/customs	High	Ensure religious sites are re-erected at the new location or still adequately maintained and reachable in the old place Consult what needs to be in place to continue with embedded customs and traditions	Low

(continued)

Table 3 (continued)

Non-economic losses as identified in literature	L&D before action to reduce non-economic losses	Action to avert, minimise or address non-economic loss	Residual L&D from non-economic loss (high, medium, low)
Social bonds/relations	High Extremely high for females	Ensure that residents who have personal or close relationships with each other move at the same time or shortly thereafter Ensure that residents who have personal or close relationships with each other stay in the same neighbourhood or house Ensure that couples with shared responsibilities move together	Low
Changed labour conditions	Extremely high	Provide education and new skills if changing occupation Provide adequate infrastructure to reach food and water resources Recognise the multiple roles of women in society for preparing, planting and gathering food as well as child minding and caring for the elderly	Low for males Medium for females

Consolidated from Fankhauser et al. (2014), Morrissey and Oliver-Smith (2013), UNFCCC (2013b), Andrei et al. (2015), Serdeczny et al. (2016a). The table shows the extent of non-economic losses and damages before and after mitigation responses have been implemented. *Source* author provides a counter action to reduce the non-economic loss and column four assesses the residual L&D.

Where appropriate, a distinction between male and female has been made to reflect the higher vulnerability status of girls and women. The non-economic loss is assumed to be felt high or extremely high for every individual if no counter-action is taken. The measures that are suggested in order to "avert, minimise or address"[5] non-economic losses as much as possible are listed. Also described are the residual L&D after implementing counter-actions. None of the non-economic losses are entirely avoided, an assumption that is also supported by Warner and van der Geest (2013) especially in the context of human mobility. Again, a distinction has been made between males and females in order to reflect a higher vulnerability status for women to residual L&D.

Table 3 can be used as a tool to integrate methods in the planning and execution of any community relocation project to reduce non-economic losses. At the same time, the costs for each action can be factored into the project budget. The concrete actions outlined in Table 3 can serve as a major guide to address strategic work stream B in the five-year rolling work plan of the WIM which calls for factoring non-economic losses into "the planning and elaboration of measures to address L&D associated with the adverse effects of climate change" (UNFCCC 2017). As non-economic losses vary from individual to individual, it is crucial in the planning process to consult with residents to identify and prioritise their needs rather than perceived needs from a project manager's point of view. Using Table 3 as a tool, allows firstly for assessment of non-economic losses, secondly, outlines possible mitigation options and thirdly, could be included in risk management plans. However, risks also need to be assessed and mitigated for each individual project and residual losses might vary.

The New Loss and Damage Spectrum

The analysis of literature and stakeholder interviews as well as the different definitions of L&D have shown that planned relocation is beyond an adaptation response. Even within the UNFCCC the concept of planned relocation blurs into adaptation and L&D debates. On the one hand, the Task Force on Displacement reports that many countries discuss planned relocation as a potential adaptation measure in national policies and strategies as well as NDCs, but on the other, appears in the L&D work plan of the WIM in conjunction with non-economic losses. The IPCC (2014), in its definition for adaptation, assumes a certain degree of adaptive capacity from a "human system" which prevents planned relocation to be entirely attributed to L&D. The choice of a community to move might be voluntary but only on the basis that there is no real alternative as the altered environment does not allow to stay in the current

[5]Refers to wording used in UNFCCC texts to show the applicability of counter actions to the work undertaken by the WIM ExCom.

place with an acceptable level of risk. To some extent, certain residents might consider planned relocation as an adaptive response, but others struggle with the change. McNamara et al. (2018), support this notion and describe planned relocation due to the adverse effects of climate change as being neither one nor the other. In fact, they believe that such a distinction is unconstructive.

Different literature have looked at L&D and human mobility and developed frameworks for their assessment. The most notable and relevant for this chapter are Hino et al. (2017), Boyd et al. (2017) and Dow et al. (2013). In the following, these frameworks are analysed to assess whether they are suitable to evaluate planned relocation projects and associated non-economic losses in the context of adaptation and L&D.

The framework by Hino et al. (2017) allows for measuring the success of a planned relocation project but does not look at the losses incurred by individuals. It therefore is unsuitable to determine when planned relocation is no longer an adaptive response or how much loss, may it be economic or non-economic, individuals have suffered. Boyd et al. (2017) identify typologies of L&D based on stakeholder interviews. These are placed on an axis. They describe how much these typologies go beyond adaptation and mitigation. However, only the identified "existential perspective" includes the mention of non-economic losses. In addition, the framework looks at L&D in its entirety and not in the context of human mobility and planned relocation. Similarly, Dow et al. (2013) develop an actor-centred risk-based approach. They describe limits to adaptation based on tolerable and intolerable risks to the individuals affected. An adaptation failure occurs therefore, when an individual cannot longer safeguard what they value. This definition supports the assumption that planned relocation goes beyond adaptation. However, the approach by Dow et al. (2013) also does not allow for the assessment of non-economic losses and their aversion, minimization or avoidance. As none of these frameworks is suitable to measure when a planned relocation project has more characteristics of L&D than adaptation, a new framework to undertake this assessment is required.

Figure 1 illustrates this framework and describes it as a spectrum from L&D to adaptation and vice versa. The spectrum can be applied to any planned relocation project and assumes that with increasing non-economic losses as defined in Table 1, associated L&D increases.

Starting from the top, non-economic losses increase the less community-focused they are. This is in line with Hino's (2017) assumption that the best possible outcome for a planned relocation project is achieved when there is support from the government and the willingness of the community to move. The more forced and without consent the relocation is, the more severe and greater residents will feel losses. The spectrum also assumes that if the project is undertaken within a short timeframe and rushed, important belongings that could be saved might be overlooked during the design phase of the project. Necessary infrastructure in the new location is not set up to fulfil residents' needs, which often disadvantages already marginalised and vulnerable community members. To counteract a rushed move, Nichols (2019) suggest a phased approach to relocation. Existing cases such as the relocation of the village of Vunidogoloa in Fiji have shown that even a relocation planned long-ahead

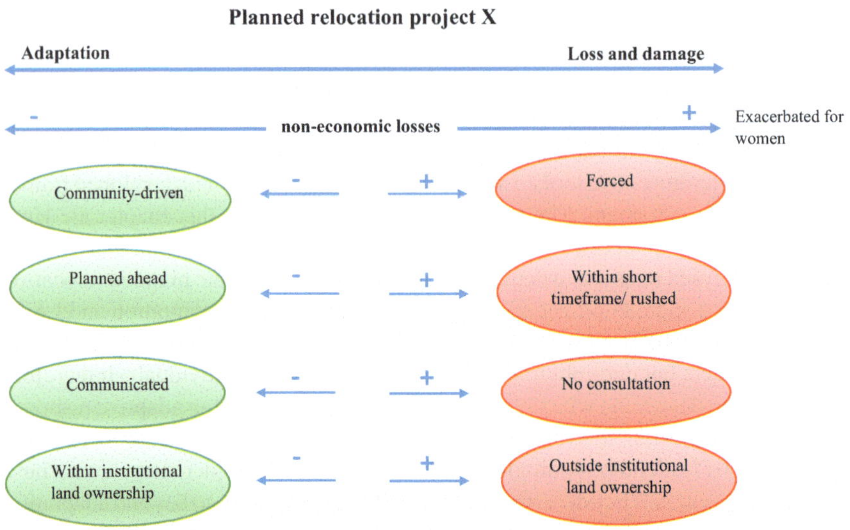

Fig. 1 Loss and damage and adaptation visualised as a spectrum for relocation projects

still results in difficulties and backlash from the community. The spectrum illustrates that barriers are even greater when no or insufficient consultation has taken place (Piggott-McKellar et al. 2019). The importance of communication is also highlighted in the case of Vunidogoloa, as some residents were still reluctant to move despite years of negotiating, consulting and the fact that the community voluntarily approached the government to relocate the village (Charan et al. 2017). A major barrier in the Pacific are landownership provisions and the connection to the land that Pacific islanders feel. Therefore, relocation should take place within customary land boundaries whenever possible to keep non-economic losses to a minimum. If undertaken within customary or institutional land boundaries, additional negotiation with neighboring landowners and potential conflict can be avoided.

The degree of L&D from planned relocation projects is gender-sensitive and women are disproportionally impacted for various reasons. They are at higher risk of experiencing L&D as well as suffering L&D to a greater extent than men (Nansen Initiative 2015; IPCC 2014; Piggott-McKellar et al. 2019). This is because, firstly, women bear most, if not all, the caring responsibilities for children and elderly family members (Campbell et al. 2005). Secondly, women are responsible for major harvest and food supply. Thirdly, they are mostly excluded from major decision-making processes and not adequately represented in community discussions. Campbell et al. (2005) conducted a focus group during the relocation of the village of Biausevu in southern Viti Levu in Fiji. Their study found that (a) paths to carry food supplies, children and firewood were inadequate, (b) setting up shelter is prioritised over domestic necessities such as cooking facilities and (c) at the beginning of the settlement stage, sanitary infrastructure and electricity are in poor supply.

Despite the aid that the L&D spectrum and the mitigation actions for non-economic losses provide in evaluating planned relocation projects, the ultimate judgement on where a relocation project is situated along the spectrum and what actions are needed, is subjective. The assessment and requirements are context-dependent and even differ between individuals. It is therefore challenging to determine when relocation of an entire community can be deemed successful or not, a view also supported by Nalau and Handmer (2018).

Regardless, to ensure that L&D to individual residents as well as the whole community are kept as minimal as possible, implementers should aim to place projects to the left end of the spectrum by using the counter actions in Table 3.

Conclusion

This chapter looked at whether planned relocation in small island developing states is an adaptation response or whether it should placed under the umbrella of L&D. The analysis of existing literature, grey and academic, revealed divergent views. The spectrum developed in this chapter provides project planners with a guide as to how much L&D from non-economic losses is incurred if no mitigation is undertaken. With current climate change projections worsening, planned relocation is likely going to be an increasingly accepted adaptation response. However, all types of movements—may they be voluntary, involuntary, planned or in haste—have associated non-economic losses. In any relocation project or program, it should be a priority to keep these non-economic losses to a minimum. Using the counter actions as a tool kit to address each non-economic loss category, allows planned relocation projects to be as much of an adaptation response as possible. The spectrum and its counter actions are the first usable "how to"-guide in order to avert, minimise and address non-economic losses in the context of L&D. The framework and counter actions go beyond discussions of the complexity of non-economic losses and the need to address them. The spectrum can be used by policy- and decision makers, the ExCom of the WIM as well as implementers for planned relocation projects . In doing so, the spectrum's suitability can be tested and shortcomings in the counter actions identified. It is important to recognise that counter actions will have to be individualised each project.

Further research is therefore required as to what people individually identify as their non-economic losses. This research should be undertaken in the form of participatory community-based fieldwork. Residents affected by planned relocation, need to be consulted to reveal how they believe their non-economic losses should be addressed. A simple compensation payment might neither be suitable, sufficient nor appropriate. Compensation could be looked at in the form of livelihood or economic improvements. However, in the context of L&D still requires research as it is explicitly excluded in the decision text of the Paris Agreement. Therefore, careful consideration needs to be given as to whether planned relocation projects should have a compensatory component. Considering planned relocation as L&D might impede funding provisions for projects that originally were considered an adaptation response.

Acknowledgements I would like to express my thanks to Dr. Anthony Weir for his advice and help in refining this chapter as well as my supervisor Professor Jamie Pittock for his continued support and guidance.

References

Adger WN, Agrawala S, Mirza MMQ, Conde C, O'Brien K, Pulhin J, Pulwarty R, Smit B, Takahashi K (2007) Assessment of adaptation practices, options, constraints and capacity. In: Parry OFCML, Palutikof JP, van der Linden PJ, Hanson CE (eds) Climate change 2007: impacts, adaptation and vulnerability contribution of working group II to the fourth assessment report of the Intergovernmental Panel on Climate Change. Cambridge University Press, Cambridge, pp 717–743

Adger WN, Arnell NW, Tompkins EL (2005) Successful adaptation to climate change across scales. Glob Environ Change 15:77–86

Adger WN, Pulhin JM, Barnett J, Dabelko GD, Hovelsrud GK, Levy M, Spring ÚO, Vogel CH (2014) Human security. In: Climate change 2014: impacts, adaptation, and vulnerability. Part A: global and sectoral aspects. Contribution of working group II to the fifth assessment report of the Intergovernmental Panel on Climate Change, Field CB et al (ed). Cambridge University Press, Cambridge and New York, pp 755–791

Albert S, Bronen R, Tooler N, Leon J, Yee D, Ash J, Boseto D, Grinham A (2018) Heading for the hills: climate-driven community relocations in the Solomon Islands and Alaska provide insight for a 1.5 C future. Reg Environ Change 18:2261–2272

Andrei S, Rabbani G, Khan H, Haque M, Ali D (2015) Non-economic loss and damage caused by climatic stressors in selected coastal districts of Bangladesh, Bangladesh

AOSIS (2008) Proposal to the AWG-LCA: multi-window mechanism to address loss and damage from climate change impacts

AOSIS (Vanuatu) (1991) Draft annex relating to Article 23 (insurance) for inclusion in the revised single text on elements relating to mechanisms (A/AC.237/WG.II/ Misc.13) submitted by the Co-Chairmen of Working Group II)

Asian Development Bank (2011) Climate change and migration in Asia and the Pacific draft edition

Barnett J, Evans LS, Gross C, Kiem AS, Kingsford RT, Palutikof J P, Pickering CM, Smithers SG (2015) From barriers to limits to climate change adaptation: path dependency and the speed of change

Barnett J, McMichael C (2018) The effects of climate change on the geography and timing of human mobility. Popul Environ 39:339–356

Birk T, Rasmussen K (2014) Migration from atolls as climate change adaptation: current practices, barriers and options in Solomon Islands. Nat Resour Forum 38:1–13

Boyd E, James RA, Jones RG, Young HR, Otto FE (2017) A typology of loss and damage perspectives. Nat Clim Change 7:723

Campbell J, Barnett J (2010) Climate change and small island states: power, knowledge and the South Pacific. Earthscan, London, Washington, DC

Campbell J, Warrick O (2014) Climate change and migration issues in the Pacific, Fiji

Campbell JR (2014) Climate-change migration in the Pacific. The Contemporary Pacific 1–28

Campbell JR, Goldsmith M, Koshy K (2005) Community relocation as an option for adaptation to the effects of climate change and climate variability in Pacific Island Countries

Charan D, Kaur M. Singh P (2017) Customary land and climate change induced relocation—a case study of Vunidogoloa Village, Vanua Levu, Fiji. Springer International Publishing, pp 19–33
Collier P (2013) Exodus: how migration is changing our world. Oxford University Press, Oxford
Connell J, Coelho S (2018) Planned relocation in Asia and the Pacific. Forced Migr Rev 46–49
Dow K, Berkhout F, Preston BL, Klein RJT, Midgley G, Shaw MR (2013) Limits to adaptation. Nat Clim Change 3:305
Durand A, Huq S (2015) Defining loss and damage: key challenges and considerations for developing an operational definition. International Centre for Climate Change and Development, Bangladesh
Fankhauser S, Dietz S, Gradwell P (2014) Non-economic losses in the context of the UNFCCC work programme on loss and damage. Centre for Climate Change Economics and Policy, Grantham Research Institute on Climate Change and the Environment. Available at: https://www.cccep.ac.uk/wp-content/uploads/2015/10/Fankhauser-Dietz-Gradwell-Loss-Damage-final.pdf
Farbotko C, Lazrus H (2012) The first climate refugees? Contesting global narratives of climate change in Tuvalu. Glob Environ Change 22:382–390
Gharbaoui D, Blocher J (2018) Limits to adapting to climate change through relocations in Papua-New Guinea and Fiji. In: Filho WL, Nalau J (ed) Limits to climate change adaptation. Springer International Publishing, pp 359–379
GIZ (2017) Adapting to climate change: a new start in Fiji. https://www.giz.de/en/mediacenter/57948.html
Government of Fiji (2018) Planned Relocation Guidelines—a framework to undertake climate change related relocation. Ministry of Economy: Suva, Fiji. Available at: https://cop23.com.fj/wp-content/uploads/2018/12/CC-PRG-BOOKLET-22-1.pdf
Hino M, Field CB, Mach KJ (2017) Managed retreat as a response to natural hazard risk. Nat Clim Change 7:364
Hoegh-Guldberg O, Jacob D, Taylor M, Bindi M, Brown S, Camilloni I, Diedhiou A, Djalante R, Ebi K, Engelbrecht F, Guiot J, Hijioka Y, Mehrotra S, Payne A, Seneviratne SI, Thomas A, Warren R, Zhou G (2018) Global warming of 1.5°C. In: Masson-Delmotte V et al (ed) An IPCC special report on the impacts of global warming of 1.5°C above pre-industrial levels and related global greenhouse gas emission pathways, in the context of strengthening the global response to the threat of climate change, sustainable development, and efforts to eradicate poverty. Geneva, Switzerland, pp 175–311
IPCC (2014) Climate change 2014: impacts, adaptation, and vulnerability. Part A: global and sectoral aspects. Contribution of working group II to the fifth assessment report of the Intergovernmental Panel on Climate Change. Cambridge University Press, Cambridge and New York
Kirsch S, Biery-Hamilton GM, Brown MF, Brush SB, Cleveland DA, Dirlik A, Dominguez V, Escobar A, Finney B, Giles-Vernick T (2001) Lost worlds: environmental disaster, "culture loss", and the law. Curr Anthropol 42:167–198
Martin M, Billah M, Siddiqui T, Black R, Kniveton D (2013) Policy analysis: climate change and migration Bangladesh
McAdam J (2014) Historical cross-border relocations in the Pacific: lessons for planned relocations in the context of climate change. J Pac Hist 49:301–327
McLeman R, Smit B (2006) Migration as an adaptation to climate change. Clim Change 76:31–53
McLeman RA, Hunter LM (2010) Migration in the context of vulnerability and adaptation to climate change: insights from analogues. Wiley Interdisc Rev Clim Change 1:450–461
McNamara KE, Bronen R, Fernando N, Klepp S (2018) The complex decision-making of climate-induced relocation: adaptation and loss and damage. Clim Policy 18:111–117
Morrissey J, Oliver-Smith A (2013) Perspectives on non-economic loss and damage: understanding values at risk from climate change
Nalau J, Handmer J (2018) Improving development outcomes and reducing disaster risk through planned community relocation. Sustainability 10:3545
Nansen Initiative (2015) Disaster-induced cross-border displacement-agenda for the protection of cross-border displaced persons in the context of disasters and climate change, vol 1

Nichols A (2019) Climate change, natural hazards, and relocation: insights from Nabukadra and Navuniivi villages in Fiji. Clim Change 1–17
Piggott-McKellar AE, McNamara KE, Nunn PD, Sekinini ST (2019) Moving people in a changing climate: lessons from two case studies in Fiji. Soc Sci 8:133
Platform on Disaster Displacement (2019) Platform on Disaster Displacement (PDD) Strategy 2019–2022
Roberts E, Huq S (2015) Coming full circle: the history of loss and damage under the UNFCCC. Int J Glob Warming 8:141–157
Roberts E, Zakieldeen A (2018) Pocket guide to loss and damage under the UNFCCC. European Capacity Building Initiative. Available at: https://ecbi.org/sites/default/files/FinalVersionLoss%26Damage_1.pdf
Serdeczny O (2017) What does it mean to "Address Displacement" under the UNFCCC? An analysis of the negotiations process and the role of research. German Development Institute. Available at: https://www.die-gdi.de/uploads/media/DP_12.2017.pdf
Serdeczny O, Pierre-Nathoniel D, Siegele L (2018) Progress on loss and damage in Katowice. Climateanalytics. Available at: https://climateanalytics.org/blog/2018/progress-on-loss-and-damage-in-katowice/
Serdeczny O, Waters E, Chan S (2016a) Non-economic loss and damage in the context of climate change understanding the challenges. Bonn, Germany
Serdeczny O, Waters E, Chan S (2016b) Non-economic loss and damage: addressing the forgotten side of climate change impacts, Germany. Available at: https://www.die-gdi.de/uploads/media/BP_3.2016_neu.pdf
Thomas A, Benjamin L (2018) Policies and mechanisms to address climate-induced migration and displacement in Pacific and Caribbean small island developing states. Int J Clim Change Strat Manage 10:86–104
UNFCCC (1992) United Nations Framework Convention on Climate Change. Available at: https://unfccc.int/resource/docs/convkp/conveng.pdf
UNFCCC (2008) Report of the Conference of the Parties on its thirteenth session. Available at: https://unfccc.int/process-and-meetings/conferences/past-conferences/bali-climate-change-conference-december-2007/cop-13/cop-13-reports
UNFCCC (2011) Report of the conference of the parties on its sixteenth session. Available at: https://unfccc.int/resource/docs/2010/cop16/eng/07a01.pdf#page=2
UNFCCC (2013a) Report of the Conference of the Parties on its eighteenth session. Available at: https://unfccc.int/sites/default/files/resource/docs/2012/cop18/eng/08a01.pdf
UNFCCC (2013b) Non-economic losses in the context of the work programme on loss and damage Technical paper. Available at: https://unfccc.int/resource/docs/2013/tp/02.pdf
UNFCCC (2014) Report of the Executive Committee of the Warsaw International Mechanism for Loss and Damage associated with climate change impacts—Annex II—Initial two-year workplan of the Executive Committee of the Warsaw International Mechanim for Loss and Damage. Available at: https://unfccc.int/sites/default/files/resource/docs/2014/sb/eng/04.pdf
UNFCCC (2017) 5-year rolling work plan—Annex I to the report of the 2017 Executive Committee. Available at: https://unfccc.int/sites/default/files/resource/Detailed%20workplan%20by%20strategic%20workstreams.pdf
UNFCCC (2018) Suva Expert Dialogue-Background: range of topic areas under loss and damage. Available at: https://unfccc.int/topics/adaptation-and-resilience/workstreams/loss-and-damage-ld/workshops-meetings/suva-expert-dialogue#eq-7
UNFCCC (2018b) Report of the Task Force on Displacement. Available at: https://unfccc.int/sites/default/files/resource/2018_TFD_report_17_Sep.pdf
UNFCCC (2019a) Katowice Climate Package-Matters relating to the global stocktake referred to in Article 14 of the Paris Agreement and paragraphs 99–101—Decision19/CMA.1 section I6. b) ii). Available at: https://unfccc.int/sites/default/files/resource/cma2018_3_add2%20final_advance.pdf#page=53

UNFCCC (2019b) Katowice Climate Package—Modalities, procedures and guidelines for the transparency framework for action and support referred to in Article 13 of the Paris Agreement—Decision 18/CMA.1 Annex IV G.115 a-c. Available at: https://unfccc.int/sites/default/files/resource/cma2018_3_add2%20final_advance.pdf#page=18

Warner K (2012) Human migration and displacement in the context of adaptation to climate change: the Cancun adaptation framework and potential for future action. Environ Plann C Gov Policy 30:1061–1077

Warner K, Afifi T, Kälin W, Leckie S, Ferris B, Martin SF, Wrathall D (2013) Changing climate, moving people: framing migration, displacement and planned relocation. Bonn, Germany

Warner K, van der Geest K (2013) Loss and damage from climate change: local-level evidence from nine vulnerable countries. Int J Glob Warming 5:367–386

Weir T, Dovey L, Orcherton D (2017) Social and cultural issues raised by climate change in Pacific Island countries: an overview. Reg Environ Change 17:1017–1028. Available at: https://doi.org/10.1007/s10113-016-1012-5

A Knowledge Network Approach to Understanding Water Shortage Adaptation in Kiribati

Rebecca Cunningham, Pierre Mukheibir, Brent Jacobs, Louise Boronyak and Pelenise Alofa

Abstract Kiribati, a small-island developing state in the Pacific, experiences a range of climate change impacts, including drought, sea-level rise, coastal erosion and salt-water intrusion to freshwater lenses. These impacts negatively affect food security and drinking water quality resulting in poor human health outcomes, particularly child morbidity. Timely warning about changes to drinking water supplies could reduce community health impacts but the existence and effectiveness of knowledge networks for water quality are unclear. This paper describes an engagement process with key stakeholders (government, community service organizations and community members) to understand how information about the impacts of climate change on potable water supplies was sought and shared using a social network analysis approach. The information networks revealed were highly fragmented and timely sharing of information was poor, which limits effective prophylactic intervention that might reduce child mortality from preventable diseases and illnesses such as diarrhoea. The main conclusion reached is that fragmented island geography and traditional forms of oral information transmission may be important factors that shape the formation and function of water knowledge networks in Kiribati. The wider application of these findings to other Pacific Island contexts requires further research to fully understand how knowledge flows could be optimized in the future.

R. Cunningham (✉) · P. Mukheibir · B. Jacobs · L. Boronyak
Institute for Sustainable Futures, University of Technology Sydney, Building 10, Level 10, 235 Jones St., Ultimo, Sydney 2007, Australia
e-mail: Rebecca.Cunningham@uts.edu.au

P. Mukheibir
e-mail: Pierre.Mukheibir@uts.edu.au

B. Jacobs
e-mail: Brent.Jacobs@uts.edu.au

L. Boronyak
e-mail: Louise.Boronyak@uts.edu.au

P. Alofa
Kiribati Climate Action Network, Eita Village, PO Box 110, Tarawa, Kiribati

W. Leal Filho (ed.), *Managing Climate Change Adaptation in the Pacific Region*,
Climate Change Management, https://doi.org/10.1007/978-3-030-40552-6_8

Keywords Social network analysis · Kiribati · Knowledge networks · Water · Climate change

Introduction

Small island developing states (SIDS) are traditionally considered among the most vulnerable to the impacts of climate change (Pelling and Uitto 2001; Kelman and West 2009). However, as recent authors have suggested, climate change may not be the most immediate issue that SIDS face (Kelman 2014) and framings of vulnerability that fail to account for an island-centred view of the capacity to adapt may be overly simplistic (Barnett and Waters 2016). Furthermore, while many islands are similarly exposed because of their location in the tropics and sub tropics (Méheux et al. 2006) and their high coast to land area ratios (Nurse et al. 2014), SIDS vary in their sensitivity to climate impacts. In particular, Kiribati's low elevation increases its vulnerability to water shortages, because atolls, unlike volcanic islands, lack high peaks that generate orographic rainfall (Houghton 1979; Campbell 2018).

Ecosystems that provide freshwater supplies of coral atolls, such as Kiribati in the South Pacific, are among those most sensitive to disruption from extreme weather events and sea level rise (Woodroffe 2008). In the absence of natural surface resources, groundwater, as freshwater lenses, forms the basis of potable water supplies for these islands. However, modelling suggests that annual salinization of freshwater lenses from increased wave height and inundation alone will force abandonment of these islands long before sea level rise renders them uninhabitable (Storlazzi 2015). In addition, freshwater resources are already under threat from a range of non-climatic stresses such as over-extraction, poor sanitation practices and poor water governance, which will be amplified under a changing climate (White 1996, White et al. 2007, de Freitas et al. 2014). These factors combined impact on water quality leading to poor human health including outbreaks of water-borne diseases such as diarrhoea, eye and skin diseases (Singh et al. 2001) as well as poor nutrition and reduced health outcomes (McMichael et al. 2003), which undermine community resilience (Storey and Hunter 2010). Moreover, children often suffer disproportionally high rates of morbidity from poor water quality (Lawler and Patel 2012).

Societies around the globe have been working together, building communities and effective transactions for millennia (Mauss 1990 [1950]). The seminal work of Marcel Mauss in small island states has been followed by numerous others from a variety of fields including human geography and marine resource management (Malm 2008), social relationships and the relationship between knowledge and language (Bennardo 2009), the exchange of art (Kirch 1988), and health information (Kuruppu and Liverman 2011; Mavoa et al. 2012). Communities can be observed through both an ethnographic lens (Brewer 2000) and through an organizational perspective; exploring collaboration rather than exchange (Heckscher and Adler 2007). Collaborative communities are developing around the world due to economic, technological and ideological shifts (Botsman and Rogers 2011). The collaborative consumption of materials is not new, however in an era of collaborative consumption through information systems (e.g., the internet), exchange through social capital remains mostly intangible.

Network theory has long been used to understand a range of transactions between individuals and communities (Mitchell 1969), including the transmission of knowledge (Borgatti et al. 2013). Further, social network analysis is one method to understand the transaction of knowledge through knowledge brokers (Borgatti et al. 2013). Social network analysis has been used to improve communities of practice (Cross et al. 2006), and to help understand and improve environmental management systems such as fisheries (Ramierez-Sanches 2011). Information flow can move through a system like an epidemic, both good (e.g., cascades of cooperation in Axelrod 2006 {1984}; Fowler and Christakis 2010) and bad (e.g., obesity in Christakis and Fowler 2007). Knowledge flow in networks is frequently facilitated by knowledge brokers (Meyer 2010) and the topology and characteristics of knowledge brokers are becoming better understood in a range of settings including climate adaptation (Cvitanovic et al. 2017) and in improving health outcomes in the Pacific (Mavoa et al. 2012). Currently, information systems within the Pacific largely rely on oral traditions and the 'lived experience' (Finnegan and Orbell 1995). Through the lived experience crucial tacit knowledge may be gained and retained; yet may also remain un-transmitted and thus unknown beyond the individual, household or community. Often understanding the process of uncovering tacit knowledge and trapped social capital requires a deep understanding of place, and an ethnographic and participatory approach to unlock and translate this knowledge (Brewer 2000; Johnson et al. 2003; Kothari et al. 2011).

Previous research on policy applications of social network analysis (SNA) for climate adaptation has demonstrated that (Cunningham et al. 2014; Harman et al. 2016):

1. Information flow through communities is often fractured;
2. Effective knowledge-brokers enhance and expedite the flow of information through social strata;
3. Visualization of information networks coupled to qualitative stakeholder analysis can reveal attributes of knowledge systems that are generally invisible to institutional decision makers (government, NGOs).

To improve communication in a range of formal and informal settings (e.g., emergency management, human health, and climate adaptation), an understanding of the structure and function of information networks is essential. For small Pacific islands, timely and effective knowledge sharing across community, NGO and government networks enables effective policy implementation and rapid operational responses with implications for community health, disaster response and loss of life. In addition, information flow is important both internally (within country) and externally (e.g., external reporting to World Health Organization (WHO) and United Nations) to the region, as effective information flow will support development aid and disaster relief, reporting and resource allocation.

This paper therefore reports on the findings from a study on the sharing of information (inter and intra flows) between government, Community Based Organisations (CBO's) and individuals on the outer islands of Kiribati. The study formed part of a larger research project *Supporting Community Adaptation to Water Shortages in Kiribati.* The overarching goal of the project was to enhance the adaptive capacity of community based water management systems in the outer western islands of Kiribati to ensure sufficient water for basic health and hygiene in a changing climate (Mukheibir and Boronyak-Vaso 2016).

Approach

Empirical data for this study was collected during a workshop held in Kiribati in September 2017. This workshop was one of a suite of workshops being a major component of the UDSAID-funded project *Supporting Community Adaptation to Water Shortages in Kiribati.* Both the funder and previous project work influenced the site of this project. Further information about the broader project is available in the book chapter "Dynamic Adaptive Management Process—Supporting Community Adaptation to Water Shortages in Kiribati" (Mukheibir and Boronyak-Vaso 2016).

A social network analysis (SNA) approach was undertaken in a one day workshop with 31 participants to uncover their understanding of how information flows to and from various stakeholder groups. Workshop participants were drawn from government (policy makers, including officials from the Ministries of Public-Works and Utilities (MPWU); Health and Medical Services; Finance and Economic Development and the Environment, and Agricultural Development), Community Service Organisations (Kiribati Climate Action Network (KiriCAN), Live and Learn and the Kiribati Red Cross) and members from local Village Water Committees (VWCs) from North Tarawa (also called Buariki) and an outer island of Abaiang. Understanding of how information is shared within and between the various groups is of relevance to the participants and policy makers to aid enhancements of information flow, especially during both rapid and slow onset extreme climate events.

A range of workshop activities were designed to explore:

- The formal and informal processes by which community members receive information related to climate change adaptation and water supply options;
- The understanding of various groups of the processes for making decisions about planning and implementing water supply options;
- The linkages between formal and informal networks to better-target engagement options and outcomes;
- The linkages between formal and informal networks leading to more transparent and informed decision-making around climate change adaptation; and
- How policy makers can better engage with communities to improve the acceptance and uptake of climate adaption and water supply policies/programs/strategies.

The workshop was run in English with translation into i-Kiribati. In total, 38 people participated in the workshop (15 Female, 23 Male) representing Government (N = 10), Community Service Organisations (NGOs) (N = 11) and community members from the Village Water Committees (N = 17) from Abaiang and North Tarawa (Buariki). Unfortunately, due to timings of arrivals of some members, five attendees did not complete the survey, as such 31 participants completed the social network analysis (N = 31).

Method

A mixed methods approach was employed, using both quantitative and qualitative techniques. The quantitative component used social network analysis (SNA) to map social networks in the region related to 'where participants access climate adaptation information' and 'who they share their information with regarding local changes in their landscape and water supplies'. SNA has been used in a range of instances including mapping community knowledge networks for climate adaptation policy (Harman et al. 2015a, b, 2016), participatory scenario planning (Krupa et al. 2018) and identification of transformation change makers (Dowd et al. 2014) among others (Borgatti et al. 2013). The researchers referred to information collected during previous project workshops to provide background data for the scenarios (Mukheibir and Boronyak-Vaso 2016).

Three scenarios were used:

1. Low Rainfall: Low or no rainfall results in dry or drought conditions with impacts to water quality and quantity (e.g., water shortage)
2. Storm Surge: Full moon and high tide with a big storm leading to coastal inundation possibly resulting in salty and contaminated water
3. Poor water quality: people getting sick from bad water causing health impacts.

Participants considered these scenarios in small groups made up of:

1. Non-Government Organisations (NGOs) and Community service organisations (CSO's) (Red Cross, Live and Learn, KiriCAN) (N = 7)
2. Government representatives—Senior managers and decision-makers from the Ministry of Public-Works and Utilities (MPWU) responsible for delivering clean water to the communities and staff from the Ministries of Health and Medical Services; Finance and Economic Development and Environment, Lands and Agricultural Development (N = 9)
3. Village/community representatives, village water committee members (VWC) from Abaiang (N = 7) and North Tarawa (Buariki) (N = 8).

There was a three step process to these activities. Firstly, participants individually completed a survey for each scenario to provide input for the SNA (Fig. 5). Within this survey participants were asked if they accessed and shared information in each of the scenarios. If they did access/share information, they were asked to describe how they both share and/or receive this information (e.g., name of a person, a newspaper, a notice board etc.). The results of the individual SNA questions regarding the knowledge network allow researchers to create knowledge flow maps using quantitative SNA. The outputs of the SNA are outlined in the results section below.

After the individuals completed the survey for each scenario, they mapped out each scenario as a sectoral group (Figs. 6, 7 and 8 offer examples of group maps). This element was the qualitative component. The scenario mapping undertaken by sectoral groups allowed for the discussion to develop around the current successes of communication, and opportunities for improvement.

Thirdly and finally, in a plenary session the sectoral maps were shared and potential next steps discussed (Figs. 6, 7 and 8). The discussion focused on where information was sourced, how information was shared about these impacts on water, and if there was a way to optimize the current information system.

Although this method within the SNA approach is considered best practice, particularly when data is collected directly from participants (Borgatti et al. 2013) there is a limitation in that only those who attended the workshop completed the SNA survey, rather than the optimum completion rates wherein every node that is mentioned in the network completes the SNA survey.

Results

The responses to the participant SNA surveys (N = 31| CSO = 7, Buariki = 8, Gov = 9, Abaiang = 7) were used to create directional symmetric matrixes for the three scenarios, thus creating 6 networks (Low Rainfall Get, Low Rainfall Share, Storm Surge Get, Storm Surge Share, Poor Water Quality Get, Poor Water Quality Share). Every nomination from a participant became an individual node in the network. For example, if a participant listed "radio" as an information source, "radio" would be a node within the network. As there are limited radio stations in the region, it was possible to qualitatively estimate and aggregate all participant answers of "radio". In order to maintain anonymity of both the participant and the information source and/or share point, each response was aggregated into a category being Community (e.g., family member, neighbour, local meetings, local wise people "Tani Bouru"), Local Civil Society (e.g., Church, School, work colleagues, Village leaders), Government (Department of Health, OB, etc.), Health (Clinic, Doctor, Nurse), High Tech Media (internet, facebook), Low tech media (Newspaper, Radio, CB radio, telephone),CSO (other CSO events, or individuals), CSO International, Observations (personal observations of weather or health symptoms), Office Bearers (Council, Mayor, Village Water Committee, Policeman, Island leaders), and Water (Water Technicians). To ensure the anonymity of the participants the respondents were aggregated and grouped into Abaiang, Government, CSO and Buariki. Within the analysis, Buariki self-nominated to be North Tarawa so is coded as NT in this analysis.

Figure 1 outlines the network nodes types and the number of each node type per network. For example, during scenario 1 (Low Rainfall) participants reported obtaining information from 12 discrete government sources, and shared with 2 government sources. Within scenario 2 (Storm Surge), participants reported obtaining information from an equal number of sources including community, government and low tech media. However, as Fig. 3 shows (storm scenario top), the low tech media (Radio), received far more nominations than the government node (MET office). Table 1 further demonstrates that participants had larger Access/Get networks in low

Table 1 Network centrality and whole network measures for each scenario

Measure	Low rain get	Low rain share	Storm get	Storm share	Poor water get	Poor water share
Average degree	1.300	1.446	1.467	1.600	1.436	1.448
Indeg H-index	5	5	4	5	5	5
Deg centralization	0.070	0.067	0.079	0.072	0.068	0.101
Out-Central	0.069	0.066	0.078	0.071	0.067	0.099
In-Central	0.187	0.362	0.354	0.383	0.218	0.260
Density	0.019	0.026	0.025	0.033	0.027	0.025
Components	70	56	60	50	55	58
Component ratio	1	1	1	1	1	1
Connectedness	0.019	0.026	0.025	0.033	0.027	0.025
Fragmentation	0.981	0.974	0.975	0.967	0.973	0.975
Ave distance	1	1	1	1	1	1
SD distance	0	0	0	0	0	0
Diameter	1	1	1	1	1	1
Wiener index	91	81	88	80	79	84
Dependency sum	0	0	0	0	0	0
Breadth	0.981	0.974	0.975	0.967	0.973	0.975
Compactness	0.019	0.026	0.025	0.033	0.027	0.025
Mutuals	0	0	0	0	0	0
Asymmetrics	0.038	0.053	0.050	0.065	0.053	0.051
Nulls	0.962	0.947	0.950	0.935	0.947	0.949
Arc reciprocity	0	0	0	0	0	0
Dyad reciprocity	0	0	0	0	0	0

Source Authors

rain and storm surge scenarios (Low Rainfall Get N = 70/Share N = 56); Storm Surge Get N = 60/Share = 50) whereas in the poor water quality scenario, participants reported to have larger access than share networks (Poor Water Quality Get N = 55/Share N = 58).

The outputs from the SNA survey were quantitatively analyzed using UCinet and Keyplayer; the visualizations were made using Netdraw (Borgatti et al. 2002). Within the visualizations, the layout is arranged so that nodes that are closer together are more similar and those on the periphery are more isolated. Within Figs. 2, 3 and 4 the size of the node also demonstrates the number of nominations that node has received; for example, the larger the node, the larger the number of participants that have listed it as a source or share point of information (Fig. 1).

A visualisation map was made for each of the scenarios:

(1) **Low rain**—Participants reported that they get information from low technology information sources and some formal government channels such as the Ministry of Public Works. Both a VWC and a Government sector participant reported to received information from local traditional holders of knowledge such as the "Tani Borau". All participants reported sharing information with their community informally with some Government and NGO participants reported to use high technology information streams (e.g., Facebook) to share information, and there were Community warnings through the Radio and local village boards (TTM boards).

(2) **Storm surge**—Participants reported that they get information from low technology information sources, such as radio (in particular those in VWCs) and participants who were from CBOs and Government also sought information from government sources (e.g., MET and Ministry of Health). All participants reported to share primarily with community members informally, with some government and NGOs noting to share information with Island leaders and newspapers.

(3) **Poor Water Quality**—Participants from VWCs reported that they primarily get information about water quality through symptoms, either as they or members of their family became sick. Whereas, NGOs and Government participants were receiving information from low tech information sources, as well as formal channels such as clinics and water technicians. All participants reported to share within their informal networks of family, friends, neighbours and church members.

Table 1 outlines the network centrality measures for each network. These quantitative measures show very low average degree numbers, that is, the average number of nodes that each node is linked to. Furthermore, the density of the network remained low in each scenario. Density is determined by calculating the number of ties divided by the possible number of connections (Borgatti et al. 2002). Connectedness is also very low in each network, with connectedness being the proportion of nodes that can reach each other (Borgatti et al. 2002). For a full explanation of these measures, please refer to Cunningham et al. (2017).

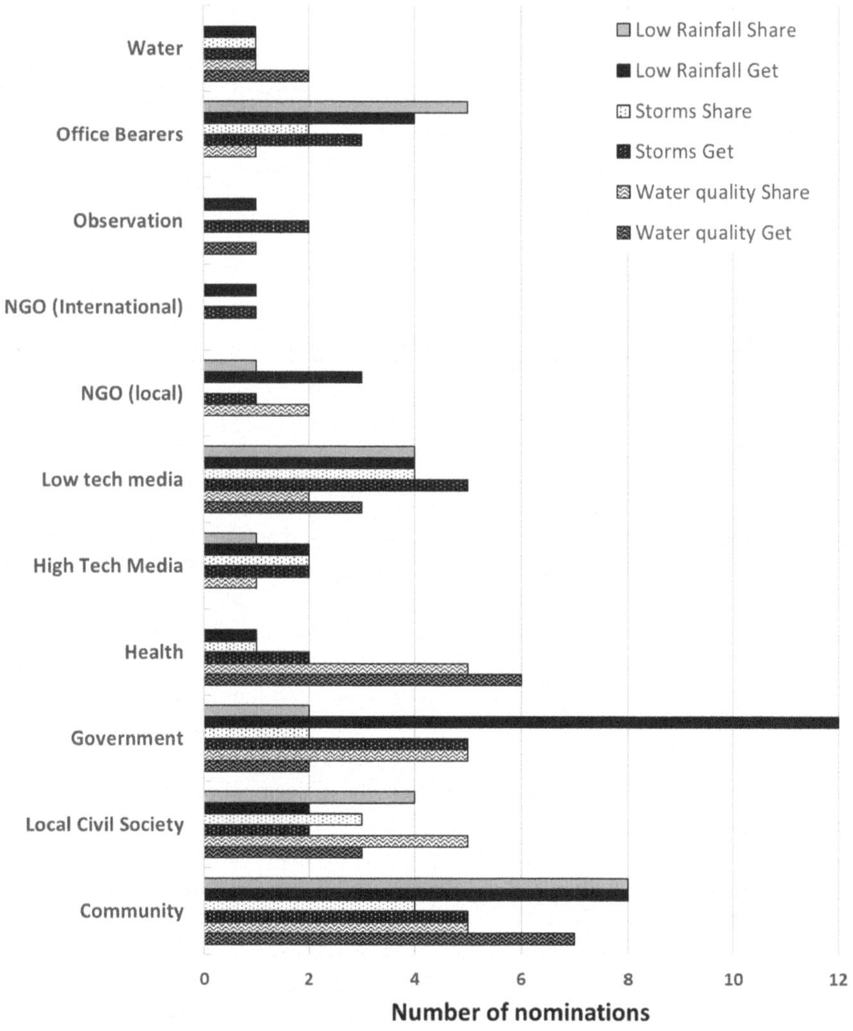

Fig. 1 Numbers of nodes nominated per sector and scenario. *Source* Authors

Table 2 outlines the findings from Keyplayer wherein the keyplayer diffusion algorithm tries to find the 3 most highly connected nodes to connect with every other node in the system. The "reach" nominates the percentage of the network connected through these three "keyplayers" (the table is to be read along each row from left to right) (Borgatti 2006). The reach of these key players for each network is low with each scenario, with the exception of "Poor Water Share" reaching less than a third of the network. Both Government and NGOs dominated the key players occupying many of the prominent positions within the network. In the "Storm Get" scenario, the algorithm found three distinct combinations of keyplayers that could reach 24.6% of the network. The low reach demonstrates the fractured nature of these networks.

Table 2 Keyplayer findings for each network

Scenario	Key players			Reach (%)
Low rain get	GOV 3	GOV4	NGO4	22.4
Low rain share	NGO4	NGO5	NT3	24.5
Storm get	GOV6	GOV8	NGO6	24.6
	GOV6	NGO6	NT1	24.6
	GOV6	NGO6	NT6	24.6
Storm share	NGO4	NGO5	NGO6	27.7
Poor water get	GOV4	NGO3	NGO5	25.0
Poor water share	GOV4	NGO4	NT3	32.7

Source Authors

Discussion

An SNA approach to information sharing among participants about aspects of water revealed knowledge networks associated with each scenario (Low Rain, Storm Surge and Water Quality). The findings demonstrate that these networks varied considerably between scenarios. In times of low rain and storm, participants reported to access information more readily from formal sources (such as Government Department of Meteorology) and share information informally (with family and friends). This was encouraging as the formal sources should have the most accurate information regarding the incidence of low rain and storm events which may impact water quality. However, within the water quality network, the majority of participants reported the appearance of symptoms of illness as the trigger to know that their water source was not of potable quality. Subsequently, participants then shared information about the quality of the water with informal information sources. This finding is important because it limits the capability of information exchange to initiate disease prevention measures. Furthermore, it emphasizes the need for improved ground water well testing and timely communication of the results as water quality is declining to reduce the impact of water borne illness. Actions to improve this situation have commenced, such as the convening of Village Water Committees and training for water technicians in regular well water testing. However, as demonstrated by the network visualizations (Figs. 2, 3 and 4), the knowledge connections are not yet in place to ensure the flow of information from water technicians through the Village Water Committees back to citizens. Establishment of this knowledge exchange represents an important opportunity for intervention to ensure that information regarding water quality is shared quickly and effectively with relevant community members and health practitioners across Kiribati.

The statistical measures each of these six networks indicated that they were highly fragmented with values of fragmentation higher than 0.95. This finding is reflected in the Keyplayer output (Table 2), which shows that utilization for information exchange of the three most highly connected nodes in each of the six networks, enables reach

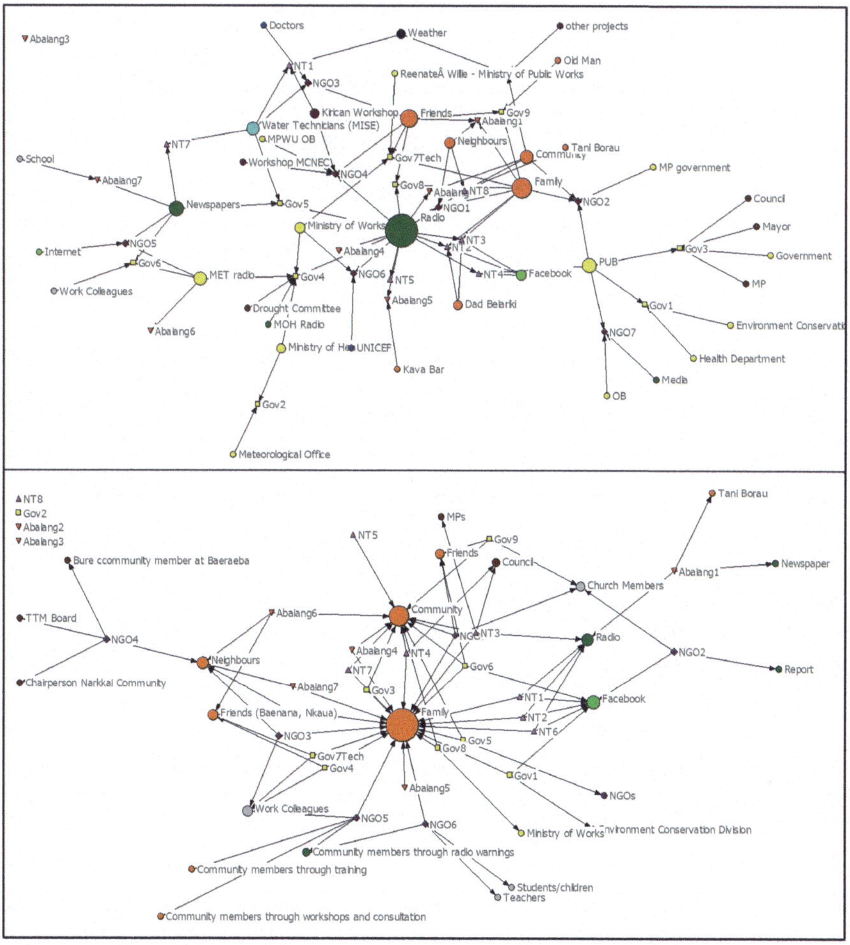

Fig. 2 Information networks for the low rainfall scenario showing get (top) and share networks (bottom). Symbols are stakeholders: Abaiang (▽), government (□), NGO (◊), NT (Δ), NA (○). Colours are categories: Abaiang (■), community (■), government (■), health (■), high tech media (■), international NGO (■), local civil society (■), low tech media (■), local NGO (■), NT (■), observation (■), office bearers (■), water (■). *Source* Authors

to less than one third of the network actors. This demonstrates high levels of network disconnectedness, and presents significant opportunity for network growth and intervention to strengthen its interconnectedness and information flows. If the networks were to become more closely connected then this would allow for improved information to flow among the stakeholder types. Deliberate insertion into these networks of trained knowledge brokers, to play a specific role in routine communication of water quality information and illness prevention may better connect these networks with

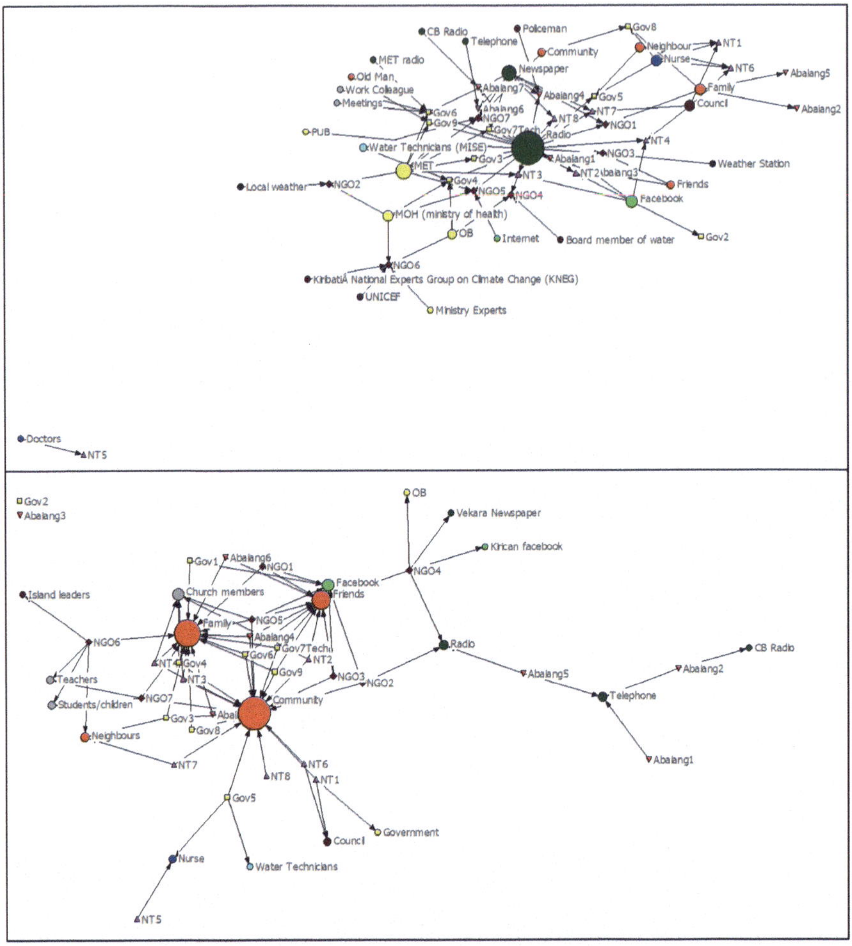

Fig. 3 Information networks for the storm scenario showing get (top) and share networks (bottom). Symbols are stakeholders: Abaiang (▽), government (□), NGO (◊), NT (Δ), NA (○). Colours are categories: Abaiang (■), community (■), government (■), health (■), high tech media (■), international NGO (■), local civil society (■), low tech media (■), local NGO (■), NT (■), observation (■), office bearers (■), water (■). *Source* Authors

associated improvements in health outcomes through timely intervention (Meyer 2010).

It was encouraging to see the Ministry of Public Works and Utility water technicians (who are based on the outer islands) feature in many of these networks, albeit mostly on the periphery of the network with often only one or two connections. They, together with the locally based nurses, have a significant role in keeping the outer island communities connected to government information sources, and vice versa,

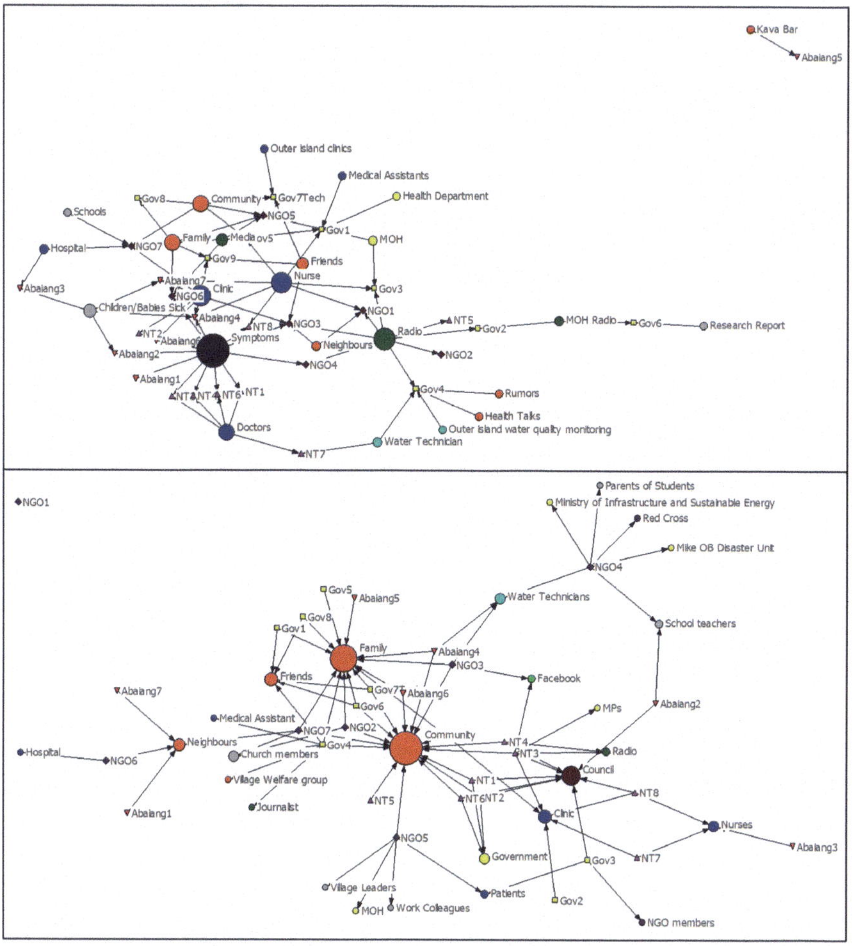

Fig. 4 Information networks for the poor water quality scenario showing get (top) and share networks (bottom). Symbols are stakeholders: Abaiang (▽), government (□), NGO (◇), NT (△), NA (○). Colours are categories: Abaiang (■), community (■), government (■), health (■), high tech media (■), international NGO (■), local civil society (■), low tech media (■), local NGO (■), NT (■), observation (■), office bearers (■), water (■). *Source* Authors

by acting as a more centralised node with potentially many inter-connections. Recognition of their role in information sharing will be key for improving the amount and rate of information flow under a changing climate. Without this improvement, there is little expectation that morbidity rates due to contaminated water will improve.

While the study revealed interesting and useful information on which to design interventions to improve the knowledge transfer, there were some limitations and areas for improvement. It is important to note that the region is geographically fragmented, Kiribati being made up of a series coral atolls, and there is an historical

USAID FROM THE AMERICAN PEOPLE
Pacific-American Climate Fund (PACAM)
UTS:ISF KiriCAN

Supporting community adaptation to water shortages: Mapping Information Flow

Access Information **Share Information**

Your full name	
Organisation/s that you represent	
Gender (Circle)	Male Female

STORM SURGE:

5. When there is a storm surge do you get any information about water supply quality?

Yes		No	

6. If yes:

Where do you get information about drinking water / water quality from?

	Who or How? (e.g., name of person, newspaper, notice board etc)
1.	
2.	
3.	

7. When there is a storm surge do you share any information about water supply quality?

Yes		No	

8. If yes:

Where do you share information about drinking water / water quality from?

	Who or How? (e.g., name of person, newspaper, notice board etc)
1.	
2.	
3.	

LOW RAINFALL:

1. When it is dry do you get any information about water supply quantity & quality?

Yes		No	

2. If yes:

Where do you get information about drinking water / water quality from?

	Who or How? (e.g., name of person, newspaper, notice board etc)
1.	
2.	
3.	

3. When it is dry do you share any information about water supply quantity & quality?

Yes		No	

4. If yes:

Where do you share information about drinking water / water quality from?

	Who or How? (e.g., name of person, newspaper, notice board etc)
1.	
2.	
3.	

POOR WATER QUALITY:

9. When people get sick from bad water, how do you know?

	Who or How? (e.g., name of person, newspaper, notice board etc)
1.	
2.	
3.	

10. If you know people are sick from bad water, who do you tell?

	Who or How? (e.g., name of person, newspaper, notice board etc)
1.	
2.	
3.	

Thank you / Ko rab'a

Fig. 5 SNA survey instrument. *Source* Authors

tradition of information being shared orally (Mauss 1990 [1950]). These factors clearly continue to play a role in shaping the effectiveness of water quality knowledge networks. However, our analysis was limited to survey responses from the participants in the workshop process rather than from the whole information system

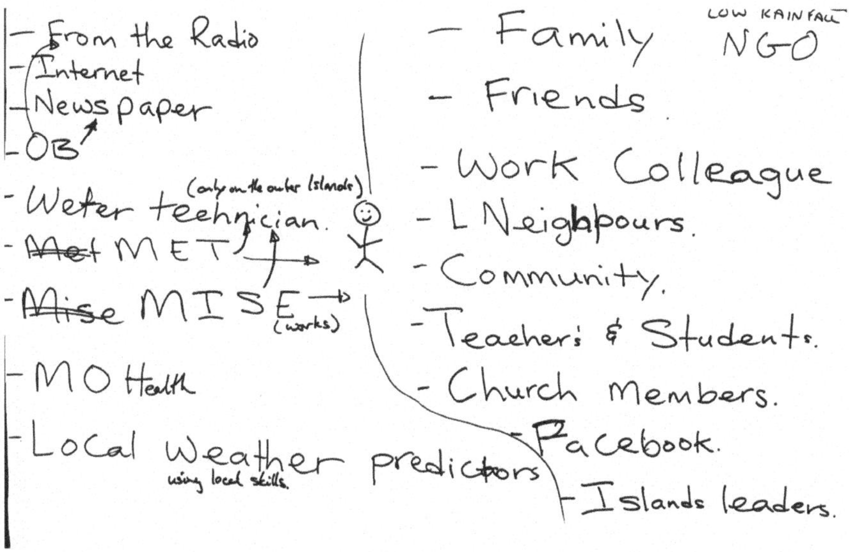

Fig. 6 Information Flow scenario 1: low rainfall—NGO/CBOs. *Source* Authors

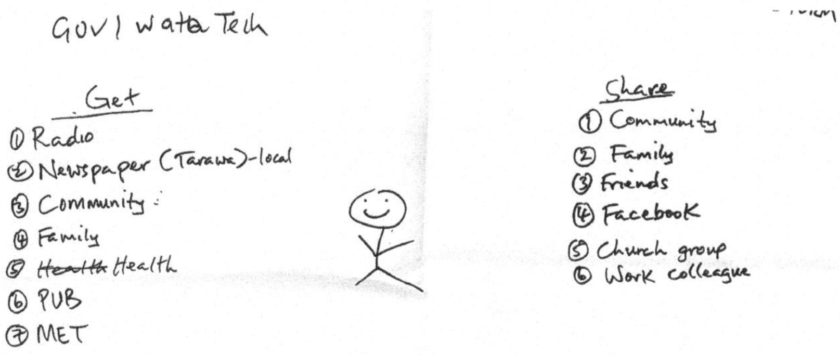

Fig. 7 Information flow scenario 2: Storm: Government. *Source* Authors

and, further, the layout of the survey limited participants' responses to three connections in the network (Fig. 5). These factors suggest that the networks may have been more extensive than our visualisations suggest. While the research presented in this paper provides a snapshot of the information flows at the time of this study, further research is required to better understand how knowledge flows could be optimized in the region for the future (Figs. 6, 7 and 8).

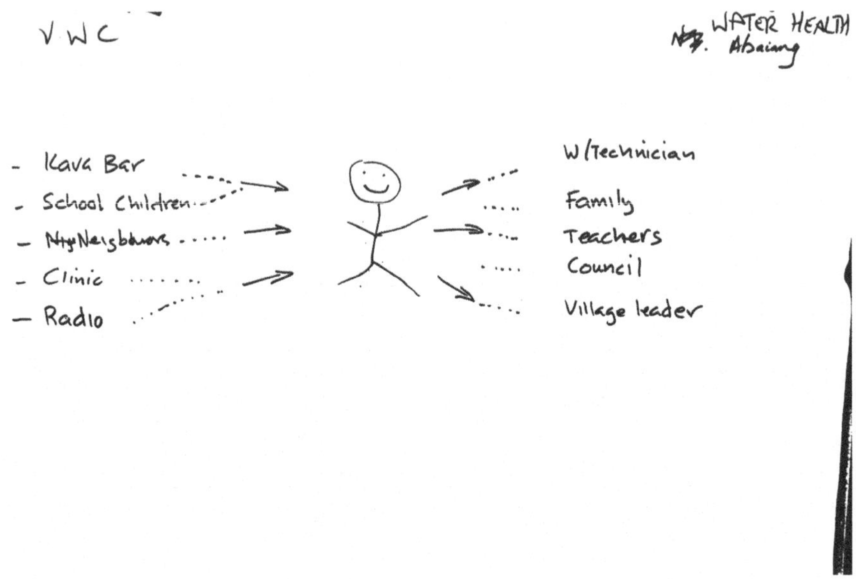

Fig. 8 Information flow scenario 3: water quality health: Village Water Committees. *Source* Authors

Conclusions

The variation of network structures between scenarios demonstrates the importance of formal and informal networks for both accessing and sharing water supply information in Kiribati.

Networks derived from the project were not definitive, however there is sufficient evidence that further enhancement of knowledge networks may deliver improved health outcomes in Kiribati, and in doing so may have significant benefits to communities (e.g., effective communication of water quality in reducing child morbidity). These options are relatively inexpensive within a developing country context as it is investment in social capital.

The approach of SNA for revealing knowledge networks may be applied to alternate settings within the Pacific region to benefit atoll communities. Finally there are broader lessons regarding the influence of geography and culture on knowledge networks both within the Pacific region and further afield.

Acknowledgements This research was funded by USAID through their Pacific-American Climate Fund. The research team acknowledge the important partnership of the Kiribati Climate Action Network (KiriCAN) staff. The authors would also like to thank all who participated in this research and each of the reviewers.

References

Axelrod R (2006{1984}) The evolution of cooperation. Basic Books, New York

Barnett J, Waters E (2016) Rethinking the vulnerability of small island states: climate change and development in the Pacific Islands. The Palgrave Handbook of International Development. J. Grugel and D. Hammett. Palgrave Macmillan, London, pp 731–748

Bennardo G (2009) Language, space and social relationships: a foundational cultural model in Polynesia. Cambridge University Press, Cambridge

Borgatti SP (2006) Identifying sets of key players in a social network. Comput Math Organ Theor 12:21–34

Borgatti SP, Everett MG, Freeman LC (2002) Ucinet for windows: software for social network analysis. Analytics Technology, Harvard, MA

Borgatti SP, Everett MG, Johnson JC (2013) Analyzing social networks. Sage Publications, London

Botsman R, Rogers R (2011) What's mine is yours: how collaborative consumption is changing the way we live. Collins, London

Brewer JD (2000) Ethnography. Open University Press, Buckingham

Campbell JR (2018) Migration and climate change in the Pacific. Handbook of migration and globalisation. A. Triandafyllidou. Edward Elgar Publishing, pp 379–396

Christakis N, Fowler J (2007) The spread of obesity in a large social network over 32 years. N Engl J Med 357:370–379

Cross R, Lasester T, Parker A, Velasquez G (2006) Using social network analysis to improve communities of practice. Calif Manage Rev 49(1):32–61

Cunningham R, Cvitanovic C, Measham T, Jacobs B, Dowd A, Harman B (2014) A preliminary assessment into the utility of social networks for engaging local communities in climate adaptation policy: working paper prepared for NSW Office of Environment & Heritage. Sydney, Australia, Institute for Sustainable Futures

Cunningham R, Jacobs B, Measham T, Harman B, Cvitanovic C (2017) Social network analysis: a primer on engaging communities on climate adaptation in New South Wales, Australia. UTS:ISF, Australia

Cvitanovic C, Cunningham R, Dowd A-M, Howden SM, van Putten EI (2017) Using social network analysis to monitor and assess the effectiveness of knowledge brokers at connecting scientists and decision-makers: an Australian case study. Environ Policy Gov 27(3):256–269

de Freitas CR, Helbig M, Matzarakis A (2014) Hydroclimatic assessment of water resources of low Pacific islands: evaluating sensitivity to climate change and variability. Int J Climatol 34:881–892

Dowd A, Marshall NA, Fleming A, Jakku E, Gauillard E, Howden M (2014) The role of networks in transforming Australian agriculture. Nat Clim Change 4(558–563)

Finnegan RH, Orbell MR (eds) (1995) South Pacific oral traditions. Indiana University Press, Bloomington, Indiana

Fowler J, Christakis N (2010) Cooperative behavior cascades in human social networks. Soc Sci. https://doi.org/10.1073/pnas.0913149107

Harman B, Cunningham R, Cvitanovic C, Jacobs B, Measham T (2015a) Community based perspectives on climate change and adaptation in the Shoalhaven region. CSIRO, New South Wales, Australia

Harman B, Cunningham R, Jacobs B, Measham T, Cvitanovic C (2015b) Engaging local communities in climate adaptation: a social network perspective from Bega Valley. CSIRO, New South Wales, Australia

Harman B, Rylandce K, Brown P, Cunningham R, Jacobs B, Measham T (2016) Engaging local communities in climate adaptation: a social network perspective from Orange Valley, New South Wales, Australia. CSIRO, Australia

Heckscher C, Adler P (2007) The firm as a collaborative community: reconstructing trust in the knowledge ecomony. Oxford University Press, USA

Houghton JG (1979) A model for orographic precipitation in the North-Central Great Basin. Mon Weather Rev 107:1462–1475 (American Meteorological Society)

Johnson JC, Boster JS, Palinkas LA (2003) Social roles and the evolution of networks in extreme and isolated environments. J Math Sociol 27(2–3):89–121
Kelman I (2014) No change from climate change: vulnerability and small island developing states. Geogr J 180(2):120–129
Kelman I, West JJ (2009) Climate change and small island developing states: a critical review. Ecol Environ Anthropol 5(1):1–16
Kirch PV (1988) Long-distance exchange and Island colonization: the Lapita case. Norw Archaeol Rev 21(2):103–117
Kothari AR, Bickford JJ, Edwards N, Dobbins MJ, Meyer M (2011) Uncovering tacit knowledge: a pilot study to broaden the concept of knowledge in knowledge translation. BMC Health Serv Res 11
Krupa M, Cenek M, Powell J, Trammell EJ (2018) Mapping the stakeholders: using social network analysis to increase the legitimacy and transparency of participatory scenario planning. Soc Nat Resour 31(1):136–141
Kuruppu N, Liverman D (2011) Mental preparation for climate adaptation: the role of cognition and culture in enhancing adaptive capacity of water management in Kiribati. Glob Environ Change 21(2):657–669
Lawler J, Patel M (2012) Exploring children's vulnerability to climate change and their role in advancing climate change adaptation in East Asia and the Pacific. Environ Dev 3:123–136
Malm T (2008) Outriggers lost in the sea of time: an overlooked aspect of cultural change and conditions for sustainable development in Oceania. Tradit Mar Resour Manage Knowl Inf Bull 23:3–12
Mauss M (1990 [1950]) The gift: the form and reason for exchange in archaic societies. W.W. Norton, New York, London
Mavoa H, Waqa G, Moodie M, Kremer P, McCabe M, Snowdon W, Swinburn B (2012) Knowledge exchange in the Pacific: the TROPIC (Translational Research into Obesity Prevention Policies for Communities) project. BMC Public Health 12(1):552
McMichael AJ, Woodruff R, Whetton P, Hennessy K, Nicholls N, Hales S, Kjellstrom T (2003) Human health and climate change in Oceania: a risk assessment. Commonwealth Department of Health and Ageing, Canberra, Australia
Méheux K, Dominey-Howes D, Lloyd K (2006) Natural hazard impacts in small island developing states: a review of current knowledge and future research needs. Nat Hazards 40(2):429–446
Meyer M (2010) The rise of the knowledge broker. Sci Commun 32:118–127
Mitchell CJ (1969) Social networks in urban situations: analyses of personal relationships in Central African Towns. Manchester University Press
Mukheibir P, Boronyak-Vaso L (2016) Dynamic adaptive management process—supporting community adaptation to water shortages in Kiribati. P. f. U. (PACAM). Institute for Sustainable Futures, University of Technology Sydney, Sydney, Australia
Nurse LA, Mclean RF, Agard J, Briguglio LP, Duvat-Magnan V (2014) Small Islands. Climate change 2014: impacts, adaptation, and vulnerability. Part B: regional aspects. Contribution of working group II to the Fifth assessment report of the Intergovernmental panel on climate change. Barros V.R., C. B. Field, D. J. Dokken et al. Cambridge University Press, Cambridge, pp 1613–1654
Pelling M, Uitto J (2001) Small Island developing states: natural disaster vulnerability and global change. Environ Hazards 3:49–62
Ramierez-Sanches S (2011) Who and how: engaging well-connected fishers in social networks to improve fisheries management and conservation. In: Bodin O, Prell C (eds) Social networks and natural resource management. Cambridge University, New York, pp 119–146
Singh RB, Hales S, De Wet N, Raj R, Hearnden M, Weinstein P (2001) The influence of climate variation and change on diarrheal disease in the Pacific Islands. Environ Health Perspect 109(2):155
Storey D, Hunter S (2010) Kiribati: an environmental 'perfect storm'. Aust Geogr 41(2):167–181

Storlazzi CD, Elias EP, Berkowitz P (2015) Many atolls may be uninhabitable within decades due to climate change. Sci Rep 5(14546)

White I (1996) Preliminary report: freshwater lens recharge at Bonriki, Kiribati. IHP-V Technical Documents in Hydrology No. 5. Paris, France, UNESCO, pp 1–38

White I, Falkland T, Perez P, Dray A, Metutera T, Metai E, Overmars M (2007) Challenges in freshwater management in low coral atolls. J Cleaner Prod 15(16):1522–1528

Woodroffe CD (2008) Reef-island topography and the vulnerability of atolls to sea-level rise. Glob Planet Change 62:77–96

Climate Change and Peri-Urban Household Food Security—Lessons from West Taraka, Morobe Province, Papua New Guinea

Zina Bird and Linda Yuen

Abstract Climate change has become a major concern towards the stability of global food production due to long and short-term climate related events. This paper will incorporate climate data to build on the existing data on the status of household food and nutrition security in one of Lae's peri-urban settlement, West Taraka in Morobe Province, Papua New Guinea. Two data sets were collected: household dietary patterns and changes in food production, and socio-economic characteristics, using stratified purposive sampling for selected fifty-eight (58) households in June 2016 through household survey and informal interviews (mixed method). Results show no statistical relationships between socio-economic characteristics of the households and their Household Dietary Diversity Score and Food Consumption Score. However, a significant inverse relationship at 95% probability exists between the numbers of household members in school with the Household Food Consumption Score. This study also found a significant positive relationship at 99% level probability between household income and Food Consumption Score signaling that income was the main determinant of household food and nutritional security.

Keywords Climate change · Food security · Household Food Consumption Score · Household Dietary Diversity Score · Papua New Guinea · Agriculture · Urbanisation

Introduction: Global Context of Climate Change and Agriculture

Climate change is a global phenomenon altering climate systems and having a wide range of impacts on human and natural systems. However, climate-related risks for

Z. Bird · L. Yuen (✉)
Pacific Centre for Environment and Sustainable Development (PaCE-SD), The University of the South Pacific, Lower Marine Campus, Laucala Campus, Suva, Fiji
e-mail: yuen.bk.linda@gmail.com

Z. Bird
e-mail: zina.bird@gmail.com

W. Leal Filho (ed.), *Managing Climate Change Adaptation in the Pacific Region*, Climate Change Management, https://doi.org/10.1007/978-3-030-40552-6_9

countries are unevenly distributed, with geographic location largely influencing the direction and rate of climatic change (IPCC 2018; Niles et al. 2017).

In agriculture or food system, it is recognized that climate and weather play a critical part towards food production and livelihoods (Niles et al. 2017). Various food crops require different stages or variation in weather conditions (max and min temperature, rainfall/precipitation) and physical conditions (soil characteristic) for development, growth and yield. Because of the dependence on climate and the physical environment, it was stated by Rauff and Bello (2015), as temperature and carbon dioxide (CO_2) increases, it may affect the biological process of plants such as photosynthesis, growth, respiration, reproduction and the nutrient content in the soil. Although increase in CO_2 and temperature can be beneficial for photosynthesis, it can also be harmful once the optimum range is exceeded and can reduce the amount of water content through the process of transpiration hindering the pores of the plants to open thus leading to a decline in yield or harvest (Rauff and Bello 2015).

For developing countries, including those of the Pacific islands region, that depend on agricultural production for socio-economic stability, declining crop production due to increase in temperature, change in precipitation pattern and introduction of new pest and diseases may have widespread negative impacts on food quality and safety (Niles et al. 2017). These biophysical factors also impacts the marine (aquaculture) and forestry systems.

For vulnerable Pacific Island countries, the impact of climate variability and extreme weather events such as flooding, drought, increase frequency of rain and cyclone result in significant loss of agricultural production (FAO 2018). Although farmers have developed and adopted potential methods to minimize climate related risk, the challenges to cope with the changes will still not be eliminated (Taylor et al. 2016). This increases the climate risk leading to increase in poverty in some population as global warming continues to increase affecting food security. Limiting warming to 1.5 °C can reduce the number of people exposed to poverty and climate related risk by 2050 (IPCC 2018). Bogard et al. 2018 found relevant indicators to measure the strength and limitations of agricultural production, analyzing the nutritional yields, potential nutrient adequacy and Rao's quadratic entropy by capturing the ability of the production system that can nourish most people. These serve as useful tools for prioritizing agriculture-focused decision-making and policy-making in the public, private and civil society sectors.

Geography and Climate of PNG

Papua New Guinea (PNG) is the largest of the Pacific small island developing states (PSIDS) and comprises the eastern half of the island of New Guinea and about 700 islands between the equator and 12° S, and 140° E to 160° E, in a region which is particularly prone to seismic activity. The population of 8.3 million is dispersed throughout a total landmass of 463,000 km^2, 74% of which is under forest cover (UNDP 2018). PNG's diverse terrestrial landscapes range from the montane rainforests of the Western Highlands to the Trans-Fly savanna and grasslands and from the mangrove swamps of the estuaries to the low-lying atolls. With over 800 distinct spoken languages, PNG is well known for being the most linguistically and culturally diverse place in the world, mainly attributed to geographical barriers hindering inter-tribal contact and interactions (Foley and Foley 1986).

Due to the proximity of PNG to the equator, there is little seasonal variation in atmospheric temperature across PNG and any significant variation is driven by changes in temperature of the sea surrounding the landmass. The southern regions experience a distinct wet season from November to April and a dry season from May to October. For example, Port Moresby receives an average annual rainfall of 1190 mm, 78% of which falls during the wet season. In comparison, seasonal rainfall variation is much weaker in the northern regions, where places like Kavieng receives as much as 3150 mm rainfall annually—almost three times more than in the southern regions. This is due to the northern region's proximity to the Western Pacific Warm Pool, which produces consistent rainfall throughout the year (BoM and CSIRO 2011). The montane rainforests of the Western Highlands receive an annual average of 7000 mm of rainfall (World Climate Guide 2019), further reflecting the uneven spatial distribution of rainfall across different regions of PNG.

The wide range in climatic conditions and physical landscapes have allowed PNG to support some of the world's most biologically diverse eco-regions. These conditions are also conducive to enabling the people of PNG to grow a wide range of food crops to meet their daily nutritional needs. Climate variables such as rainfall, temperature (air and sea), soil moisture and sea level directly affect plants development and yield. The increasing frequency and intensity of extreme climate events like frost, tropical cyclone and drought under current anthropogenic climate change is becoming a growing threat to crop production and food security (Wairiu et al. 2012; Weber 2014).

Although climate change is a global phenomenon, the impact in food security differs between regions. PNG is vulnerable to climatic extremes, such as frost, drought and flooding, which could cause poor crop harvests and increase food insecurity, poverty and disease (Bourke 2001).

Subsistence Agriculture

Papua New Guinea's local economy is driven mainly by two sectors—(i) the mineral and energy extension sector, accountable for national export earnings, and (ii) the agriculture, forestry and fishery sector, mainly supporting the domestic subsistence economy (Gwatirisa et al. 2017).

Globally, 800 million people practice urban agriculture for basic consumption purposes (Game and Primus 2015; Lwasa et al. 2014). Similarly, for most PSIDS, agriculture is one of the most important sectors contributing towards the livelihoods, food security and gross domestic product (GDP) (Rosegrant et al. 2015; Taylor et al. 2016).

In PNG, the local agriculture sector supplies 83% of food energy and 76% of protein towards the country's nutritional needs (Bourke and Harwood 2009). It provides a safety net to rural and peri-urban communities in PNG, employing about half of the labour force and contributing 15% towards GDP (ACIAR 2018). Most of the rural population is involved in producing most of their own staple foods and are also engaged in other forms of income earning activities to enable them to purchase foods which they do not produce themselves. Therefore, PNG can be generally considered as being food secure. However, some studies done in various locations in PNG (Bourke 2001; Bue 2013; Gwatirisa et al. 2017) have found that certain locations in the country are facing threats in household food and nutritional security. One such setting is the informal or peri-urban settlements which surround major urban centres of PNG. Vulnerable populations such as these have contributed to PNG's classification by FAO as a Low-Income Food-Deficit Country (LIFDC) (FAO 2015). Although PNG, on a national scale, produces enough of its own food, it is still facing malnutrition issues, exacerbated by social inequality and inadequate awareness on nutritional security and crop resilience.

In most of the informal peri-urban settlements that exist on the outskirts of PNG's major urban centres, public infrastructure and income-generating opportunities are often rudimentary, if available at all. The livelihoods of majority of the populations in these communities are centred on low wage employment and the informal sector (Umezaki and Ohtsuka 2003). With limited or no access to land for food gardening, households are heavily dependent on fresh food purchased from the local produce markets and store-bought processed food. The high cost of living in urban areas often force low income-earners to compromise their children's welfare and education in order to acquire just the basic staple foods for the household.

Like most Melanesian countries, most (97%) of the land in PNG is under customary ownership (AusAID 2008). The State owns 2.5% as public land while the remaining 0.5% is freehold land which can be privately owned by individuals. The development of customary land for commercial purposes is usually managed through Incorporated Land Groups (ILG), which in essence serves as a trust for individual landowning groups.

At 13.1%, PNG currently has the lowest proportion of urban population among PSIDS, however, this is projected to rise to 24% by 2050 (UN DESA 2018). Rapid urbanisation, compounded by the impacts of climate change will make it increasingly difficult for urban and peri-urban populations meet basic household nutritional needs.

There is currently limited information on the relationship between climate change and household nutrition in PNG, particularly for informal settlements. This study aims to understand the status of household food and nutritional security of the West Taraka peri-urban settlement by profiling the selected socio-economic characteristics of households, examining the household dietary patterns and analysing the relationship between the two. The impact of climate change is also taken into consideration for this benchmark West Taraka study, which will potentially be a useful contribution to filling in the existing knowledge gap and to inform future policy-making processes from the community level up to the national level.

The Study Site

This study focuses on a typical peri-urban settlement—West Taraka, situated on the fringes of Lae City in the Morobe Province (Figs. 1 and 2).

The site was selected mainly for two reasons. Firstly, it is one of the project site of the Agriculture Department of the PNG University of Technology (Unitech) which conduct extension programmes to the community who farm the state agricultural land. Secondly, the site is close to Unitech and allowed easy access for the researcher to conduct the field survey.

West Taraka was initially established in 1974, on state residential land, by the PNG National Housing Commission (NHC) under the West Taraka Housing Scheme to provide low cost housing for civil servants, who later received the land titles. However, soon after the allocation of titles, titleholders began to subdivide and lease land to other people. Eventually, more people continued to move into the area and started illegally occupying the land around the demarcated Housing zone, which consists of a mixture of state reserved land, state agricultural land and customary land. Since its establishment, urban migration into West Taraka, mainly by rural migrants from the Highlands seeking employment opportunities in Lae City, has resulted in population growth rates even exceeding that of the main Lae City (Tapulu et al. 2014; Walsh 1987).

Although not officially categorised, West Taraka is generally considered a peri-urban settlement because it is located in the transition area between the urban boundary and the customary land. Urban agriculture is common practice for the community, utilising whatever little space available in the backyard of their homes to grow food for household consumption. Any excess produce is shared with neighbours and family or taken to be sold at the roadside markets. Sometimes people in the community are able to lease parcels of customary land from traditional landowners for a small fee or through informal arrangements. In this way, some household in West Taraka is able to have access to more space for crop production, both for household consumption and for selling at the local market.

Fig. 1 Map of Papua New Guinea showing Lae City. *Source* http://www.justmaps.org/maps/images/papua/lae-map1.gif

Lae's Nadzab Airport is the closest weather station to the site of this study. Rainfall records show that Lae receives almost 4500 mm of rainfall annually (Fig. 3)—an extraordinary amount which is distributed with only slight seasonal variation (PNG National Weather Service 2019). While high annual rainfall may provide favourable growing conditions for some crops, the increasing frequency of extreme rainfall events leading to severe flooding and soil erosion is problematic to the consistency of crop production.

Extreme climate events, such as tropical cyclone, flooding, drought and frost, affect squatter settlements disproportionately (United Nations Population Fund

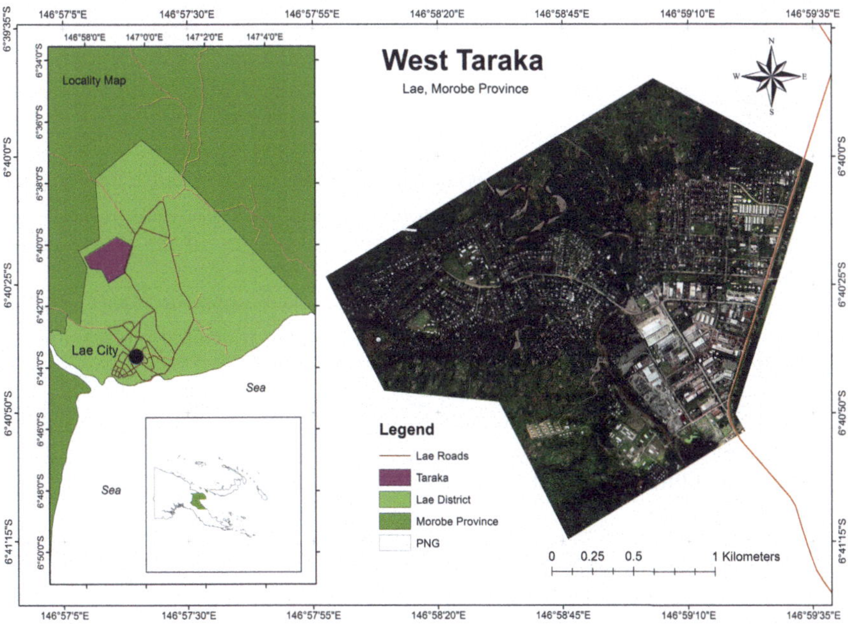

Fig. 2 Map showing location of the study site—West Taraka settlement (Tarutia 2019)

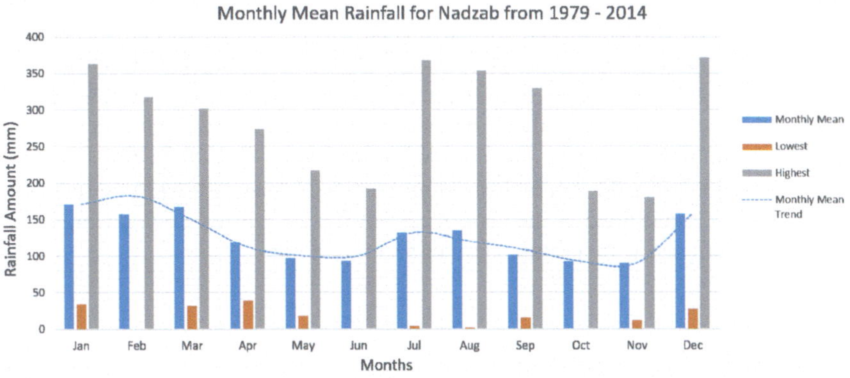

Fig. 3 Monthly rainfall for Lae (Nadzab Airport Station) (PNG National Weather Service 2019)

2007). The rapid growth and unmanaged land use (including unauthorized land clearance for backyard gardening) since the establishment of West Taraka were major factors which led to the disastrous September 1983 flooding event on the Bumbu River where 8 people were killed and hundreds of houses located along the river banks were destroyed (Atkins 2013). However, the scale of this disaster did not seem to deter the community from resuming its home gardening practices in order to produce food to meet household consumption needs.

Method

Sampling

The 2-week field survey was conducted in June 2016. However because limited data was collected on climate change impact on food productions, literature review was used to relate how climate change influences the livelihoods of the West Taraka community.

Two sampling methods were used to select the sample population. First, stratified sampling was used to demarcate households based on the four land types found in West Taraka:

(i) state reserve land;
(ii) state residential land;
(iii) state agricultural land, and
(iv) customary land.

Stratification by land type was selected to assess the impact of accessibility to land for subsistence farming on household food consumption.

Research observation was also used during the survey, where the researcher recorded personal observation on the living conditions and the classification of the settlement. Peri-urban settlement was not a category in the 2011 PNG National Census, however it is classified as such in this survey because of the location of the settlement and its extension into other land areas.

Purposive sampling was then applied to select 4% (n = 60) of 1500 households occupying West Taraka. While the sample size was initially 60 households, two of the households became unavailable halfway through the survey. Introducing new replacement households would increase the probability of bias created by the inaccuracy of results since the respondents would inevitably have difficulty in recalling food and drinks consumed in the prior week for the 24-h recall component. Thus, the survey continued with a sample size of 58 households.

Assessment of Dietary Patterns—24-h Diet Recall

Dietary patterns are the "quantities, proportions, variety or combinations of different foods and beverages in diets, and the frequency with which they are habitually consumed" (Rodgers 2015). At the household level, it is the consistent consumption of certain meal ingredients from different food groups by people living within the same dwelling.

The 24-h diet recall method was used previously in Bue (2013) and Koczberski et al. (2012) to capture the food security level in different regions in PNG. The meal components (ingredients, source and meal number) are measured using two proxy, HDDS and FCS, which were developed for non-nutritionist to assess the status of food security without using anthropometric measurements (the size, shape and composition of the human body). These proxies have been found to be an efficient application to measure household nutritional quality in studies in Belgium, Burkina Faso, Mali, PNG and Bangladesh (Bue 2013; Savy et al. 2007; Torheim et al. 2004; Thorne-Lyman et al. 2009; Vandevijvere et al. 2010) and was therefore adopted in the survey.

(a) Household Dietary Diversity Score (HDDS)

The HDDS (scores ranging from 0 to 13) is the sum of food groups consumed within the last 24-h (FAO and USAID 360 2016; Swindale and Bilinsky 2006) and calculated for the duration of the 14-day survey. Instead of 12 food groups, 13 food groups was used. The 13 standard food groups (Bue 2013) used corresponds to specific meal ingredient, meal source and total meal consumed.

HDDS is also used as an indicator to measure the economic stability of a household. The number of food groups consumed is an indicator of household's purchasing power to access various food groups for a quality diet and nutritional security, thus, the higher the HDDS, the more money the household has to purchase different foods.

(b) Food Consumption Score (FCS)

FCS was calculated using the 24-h diet recall for the 14-day survey. Instead of using the normal 7-day frequency, a 14-day food recall was conducted to observe the diet trend for the households to establish if their diet is stable or fluctuating.

The FCS contains 9 food groups, however, 10 food groups were used for this survey (Bue 2013) which combines two food groups in the HDDS according to the given standard weight allocation based on its nutritional properties. The FCS measures the frequency of food over a period of time (7 days). The frequency of the food groups (per week) are multiplied by their given weights to calculate their nutritional value to obtained the FCS. The FCS is then used to determine the threshold of the households to indicate their level of food security. This proxy also allows the researcher to observe which food groups are consumed at a daily basis, giving the research an indication of the household's main staple food. In order to calculate the FCS, the formula used is: $\text{FCS} = \text{food groups} \times \text{weight}$. Each of the food groups (10) are multiplied with their own nutritional score to obtained the FCS.

People in West Taraka are engaged in a range of economic activities, including formal employment in public or private sectors, industries and self-employment. Household food consumption will inevitably be linked to household earnings, frequency of income and amount of harvest from food gardens. Therefore comparing the 14 days will give a good indicator of their diet pattern by analyzing the total frequency of foods consumed each day over the duration of the survey.

(i) FCS thresholds

The FCS threshold was calculated to determine the nutritional profile of the households. Under this method, the researcher has the flexibility to use their own background knowledge on the type of staple foods consumed in the site and the consumption pattern to modify the FCS thresholds in order to produce the household nutritional profiles (Swindale and Bilinsky 2006).

The first profile for West Taraka was calculated to be the FCS range of 23–32, based on evidence that meals consisted mainly of cereal/tuber (rice and/or root crops), banana with green leafy vegetables cooked in coconut milk or oil, and tinned fish/meat and sugared tea. These are food groups consumed as the households' main staple. This finding agreed with Bue (2013) and Gwatirisa et al. (2017) who identified similar dietary patterns. The consumption of food groups for 3 or more days was considered as the main staple food to calculate the first threshold.

The second threshold of 32.5–40, consisted of all food groups in the first threshold, with the addition of fresh fish/chicken/meat in the calculation. Finally, the third threshold consisted of all 10 food groups, producing scores of over 40.

These thresholds are calculated based on the observed diet pattern during the 2-week survey.

Assessment of Socio-Economic Level

Eight household socio-economic characteristics were surveyed through interviews in West Taraka. These are:

(1) Age of head of household;
(2) Highest education level of head of household;
(3) Current occupation of head of household;
(4) Household size;
(5) Number of dependents attending school;
(6) Number of household members employed in the formal sector;
(7) Number of household members employed in the informal sector; and
(8) Total household income per fortnight.

Households occupying informal settlements around major urban centres generally have a lower level of engagement in the formal economy, a lower number of income earners and generate lower combined household income (Barber 2003), which all affect the ability to meet basic nutritional security. This assessment is used to determine whether there is a similar relationship between household socio-economic level and dietary patterns in West Taraka.

Data Analysis

Description analysis method was used to analyze the percentage, mean, standard deviation, frequency and correlation coefficient test to explore the relationships between basic data collected during the survey and interviews. The analysis involves the use of Microsoft (MS) Excel and Statistical Package for the Social Sciences (SPSS).

Results and Discussion

Household Characteristic Profile

Eight socio-economic characteristics of the study site are presented in Table 1. It was found that the mean age for household heads was 48 years. Over half of the household heads are literate, 29% have reached secondary school, 23% completed their tertiary levels while 31% have completed primary school to Grade 8 level. Primary school completion is the most common education level attained in the rural community in the highlands of PNG (Schmidt et al. 2019). The level of literacy also corresponded with the engagement in wage employment by 59% of household heads who are involved in some form of wage/salary employment. Although 17% of the household heads are illiterate, only 12% are unemployed, either due to old age or unable to find employment. However, some of the unemployed household head are engaged in income generating activities such as farming on state agricultural land, backyard gardening and operating roadside stalls selling *buai* (betel nut or Areca nut), tobacco or cooked food. Majority of the households (86%) have at least one dependent child in school. Some had over 4 children attending school.

According to field observation, differences in housing structures can be associated with the household's social status, similar to findings in Khan (2014). It was found that 62% of households consisted of at least 5 members while 38% consisted 4 or less members. While the average household size of 5 members in the study site, corresponds with PNG's national average, this is lower than the average size of households in Lae District (6.8 members) (PNG National Statistical Office 2011).

From the analysis, it was found that having a large family in a low socio-economic condition has largely negative impacts on the family's well-being. The impacts are exacerbated by limited land accessibility for gardening and other social constraints. Kiran and Dhawan (2015) found that larger households tends to have a wider range of expenses (such as school fees, stationeries, and other necessities) to cater to the needs of non-working dependents. Therefore, income diversion was a problem for the households, particularly for low income earners with large family members, making it difficult to accumulated savings.

The survey found that maximum income earned by the households fluctuated during the two weeks especially for those involved in the informal sectors. The income ranged from K45.00, for those involved in non-farming activities, to K7800.00, for

Table 1 Household characteristic profile (n = 58)

Characteristic	Range	Measuring unit	Category	Frequency	%	Mean	SD
Age of Hh heads	25–80	Years	25–35	11	19	48	13
			36–50	27	47		
			Above 50	20	35		
Education level of Hh heads	0–13	Grades	Illiterate	10	17	8	5
			Primary (3–8)	18	31		
			Secondary (9–12)	17	29		
			Tertiary (Above 12)	13	23		
Occupation of Hh heads	–	Employment	None	7	12	–	–
			Formal	34	59		
			Informal	17	29		
Family size	2–10	Number of members	Small (Up to 4)	22	38	5	1
			Medium (5–7)	30	52		
			Large (Above 7)	6	10		
Dependents in school	0–6	Number of dependents	None	8	14	2	2
			Up to 2	31	54		
			3–4	13	22		
			Above 4	6	10		
Hh members employed in the formal sector	0–4	Number of members	None	29	50	–	–
			Employed	29	50		
Hh members employed in the informal sector	0–2	Number of members	None	31	53	–	–
			Employed	27	47		
Total Hh fortnightly income	45–7800	Kina	Up to 500	22	38	1160	1460
			501–1000	17	29		
			1001–1500	8	14		
			Above 1500	11	19		

households with small businesses. It was also noted that 1 member in 7 households earned income for the whole family, this shows the economic constraints families faced. This data confirms other findings like Barber (2003) that the higher the number of household members working, the higher the combined household income, thus determining the HDDS and FCS. However, the 38% of households that earn below K500 has to improvise or prioritise spending, which often leads to sacrificing main meals per day.

Change in Food Production

During the 2015 frost and drought event in the Highlands region, main staple crops such as the sweet potato (*kaukau*), taro and yam took take up to nine months of growth before reaching maturity, compared to the usual six months. Inadequate technology in agricultural recovery after extreme climate events have resulted in food insecurity remaining as a major concern in the country. The frost event was the worst to affect the PNG Highlands region in 40 years and caused food severe food shortage for about 300,000 people (Cobon et al. 2016). International emergency response was activated to distribute food supplies to meet the basic nutritional needs of the communities affected (IOM 2016), illustrating the severity of the impacts of climate change and extreme climate events on food security to the vulnerable communities.

As peri-urban settlements expanded, farming activities decline. Majority of the foods consumed in PNG are locally produced, contributing to the high food security in rural areas due to availability of land resources for food production (Bourke 2001; Schmidt et al. 2019). However, it was reported that many villagers and settlements in PNG still remain vulnerable to food shortage due to extreme drought (El Niño) and severe frosting in the high altitudes (Kanua et al. 2016). Decline in crop production also disrupts supply to various markets in PNG, including Lae Market Centre, and may lead to temporary food shortage.

Agricultural productivity is sensitive to physical variables such as local weather conditions, soil nutrient status, moisture level and temperature. PCCSP climate projections for PNG, indicate with very high confidence, that sea surface temperature, atmospheric temperature, seasonal mean rainfall and annual rainfall are all likely to continue increasing throughout the 21st century (BoM and CSIRO 2011). Increase in temperature reduces the efficiency of photosynthesis, and as the long-term average local temperature increases beyond the upper threshold (25 °C) of the optimum range, most tropical crops like sweet potato, taro, cassava, and yams are directly affected (SPREP, n.d.). Survey respondents indicated that their food consumption level either declined or remain unchanged during the year (2015–2016). There was no incidence of increased household food consumption.

A total of 34 households (58.6%) who are involved in home gardening also experience a decline in production from both social and environmental impacts which led to increased dependence on processed food. Table 2 represents a summary of the respondents' perceptions to explain the experienced decline in food production.

Table 2 Household response on the impact on crop production

Social impact	Environmental/climate change impact
Customary land Hh has to ask for permission to cultivate in their land and pay certain fees given by the owners	Drought Loss of crops, water/irrigation
Thief	*Rainfall* Flooding due to the location of the settlement and poor drainage system
Agricultural land Was assigned to certain person thus not all have the accessibility	*Frost* For Hh's with farms in the interior part of PNG
Labour force Most Hh members are working either in the formal or informal sector	*Temperature* Increase in heat affecting the time spend in the gardens
Hh heads Hh heads (female) have limited support to help out in the garden due to old age or members are children	
Overcrowding in the area	

Majority of households attributed crop decline mainly to a range of social impacts, with only a few highlighting the impacts of environmental changes. A recent study on the impacts of the 2015 drought in PNG (Gwatirisa et al. 2017), found that while many participants were unable to give a clear recollection of when the drought started, they agreed that it would have a lasting effect on their livelihoods. It indicated that while local farmers were able to identify the impacts on crops, they have little adaptation mechanism to cope. In comparison, respondents in West Taraka mainly attributed social factors rather than environmental and climate change on declining crop production.

Household Dietary Diversity and Food Consumption Scores

HDDS

Analysis on meal characteristics (Table 3) indicate that majority of households only consumed 2 meals (breakfast and dinner) on a daily basis, which reflects similar patterns for oil palm farmers in Madang Province and West New Britain Province (Bue 2013; Nahuet and Bue 2015).

Two main factors that influence household diet were income and family size. This indicates that the households reduced their meal spending to cater for the children's school fees.

Analysis of the households' meal sources (Fig. 4) found that 59% of foods were bought from the store, 23% from the market, 14% from own garden, 3% were shared among households and 1% from friend's garden. This indicates that majority of the households depended on processed foods which also confirms previous studies (Bourke 2001; Gwatirisa et al. 2017; Koczberski et al. 2012; Nahuet 2014; Nahuet and Bue 2015; Yamauchi et al. 2001). This shows that while store-bought food usually makes up a larger portion of household food sources in urban centres, this trend is becoming more common in the rural and peri-urban areas as well.

Majority of the households' HDDS range from 1 to 8 food groups consumed and a mean of 5 food groups consumed daily, which technically meets the daily food consumption needs based on international guidelines (WHO 2003). However, the foods most commonly consumed in West Taraka are high in carbohydrate and low in other nutritional value. There is insufficient consumption of high vitamin and protein food groups, subjecting many households to the risk of malnutrition.

The average frequency of consumption of different food groups during a typical week is presented in Table 4. It was found that rice was consumed most frequently in West Taraka, with households consuming it 6 out of 7 days. In comparison, root crops such as kaukau (sweet potato) was only consumed 3 times a week, banana twice a week and cassava once a week. Tin fish (consumed 4 times a week) was substituted as a protein cooked with noodles (3 days per week), green leafy vegetables (6 days) cooked in coconut milk or oil (4 days per week).

Table 3 Number of meals consumed per day (n = 58)

Characteristic	Range	Measuring unit	Category	Frequency	Percent	Mean	SD
Number of meals per day	1–3	Meal	1	3	5	2	0.6
			2	36	62		
			3	19	33		

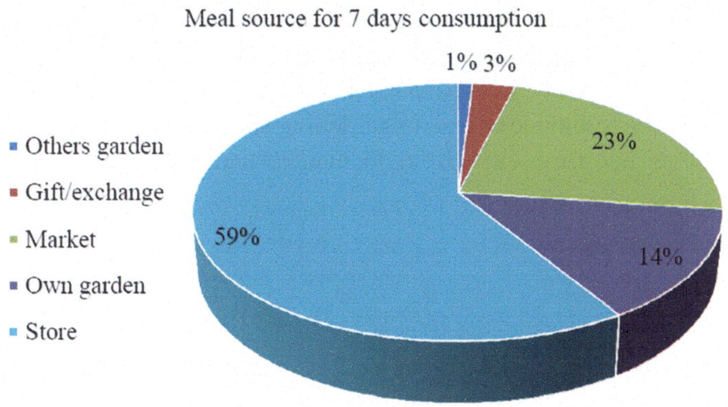

Fig. 4 Household meal sources for seven days

Table 4 Average consumption of the specific foods and the collapsed food groups over a 7 day period

Food Items	Average consumption of specific foods per week	Average consumption of collapsed food groups per week
Banana	2	
Kaukau	3	
Cassava	1	7
Rice	6	
Noodle	3	
Tin fish	4	
Tin meat	1	4
Pork	0	
Beef	0	
Chicken	1	
Sausage	0	2
Fish	0	
Lamb	0	
Egg	0	
Vegetables	6	6
Coconut milk	4	4
Condiments	4	4

Households in West Taraka are highly dependent on imported processed foods to make up their daily nutritional requirements. Low cost protein and fat substitutes such as tinned fish/meat and coconut cream are often consumed instead of the fresh options (Gwatirisa et al. 2017). 62% of households consume 4–6 food groups daily, 14% consumed more than 6 groups while 24% had less than 3 food groups—are falling below the basic daily food requirement leading to malnutrition. The trend of their diet within the 14 days shows little to no change in their dietary pattern. Results from this study supports studies done by Bourke (2001), Hodge et al. (1996), Yamauchi et al. (2001), Nahuet (2014), Koczberski et al. (2012) and Bue (2013). These studies all agreed that, due to limited access to gardening space, urban households tend to purchase most of their fresh foods from the markets and main staples from stores.

FCS Threshold

Based on the analysis, we calculated the FCS according to the observation and data collected for the 2 weeks based on which food groups were consumed more frequently for 7 days. From the frequencies, we identified the main staple foods which we used to calculate the threshold which was adjusted from the WFP guidelines.

Although the FCS, which range from 23 to 61.5, did not present the household caloric intake of micro and macro nutrients, it was used to measure the household nutritional security or the nutritional profile. At the end of Week 1 of the survey (Fig. 5), 60% of the households have high FCS and are nutritionally secure. The 26% of households on the borderline have either the potential to improve their diet or are risk of diet declining leading to lack of nutritional security. The 14% who have low FCS have a high risk of nutritional deficiency.

The analysis for Week 2 (Fig. 6) shows that there was only a slight difference in the FCS. 62% of the households showed high dietary consumption, however these are not the same households represented in Week 1. Some of the households that were on the borderline in Week 1 actually improved their nutritional intake and moved

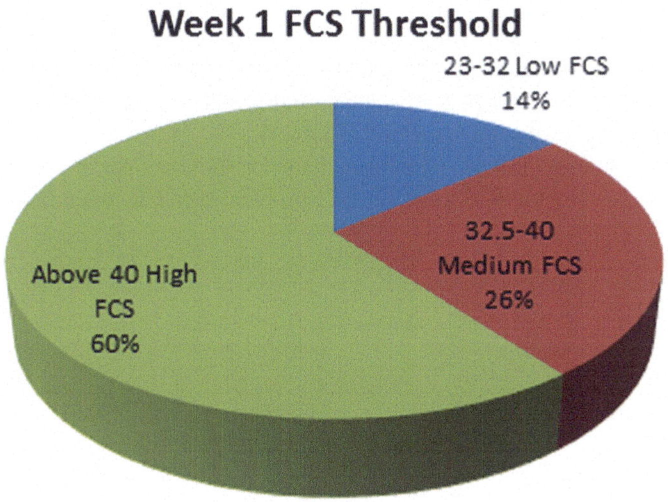

Fig. 5 Household FCS threshold for week 1

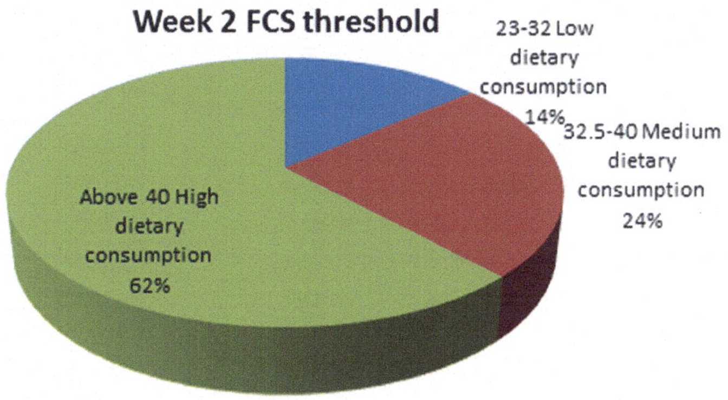

Fig. 6 Household FCS threshold for week 2

up from the medium threshold to the high threshold in Week 2. The reason was that households not only depend on their income but also diversification of income earning activities to provide food. The finding also supports Bue (2013) and Koczberski et al. (2012) that fortnight and non-fortnight earnings and the fluctuation in oil palm price influenced the dietary patterns of the oil palm smallholder households.

Correlation Test Between the Selected Socio-Economic Characteristics of the Households and Their FCS and HDD

Previous studies have established a relationship between socio-economic characteristics and household dietary patterns (Drewnowski and Darmon 2008; Giskes et al. 2007; Murakami et al. 2009; Mirmiran et al. 2002; Rezazadeh et al. 2010; Lenz et al. 2009). Therefore, the 8 selected household socio-economic characteristics were correlated against HDDS and FCS to evaluate the presence of any similar statistical relationships. The results are presented in Table 5.

Out of the 8 household socio-economic characteristics, income showed the largest positive correlation with FCS (0.361 at 0.01 probability), indicating that as income

Table 5 Correlation coefficient between the socio-economic characteristics of the households and their FCS and HDDS (n = 58)

Socio-economic characteristics of the households	Independent variable	Correlation coefficient
Age of Hh heads	FCS	−0.051
	HDDS	−0.046
Education level of Hh heads	FCS	0.19
	HDDS	0.128
Occupation of Hh heads	FCS	0.201
	HDDS	0.053
Family size	FCS	−0.214
	HDDS	0.023
Dependents in school	FCS	−0.281*
	HDDS	−0.124
Hh members employed in the formal sector	FCS	0.063
	HDD	0.063
Hh members employed in the informal sector	FCS	0.085
	HDDS	−0.005
Total Hh fortnightly income	FCS	0.361**
	HDDS	0.112

*Significance at 0.05 level of probability
**Significance at 0.01 level of probability

increases, households were able to consume more food groups. The high income level is linked to formal employment opportunities and other income-generating activities that households are engaged in, thereby increasing purchasing power to access quality foods such as fresh meats (chicken, meat, fish egg and dairy products). In addition, households are able to consume more than one meal per day thus contributing to their high FCS. This significance was also found in other countries (González et al. 2008; Butt and Iram 2004; DeWalt 1983; Kirkpatrick and Tarasuk 2003) and in selected locations in PNG (Bue 2013; Nahuet and Bue 2015; Koczberski et al. 2012; Shack et al. 1990). The education level of household heads and their current occupation also have a positive correlation with FCS (0.190 and 0.201, respectively), indicating the ability of literate heads of household to engage in stable income-generating activities to provide for their families.

In comparison, a negative correlation of −0.281 at 0.05 level of probability was seen between the number of dependents who were in school and household FCS, indicating that as the number of children in school increases, the lower the household FCS. A similar inverse relationship was seen between the household size (−0.214) and FCS. It is often more difficult for larger households to meet basic nutritional needs due to the diversity of expenses incurred to cater for the needs of non-income generating dependents such as school-aged children.

While this study did not separately evaluate the characteristics of mothers in the household, studies in Mozambique (Sahn and Alderman 1997; Garrett and Ruel 1999; Burchi 2012) have found significant positive relationships between FCS and the mothers' education level and knowledge in meal preparation. This shows mothers' awareness of the nutritional value of the meals they prepare that promotes nutritious benefit for their families.

Overall, this study confirms studies done by Amaza et al. (2008), Mueller et al. (2001), Drewnowski and Darmon 2008, Gwatirisa et al. (2017): Giskes et al. (2007), Murakami et al. (2009), Mirmiran et al. (2002); Rezazadeh et al. (2010) and Lenz et al. (2009) that the socio-economic factors of the households influences the households' dietary patterns, hence the household food and nutritional security.

Conclusion, Limitations and Recommendations

The population of the West Taraka peri-urban settlement has continued to grow as urban migration continue to increase, resulting in unauthorized settlements on both state and customary land. Few households are able to gain access to both customary and state land for food gardening unless they can afford to pay for a formal lease. This study was conducted to understand the status of household food and nutritional security in peri-urban settlements by profiling eight selected socio-economic characteristics, examining the household dietary patterns and analysing the relationship between the two. The sample population was identified through stratification by the four land types found within the study site and purposive sampling. Assessment of dietary patterns using the 24-h diet recall method from which two quantitative indicators, i.e. HDDS and FCS, were calculated for the 2-week survey. Pearson's correlation analysis was used to examine the relationship between the socio-economic characteristics and household food consumption.

The study found that 83% of the household heads are literate, corresponding with the 88% engagement in some form of employment—59% in formal employment and 29% in the informal sector. Only 12% are unemployed and unable to generate any form of income, due to old age or retirement. Despite this, other household members are able to support the household in other ways, notably through crop gardening, which is practiced by 58.6% of the households.

Most households consume only 2 meals daily—breakfast (before members leave the house) and dinner (when all members return home). Most of the meal ingredients are purchased either from the store (59%) or from the local fresh produce market (23%), while 14% were obtained from their own gardens. The remaining 4% were obtained through gifting and sharing of food, most commonly given by households that have access to farming space to those who do not. The food consumption score (FCS) for households in West Taraka ranged from 23 to 61.5, with 60% households found to be nutritionally secured, 26% on the borderline, and 14% at risk of nutritional insecurity.

Pearson's correlation analysis shows that there is a significant positive relationship between the FCS and income, indicating that income drives the consumption pattern of a household. Education level of household heads and their current occupation also have a positive correlation with FCS, indicating the ability of literate heads of household to engage in stable income-generating activities to provide for their families. While some households were faced severe economic constraints, they were able to maintain a basic diet. In comparison, a significant inverse relationship was found between FCS and the number of household members in school, indicating that expense diversification compromised the consumption of quality diets.

Previous studies have also established linkages between household diet and agriculture and highlighted the threat that climate change poses on community livelihoods (Cobon et al. 2016; Gwatirisa et al. 2017; IOM 2016). The 2015 drought and frost event which affected large populations in the PNG Highlands regions illustrated the threat of climate extremes to crop production and food security, particularly for vulnerable rural and peri-urban populations without adequate access to housing, land resources, income-generating opportunities and basic awareness of nutritional diets and climate change risks.

It may be necessary for people in West Taraka to diversify their income-generating activities in order to be able to meet their basic household nutritional needs. The growing influence of urbanization and limitations in access to farming space has resulted in transformation of the community's dietary patterns to a more processed-food diet. 90% of the local rural population in PNG derive their income from selling fresh foods (Bourke and Harwood 2009). Both temperature and rainfall are projected to increase over the 21st century in PNG, bringing increased frequency of extreme heat, extreme frost and severe rainfall events (BoM and CSIRO 2011). Crop production, both on a commercial and subsistence scale, have already been experiencing significant decline in PNG due to climate-related extreme events (IOM 2016). Rural and peri-urban communities will find it increasingly difficult to rely solely on agriculture-based activities to generate a stable level of household income. Households that rely on home gardening to meet their nutritional needs will face difficulty in producing enough food due to the increasingly hazardous climatic conditions, further jeopardising household food security.

Many PSIDS have shared development challenges and lessons learnt from country is often applicable to another. The challenges associated with urban sprawl, compounded with the pressures of complex land tenure systems and increasing threat of climate change inevitably affects the long-term food security of local communities. This research in PNG provides valuable lessons to guide other PSIDS in carrying out similar studies to build community resilience again climate change impacts.

Limitations and Implications of the Study Site

As the first household nutritional security study to be conducted in West Taraka, inferences and conclusions were drawn based on 58 households. The scope of this study was limited by the amount of research funding, duration of study, availability of households in the community, field assistance, challenges to personal safety at the study site and lack of recent nutritional data to support the findings in the study site.

There is a clear lack of awareness in climate change and household food security among the West Taraka community. The researchers recommend the prioritisation of awareness programmes for rural and peri-urban communities to empower people in making informed decisions for their household nutritional security under the exacerbating threats of climate change on crop production. The findings also contribute valuable information to the Ministry of Health's efforts in making effective policies to enhance national nutritional status for PNG.

The researchers recommend further studies to be conducted in the future to confirm the current findings, with particular focus on the impacts of environmental and climate change on food security and household diet. It would also be valuable to expand the sample size and survey duration, if adequate time, community assistance and financial resources are available. The enhanced study would contribute to filling the knowledge gap in understanding the perspectives of PNG peri-urban communities on the impacts of climate change on their livelihood, given the socio-economic and environmental constraints they face.

References

Australian Centre for International Agricultural Research [ACIAR] (2018) ACIAR. Annual Repot 2017–18. Canberra, Australia. ISBN 978-1-925746-53-2 (PDF/online)

Amaza PS, Adejobi AO, Fregene T (2008) Measurement and determinants of food insecurity in northeast Nigeria: some empirical policy guidelines. J Food Agric Environ 6(2):92–96
Atkins GP (2013) Erosion and drainage problems in the environs of Lae: a disaster waiting to happen–again. In: Proceedings of the 7th Huon seminar—achieving vision 2050 though higher education, research, science and technology (13th–14th Nov 2013, Papua New Guinea University of Technology, Lae, PNG), p 92
AusAID (2008) Making land work: volume two—case studies on customary land and development in the Pacific. Australian Agency for International Development, Canberra
Bogard RJ, Mark CG, Thilsted S, Wood HS (2018) Measuring nutritional quality of agricultural production systems: application to fish production. Glob Food Secur 16:54–64
Barber K (2003) The Bugiau community at eight-mile: an urban settlement in Port Moresby, Papua New Guinea. Oceania 73(4):287–297
BoM (Australia) and CSIRO (2011) Papua New Guinea. Climate change in the Pacific: scientific assessment and new research. Volume 2: country reports, vol 2. Australian Bureau of Meteorology and Commonwealth Scientific and Industrial Research Organisation, Aspendale, VIC, pp 169–184
Bourke RM (2001) An overview of food security in Papua New Guinea. Paper presented at the Proceedings of the Papua New Guinea food and nutrition 2000 conference, 26–30 June 2000, PNG University of Technology, Lae. https://ageconsearch.umn.edu/record/135368/files/PR099.pdf#page=19
Bourke R, Harwood T (2009) Food and agriculture in Papua New Guinea
Bue VG (2013) The role of smallholder farmers in sustaining household food security at Bialla and Hoskins oil palm land settlement schemes, Papua New Guinea (Ph.D.), Curtin University
Burchi F (2012) Whose education affects a child's nutritional status? From parents' to household's education. Demographic Res 27:681–704
Butt MS, Iram U (2004) Determinants of household food security: an empirical analysis for Pakistan. Int J Soc Econ 31(8):753–766. https://doi.org/10.1108/03068290410546011
Cobon DH, Ewai M, Inape K, Bourke RM (2016) Food shortages are associated with droughts, floods, frosts and ENSO in Papua New Guinea. Agric Syst 145:150–164. https://doi.org/10.1016/j.agsy.2016.02.012
DeWalt KM (1983) Income and dietary adequacy in an agricultural community. Soc Sci Med 17(23):1877–1886. https://doi.org/10.1016/0277-9536(83)90165-X
Drewnowski A, Darmon N (2008) Does social class predict diet quality? Am J Clin Nutr 87(5):1107–1117. https://doi.org/10.1093/ajcn/87.5.1107
FAO (2015) Low-Income Food-Deficit Countries (LIFDC)—list for 2015. Retrieved 09 Sept 2016. http://www.fao.org/countryprofiles/lifdc/en/
FAO (2018) Asia and the Pacific regional overview of food security and nutrition 2018—accelerating progress towards the SDGs. Bangkok. License: CC BY-NC-SA 3.0 IGO
FAO and USAID (2016) Minimum dietary diversity for women—a guide to measurement. Food and Agriculture Organization of the United Nations and USAID's Food and Nutrition Technical Assistance III Project (FANTA), Rome, Italy, p 82
Foley WA, Foley WA (1986) The Papuan languages of New Guinea. Cambridge University Press
Game I, Primus R (2015) Urban Agriculture. State University of New York College of Forestry and Environmental Science
Garrett JL, Ruel MT (1999) Are determinants of rural and urban food security and nutritional status different? Some insights. World Dev 27(11):1955. https://doi.org/10.1016/S0305-750X(99)00091-1
Giskes K, Turrell G, van Lenthe FJ, Brug J, Mackenbach JP (2007) A multilevel study of socio-economic inequalities in food choice behaviour and dietary intake among the Dutch population: the GLOBE study. Public Health Nutr 9(1):75–83. https://doi.org/10.1079/PHN2005758
González W, Frongillo EA, Jiménez A, Madrigal G, Muñoz LM (2008) Development and validation of measure of household food insecurity in Urban Costa Rica Confirms proposed generic questionnaire. J Nutr 138(3):587–592. https://doi.org/10.1093/jn/138.3.587

Gwatirisa PR, Pamphilon B, Mikhailovich K (2017) Coping with drought in Rural Papua New Guinea: a Western Highlands case study. Ecol Food Nutr 56(5):393–410. https://doi.org/10.1080/03670244.2017.1352504

Hodge AM, Montgomery J, Dowse GK, Mavo B, Watt T, Alpers MP, Zimmet PZ (1996) Diet in urban Papua New Guinea population with high levels of cardiovascular risk factors. Ecol Food Nutr 35:311–324

IOM (2016) Papua New Guinea drought and floods response: situation report. From: https://reliefweb.int/sites/reliefweb.int/files/resources/IOM%20Sitrep%20%238%20%20PNG%20Drought%20%28May%202016%29_v2.pdf

IPCC (2018) Summary for policymakers. In: Global warming of 1.5°C. An IPCC special report on the impacts of global warming of 1.5°C above pre-industrial levels and related global greenhouse gas emission pathways, in the context of strengthening the global response to the threat of climate change, sustainable development, and efforts to eradicate poverty [V. Masson-Delmotte, P. Zhai, H. O. Pörtner, D. Roberts, J. Skea, P. R. Shukla,A. Pirani, W. Moufouma-Okia, C. Péan, R. Pidcock, S. Connors, J. B. R. Matthews, Y. Chen, X. Zhou, M. I. Gomis, E. Lonnoy, T. Maycock, M. Tignor, T. Waterfield (eds.)]. World Meteorological Organization, Geneva, Switzerland, 32 pp

Kanua M, Bourke RM, Jinks B, Lowe M (2016) Assessing village food needs following a natural disaster in Papua New Guinea. Church Partnership Program, The Australian Government Department of Foreign Affairs and Trade (DFAT), Canberra

Khan AA (2014) Analysis of housing structure in Peri Urban areas of Faisalabad, Pakistan. Asian J Multidisciplinary Stud 1(2):57–62

Kiran T, Dhawan S (2015) The impact of family size on savings and consumption expenditure of industrial workers: a cross-sectional study. Am J Econ Bus Adm 7(7):177–184

Kirkpatrick S, Tarasuk V (2003) The relationship between low income and household food expenditure patterns in Canada. Public Health Nutr 6(6):589–597. https://doi.org/10.1079/PHN2003517

Koczberski G, Curry GN, Bue V (2012) Oil palm, food security and adaptation among smallholder households in Papua New Guinea. Asia Pac Viewpoint 53(3):288–299. https://doi.org/10.1111/j.1467-8373.2012.01491.x

Lenz A, Olinto MTA, Dias-da-Costa JS, Alves AL, Balbinotti M, Pattussi MP, Bassani DG (2009) Socioeconomic, demographic and lifestyle factors associated with dietary patterns of women living in Southern Brazil. Cadernos de saude publica 25:1297–1306

Lwasa S, Mugagga F, Wahab B, Simon D, Connors J, Griffith C (2014) Urban and peri-urban agriculture and forestry: transcending poverty alleviation to climate change mitigation and adaptation. Urban Clim 7:92–106. https://doi.org/10.1016/j.uclim.2013.10.007

Mirmiran P, Mohammadi F, Allahverdian S, Azizi F (2002) Association of educational level and marital status with dietary intake and cardiovascular risk factors in Tehranian adults: Tehran lipid and glucose study (TLGS). Nutr Res 22(12):1365–1375. https://doi.org/10.1016/S0271-5317(02)00440-2

Map of Lae (n.d.) From: http://www.justmaps.org/maps/images/papua/lae-map1.gif

Mueller I, Vounatsou P, Allen BJ, Smith T (2001) Spatial patterns of child growth in Papua New Guinea and their relation to environment, diet, socio-economic status and subsistence activities. Ann Human Biol 28(3):263–280. https://doi.org/10.1080/030144601300119089

Murakami K, Miyake Y, Sasaki S, Tanaka K, Ohya Y, Hirota Y (2009) Education, but not occupation or household income, is positively related to favorable dietary intake patterns in pregnant Japanese women: the Osaka maternal and child health study. Nutr Res 29(3):164–172. https://doi.org/10.1016/j.nutres.2009.02.002

Nahuet P (2014) Perceived impact of village oil palm production on the livelihood of the growers in the Markham valley of Papua New Guinea (Ph.D.), Papua New Guinea University of Technology, Lae, PNG

Nahuet P, Bue V (2015) Perceived impact of village oil palm production on the livelihood of the growers in the Markham valley of Papua New Guinea. Niugini Agrisaiens 10:40–46

New Guinea Map (n.d.) From http://www.justmaps.org/maps/images/papua/lae-map1.gif

Niles MT, Ahuja R, Esquivel MJ, Mango N, Duncan M, Heller M, Tirado C (2017) Climate change and food systems: assessing impacts and opportunities. Meridian Institute, Washington, DC

PNG National Statistical Office (2011) National population and housing census 2011. Government of Papua New Guinea

PNG National Weather Service (2019) PNG national weather service. Retrieved 01 Apr 2019. http://www.pngmet.gov.pg/

Rauff KO, Bello R (2015) A review of crop growth simulation models as tools for agricultural meteorology. Agric Sci 6:1098–1105. https://doi.org/10.4236/as.2015.6910

Rezazadeh A, Rashidkhani B, Omidvar N (2010) Association of major dietary patterns with socioeconomic and lifestyle factors of adult women living in Tehran, Iran. Nutrition 26(3):337–341. https://doi.org/10.1016/j.nut.2009.06.019

Rodgers AB (2015) Scientific report of the 2015 dietary guideline advisory committee: advisory report to the secretary of health and human service and the secretary of agriculture. U.S department of agriculture and Department of Health and Human Services, Washington, DC

Rosegrant MW, Valmonte-Santos R, Thomas T, You L, Chiang C (2015) Climate change, food security, and socioeconomic livelihood in Pacific Island countries. Asian Development Bank and International Food Policy Research Institute, Mandaluyong City, Philippines

Sahn DE, Alderman H (1997) On the determinants of nutrition in Mozambique: the importance of age-specific effects. World Dev 25(4):577–588

Savy M, Martin-Prével Y, Traissac P, Delpeuch F (2007) Measuring dietary diversity in rural Burkina Faso: comparison of a 1-day and a 3-day dietary recall. Public Health Nutr 10(1):71–78. https://doi.org/10.1017/S1368980007219627

Schmidt E, Gilbert R, Holtemeyer B, Rosenbach G, Benson T (2019) Papua New Guinea survey report: rural household survey on food systems

Shack KW, Grivetti LE, Dewey KG (1990) Cash cropping, subsistence agriculture, and nutritional status among mothers and children in lowland Papua New Guinea. Soc Sci Med 31(1):61–68. https://doi.org/10.1016/0277-9536(90)90010-P

SPREP (n.d.) Pacific adaptation to climate change, Papua New Guinea: Report od In-Country consultation. Retrieved from: https://www.sprep.org/attachments/Climate_Change/PACC_Report_of_in-country_consultations_Nauru.pdf

Swindale A, Bilinsky P (2006) Household dietary diversity score (HDDS) for measurement of household food access: indicator guide. Food and Nutrition Technical Assistance Project, Academy for Educational Development, Washington, DC

Tapulu W, Jana KS, Sekae T (2014) Change detection study of Lae Urban. The second largest city of Papua New Guinea using multi-temporal high resolution remote sensing data. Int J Manage Soc Sci Res (IJMSSR) 3(10). ISSN: 23319-4421

Tarutia R (2019) Map of West Taraka, Lae, Morobe Province. DroneTree Mapping and Solutions, Lae (Papua New Guinea)

Taylor M, McGregor A, Dawson B (2016) Vulnerability of Pacific Island agriculture and forestry to climate change. Pacific Community (SPC), Noumea, New Caledonia

Thorne-Lyman AL, Valpiani N, de Pee S, Bloem MW, Sun K, Semba RD, Sari M (2009) Household dietary diversity and food expenditures are closely linked in Rural Bangladesh, increasing the risk of malnutrition due to the financial crisis. J Nutr 140(1):182S–188S. https://doi.org/10.3945/jn.109.110809

Torheim LE, Ouattara F, Diarra MM, Thiam FD, Barikmo I, Hatløy A, Oshaug A (2004) Nutrient adequacy and dietary diversity in rural Mali: association and determinants. Eur J Clin Nutr 58:594. https://doi.org/10.1038/sj.ejcn.1601853

Umezaki M, Ohtsuka R (2003) Adaptive strategies of highlands-origin migrant settlers in Port Moresby, Papua New Guinea. Human Ecol 31(1):3–25. https://doi.org/10.1023/A:1022881506510

United Nations Population Fund (2007) State of the world population: unleashing the potential of urban growth. United Nations Population Fund, New York

UN DESA (2018) World Urbanization prospects: the 2018 revision (dataset). Retrieved from: https://esa.un.org/unpd/wup/Download/
UNDP (2018) Human development indices and indicators—2018 statistical update. United Nations Development Programme, New York
Vandevijvere S, De Vriese S, Huybrechts I, Moreau M, Van Oyen H (2010) Overall and within-food group diversity are associated with dietary quality in Belgium. Public Health Nutr 13(12):1965–1973
Wairiu M, Lal M, Iese V (2012) Climate change implications for crop production in Pacific Islands region. Food production-approaches, challenges and tasks. IntechOpen, London
Walsh BJ (1987) The growth and development of squatter settlements in Lae, Papua New Guinea. In: Mason L, Hereniko P (eds) In search of a home. Institute of Pacific Studies of the University of the South Pacific, Suva
Weber EH (2014) Climate change, agriculture and food security in the Pacific Islands. J South Pac Agric 20
WHO (2003) Diet, food supply and obesity in the Pacific. World Health Organization, Geneva
World Climate Guide (2019) Tabubil climate guide, Papua New Guinea. Retrieved 01 Apr 2019. http://www.worldclimateguide.co.uk/guides/papua-new-guinea/tabubil/
Yamauchi T, Umezaki M, Ohtsuka R (2001) Influence of urbanisation on physical activity and dietary changes in Huli-speaking population: a comparative study of village dwellers and migrants in urban settlements. Br J Nutr 85(1):65–73. https://doi.org/10.1079/BJN2000208

Resilience in Education: An Example from Primary School in Fiji and Technical Vocational Education and Training

Peni Hausia Havea, Antoine De Ramon N' Yeurt, Apenisa Tagivetaua Tamani, Amelia Siga, Hélène Jacot Des Combes, Sarah Louise Hemstock and Johannes M. Luetz

Abstract In the Pacific, the capacity of curriculum writers for integrating the content of climate change into their curricula and/or taught Resilience [Climate Change Adaptation (CCA) & Disaster Risk Reduction (DRR)] in education is limited. This paper described the findings of a 2018 study on the integration of climate change into primary and secondary schools' curricula and taught resilience in education in TVET. It involves teachers (n = 30) from Kadavu and Levuka islands, curriculum writers and editors from the Ministry of Education, GIZ, SPC, and USP—in Fiji. An exploratory design was used to explore the curricula for Fiji and the EU PacTVET project at SPC. Information was collected from workshops and training events, interviews and project documents. Using BEKA (Benchmarking, Evidencing, Knowing, Applying) and the

P. H. Havea (✉) · A. De Ramon N' Yeurt
Pacific Center for Environment and Sustainable Development (PaCE-SD), University of the South Pacific (USP), Suva, Fiji
e-mail: ilaisiaimoana@yahoo.com

A. De Ramon N' Yeurt
e-mail: nyeurt@gmail.com

A. T. Tamani
Deutsche Gesellschaft für Internationale Zusammenarbeit (GIZ) GmbH, Suva, Fiji
e-mail: apenisa.tamani@giz.de

A. Siga
GeoScience, Energy & Maritime Division (GEMD), Pacific Community, Suva, Fiji
e-mail: amelias@spc.int

H. J. D. Combes
National Disaster Management Office, Government of Marshall Islands, Majuro Atoll, Marshall Islands

S. L. Hemstock
School of Humanities, Bishop Grosseteste University, Lincoln, UK
e-mail: sarah.hemstock@bishopg.ac.uk

J. M. Luetz
Christian Heritage College (CHC), 322 Wecker Road, Carindale, QLD 4152, Australia
e-mail: jluetz@chc.edu.au; j.luetz@unsw.edu.au

University of New South Wales (UNSW), Sydney, NSW 2052, Australia

W. Leal Filho (ed.), *Managing Climate Change Adaptation in the Pacific Region*, Climate Change Management, https://doi.org/10.1007/978-3-030-40552-6_10

concept of ako (e.g. to study or educate), a model of climate change and resilience in education was designed as part of this research to help Pacific schools with their curricula. These results indicate how behavioural changes may shape Resilience, thus placing them in a better position to achieve the UNFCCC, the SDGs, the Sendai Framework and the Framework for Resilient Development in the Pacific (FRDP) targets and objectives by 2030 and beyond.

Keywords Resilience · Climate change · Education · Pacific · Sustainability

Introduction

The Pacific is known to be the most at-risk region in the world to be affected by climate change and disasters caused by natural hazards (IPCC 2014a, b; Nunn 2013; Nunn et al. 2014; Singh et al. 2001; SPC and GIZ 2016; United Nation University 2016; United Nations University, Institute for Environment and Human Security (UNU-EHS) and Universität Bonn 2013; WHO 2015). As a result, many climate change programmes are preparing for these inevitable changes (Bell et al. 2011; EU PacTVET Project 2017; EU PacTVET Project, SPC and USP 2016a, b; Luetz and Havea 2018; Pacific Community et al. 2016; Pacific Community (SPC) 2014; Regmi 2015; UNESCO 2006; WHO 2015; Yamamoto and Esteban 2017). However, albeit this climate change foray has been a significant lacuna, the fact that it will also raise the level of awareness of the people in the Pacific, to be more self-consciousness, sustainable and resilient in education (McIver et al. 2016; Pacific Community (SPC) 2015; Rochat 2003; SPREP 2017; Taylor et al. 2016), plays a critical role in the building of resilient Pacific Islanders by 2030 and beyond.

From what is known in the Pacific, climate change education is already experienced by children and young people in the primary and secondary schools and Technical Vocational Education and Training (TVET) sectors in Fiji, Kiribati, Samoa, Tonga and Vanuatu through integration into the primary school syllabus, secondary school basic science, geography, agricultural science and TVET education (Pacific Community (SPC) 2011). Even other countries have done so, including in Europe (Cefai et al. 2014), Asia (Luetz and Sultana 2019), and Africa (Boakye 2015; Chirwa et al. 2014; Mokaya et al. 2016). Therefore, this work will generate fresh insight as to why this integration-based course of study is vital for the region.

To begin with, this is in-line with the regional SPC/GIZ programme 'Coping with Climate Change in the Pacific Island Region' (CCCPIR), which aims at integrating climate change into primary and secondary education and TVET through its European Union Pacific Technical Vocational Education and Training in Sustainable Energy and Climate Change Adaptation (EU PacTVET) project (EU PacTVET Project 2017; EU PacTVET Project et al. 2016a, b; Hemstock et al. 2018; SPC and GIZ 2016). As a result, even though this integration of climate change into the Pacific schools' curricula is unquestioning and essential, to play an active role in bolstering the next generation to come, to be more resilient by 2030 and beyond, there is a need to conceptualise thinking about Resilience (CCA & DRR) as a resolution rather than focussing primarily on how many people will be affected by climate change.

Second, the Framework for Resilient Development in the Pacific (FRDP) 2017–2030 (Pacific Community et al. 2016) goal 1 (e.g. to strengthen integrated adaptation and risk reduction to enhance resilience to climate change and disasters), and goal 2 (e.g. preparedness, response and recovery), both indicate calling to increase awareness or program visibility, and most importantly an option for training and education. All of these are significant in achieving the Pacific plan for a Climate Change Education for Sustainable Development (CCESD), by 2030 and beyond.

Third, the qualifications at levels 1–4 were developed regionally by all 15 P-ACP (Pacific African Caribbean Pacific) countries, which was facilitated by the Fiji Higher Education Commission (FHEC), thus allowing the region to bridge the lower level to the milestones for climate change or Resilience in higher education for the future. More importantly, the EU PacTVET project was given endorsement from 15 P-ACP countries (Cook Islands, Federated States of Micronesia (FSM), Fiji, Kiribati, Nauru, Niue, Palau, Papua New Guinea, Republic of Marshall Islands (RMI), Samoa, Solomon Islands, Timor-Leste, Tonga, Tuvalu and Vanuatu), to proceed with the development of these "regional qualifications" in Resilience, in the sense that it incorporated country priority needs, which were identified through the national needs and gaps analysis.

National stakeholders at the EU PacTVET Inception Meeting were keen to ensure opportunities for learning via formal qualifications available to all people affected by climate change. This call for action is vital not only for the Pacific but other cultures as well, so that the capacity building opportunities should be applicable across the board from grassroots community members, through to technicians and government and private sector managers. In this regard, vocational qualifications have been constructed around a "competency" and "skillset" approach, so that people can pick what competencies they need to "up-skill", to improve their own capacity—a menu of competencies and skill sets are available within the qualifications. The national stakeholders were also keen to ensure that aspects of "resilience" (not just climate change) were covered in the qualifications developed by the EU PacTVET project.

In Europe, some aspects of resilience and environmental responsibility has already been taught in schools, so it would be easier to integrate the concepts of climate change and disasters caused by natural hazards to the current curriculum (Boniwell et al. 2013; Cefai et al. 2014; Wosnitza et al. 2018). The Australia Pacific Training Coalition (APCT) is also offering the same resilience-based course of study in Fiji, Papua New Guinea, Samoa, Solomon Islands and Vanuatu (APTC 2018). Fortunately, Vanuatu is the first country in the Pacific to have its cohort graduated with certificates in resilience, which was supported by the EU PacTVET project. More significantly, not until recently, the Pacific TAFE at the University of the South Pacific (USP) has offered Resilience certificate levels 3 and 4 as part of its program to meet the University's targets and objectives for sustainable education (USP 2018) by 2030 and beyond. As a result, this study is also consistent with other research needs and programs serving the same mission in the region, to have sustainable education.

So, if this is the case, then this paper used Resilience in education rather than climate change per se for the TVET because it is imperative, popular and disaster-oriented. Adding to this information, its technical application and foundations for the concept of ako[1] (e.g. to educate people) about both climate change and hazards would appear to be necessary. The discovery of resilience was in the early 17th century. It was derived from a Latin word resilire, which means 'to rebound or recoil'. Since then, the usage of resilience has emerged and recognised internationally and applied in the following discipline:

(1) science (e.g. Woods 2015);
(2) ecology (e.g. ecological resilience);
(3) individual (e.g. resilience of individuals);
(4) community (e.g. community resilience);
(5) organisation (e.g. organisational resilience);
(6) economics (e.g. resilience of the economy);
(7) national level (e.g. national resilience);
(8) humanitarian (e.g. humanitarian resilience);
(9) security (e.g. resilience and security);
(10) vulnerability (e.g. resilience and vulnerability);
(11) risk (e.g. resilience and risk).

Now, resilience has also been applied to climate change (e.g. climate resilience) as well (Barrow Cadbury Trust 2012; Flinders University 2018; Vernon 2004).

In the Pacific, resilience has now featured at the forefront of climate change leadership and hazards management platform. It features not only in the FRDP for the Pacific but most importantly, it has defined and built its development to fit the concepts of resilient Pacific Islanders, as indicated by the SDGs (UNDP 2015) and the Sendai Framework (Maini et al. 2017; United Nation Office for Disaster Risk Reduction (UNISDR) 2015). As an integrated Climate Change Adaptation (CCA) and Disaster Risk Management (DRM) framework for sustainable development in education, FRDP defined resilience using the above state of affairs. As a result, it describes "development processes and actions that address the risks and impacts of disasters and climate change while progressing to stronger and resilient communities" (Pacific Community et al. 2016, p 3).

This definition provided an opportunity to advance the understanding of the capacity building in resilience or sustainable development in education because it is both climate and disaster-oriented—personalising and professionalising the resilience

[1] In the context of the Pacific, ako means to teach (e.g. educator) or to study (e.g. leaner). It is a two-way process of learning founded on reciprocity and student-based approach. The idea is that when educating students in Fiji and the Pacific, the focus is not only academic, but also physically, mentally, and spiritually. So, using this concept of ako as a guide, learners of climate change must be holistically educated. Because climate change and disasters are affecting people lives at different levels of society, therefore ako about climate change should not separate the learners from families, Vanua (land), and religions.

sectors in the Pacific Islands region (Hemstock et al. 2018). Using this modern definition as a guide, integrating climate change and using resilience in education is to be aligned with government development planned. To achieve this goal, Fiji will be used as a case study for integrating climate change and the EU PacTVET project for resilience in education. This integration of climate change into the school system has been requested by the Government of Fiji via its Ministry of Education to the GIZ office in Suva.

The basis is that since climate change is a cross-sectoral (e.g. crosscut between economic, health, tourism, fishery, energy) area, it is much cheaper and convenient to relate it and fill in gaps in the existing learning system (e.g. integrating it into social science, agriculture and science subjects) rather than having it as a stand-alone subject. Furthermore, by doing it in this way may also help the ministry to forcibly prevent polarising issues from teacher turnover in the sense that it is cheaper to prepare the current teachers and retain their jobs than to train new climate change teachers and issue of resignation and/or relocation. Based on this analysis, using the BEKA (benchmarking, evidencing, knowing, applying) framework as a guide, a model was designed to illustrate how the integration of climate change in education may be implemented for the primary and secondary education sector in Fiji and TVET sector for Resilience education in the region.

Methods

Methodology

This study used a mixed method approach named exploratory design. For the quantitative aspect of the study, obtaining information on CCE and ESD was from teachers' workshops, training and survey (Clark and Ivankova 2016; Curry and Nunez-Smith 2015; Havea et al. 2017; Nastasi et al. 2007; Parvaiz et al. 2016). For the qualitative aspect of the study, data were deducted from project documents and in-depth interviews (Brady and O'Regan 2009; Creswell and Plano Clark 2011; Curry and Nunez-Smith 2015; Parvaiz et al. 2016; Tanyanyiwa and Kanyepi 2015). This method relied heavily on the qualitative aspect of the study (Clark and Ivankova 2016; Creswell and Plano Clark 2011; Curry and Nunez-Smith 2015; Havea et al. 2017; Heyvaert et al. 2017; Nastasi et al. 2007; Parvaiz et al. 2016; Tanyanyiwa and Kanyepi 2015).

Data Analytical Strategy

Information from the study was analyzed using Benchmarking, Evidencing, Knowing and Applying (BEKA) framework. The benchmark was used to assess the clarity of integrating climate change into primary and secondary education curriculum and resilience in education as a stand-alone certificate level 1–4 in TVET. Evidencing was used to assess the curriculum content and resources to map against the benchmarks. Knowing and applying were then used for deeper mining to have a complete understanding and to corroborate the evidence collected.

To achieve this goal, an exploratory design used desktop research (e.g. printed documents), document analysis, surveying and in-depth interview to provide a more thorough understanding as to how the integration of climate change by the Ministry of Education in Fiji, SPC/GIZ from CCCPIR, EU PacTVET project and FHEC has achieved. Once this information was gathered, the BEKA process was then used to map the relevant data to the framework's content. Information on the integration of climate change was moved iteratively between code and text to derive themes related to how climate change had been integrated into the school's curriculum using benchmarking, evidencing, knowing and applying content as a guide (Hall 2014).

Limitation of the Paper

Although the paper has reached its aim, there were some unavoidable limitations. First, because there was no funding to support this work within a set time limit, this research was conducted only on a small size of the population who was working for the EU PacTEVT at SPC (Pacific Community). Therefore, to generalise the results for all the Pacific Island Countries (PICs), the study would have benefitted from involving more participants at different levels. Second, because the model developed in this study was based on Fiji and the EU PacTVET project, it may not be applicable beyond Fiji's cultural context. Even so, research designer might adapt the model to address respective issues in other countries.

Third, the data regarding the quantitative aspects of the study are limited. Because of this, the authors also used the data from a research that was conducted in five coastal communities in Tongatapu Island: Kanokupolu, 'Ahau, Tukutonga, Popua and Manuka—to help address this topic. Another important limitation worth mentioning is the issue regarding policy anticipation. Hence, the respective countries should adjust or align their policy recommendations based on what they need. For example, countries that prefer to have Resilience as a major topic of study for schools may differ from those that prefer integration. As a result, policy development should be implemented accordingly.

Results

Integrating Climate Change Adaptation and Disaster Risk Reduction—A Key for Education for Sustainable Development in Tonga

Intriguingly, a study by Havea et al. (2018) on impacts of climate change on livelihoods, health and well-being in five coastal communities in Tongatapu, Tonga: Kanokupolu, 'Ahau, Tukutonga, Popua and Manuka—found that of the 460 participants, 97.8% (450) considered integrating climate change adaptation and disaster risk reduction as a key for education for sustainable development in Tonga (Table 1).

Furthermore, when asked whether this could be one of the solutions on the impacts of climate change on people's livelihoods, health and well-being in Tonga, one participant responded, "yes" (Table 2), although many other solutions were also recommended in the study.

Most importantly, this cosmological framing of integrating climate change into the school's curriculum in Tonga seems to be relevant for the concept of fulfilling God's scripture in the Bible, although it needs further research and development in the context of Tonga. This state-of-the-art-idea is pertinent when considering Tongan communities across the Pacific, in which a high percentage are Christians. Their worldviews and cosmologies comprise a rich epistemological framework of human-nature-spiritual relationships that are entwined with evangelical faiths and comprise

Table 1 Do you think it is a good idea that this research may be used as a benchmark to initiate a course curriculum structure on climate change and its impacts for all young people in Tonga and for our future generations to learn?

	Frequency	Percent
No	10	2.2
Yes	450	97.8
Total	460	100

Source Authors

Table 2 What are the best solutions to tackle the impacts of rising sea levels, extreme rainfalls, flooding, cyclones, droughts, temperature rise and tsunami on livelihoods, health and well-being amongst Tongans aged 15–75?

	Frequency	Percent
Climate change course structure curriculum, god scriptures fulfill	1	0.22
Total	460	100

Source Authors

spirituality-based resilience. This approach is vital in crafting and/or hatching effective and efficient measures for spiritual resilience that would contribute significantly to the building of a resilient Pacific Islanders by 2030 and beyond.

Adding to the above state of affairs, as Tonga was not part of this current study, the fact that the people are also cognizant about educating their children and young people as part of the solution to the problems is significant. By building the capacity of all Pacific Islanders to be more resilient and sustainable in nature is both pertinent for their lives and also allows the message to reach out to the communities at the household levels through their children who learn climate change adaptation at schools and take the message home to their parents for implementation. This is evident in what Fiji's Ministry of Education is doing. As a result, whether climate change is integrated and/or used as a separate subject of learning like Mathematics or English, it puts the Pacific ahead of this adaptation curve.

Integrating Climate Change in Education

There are two models for merging and using climate change in education. The first one is the model for primary and secondary schools in Fiji (Fig. 1).

EU PacTVET Resilience in Education

The second model is for TVET education in the region (Fig. 2).

This model was developed by the EU PacTVET project to deliver resilience qualifications certificate levels 1–4 through eight streams: agriculture, coastal management, energy and infrastructure, fishery, forestry, health, tourism and water resource management (EU PacTVET Project et al. 2016a, b).

Discussion

The BEKA process consists of four components: benchmarking, evidencing, knowing and applying, and it has been used successfully to demonstrate how the integration of climate change into the school curriculum in Fiji and having resilience in education for TVET's would benefit the children and young people of the Pacific both now and in the future.

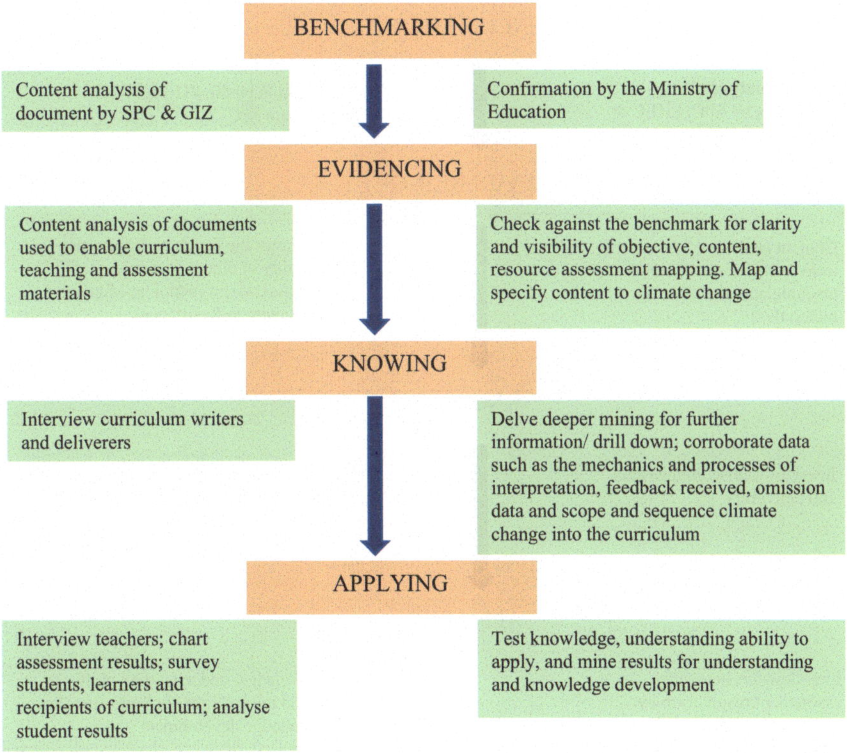

Fig. 1 Integration model for primary and secondary education in Fiji

Integration Climate Change in Education for Primary and Secondary Schools

The integration of climate change and disaster risk management topics into the school curriculum were done by the SPC/GIZ CCCPIR programme to meet Fiji Education for Sustainable Development (ESD) and Climate Change Education (CCE), and most importantly, the objectives of the FRDP framework. Fiji along with Kiribati, Samoa, Tonga, and Vanuatu cast the first countries in the Pacific to pilot this programme of integration targeting schools at the national level. The strategies that were used to integrate climate change into the school curriculum comprised of four major steps:

Step 1—Integration of climate change into curricula
Step 2—National curriculum consultations
Step 3—Development of teaching materials
Step 4—Training of teachers.

There were several strategies used to integrate climate change into the school's curricula: curriculum mapping (CAS-TVET), curriculum scoping and sequence,

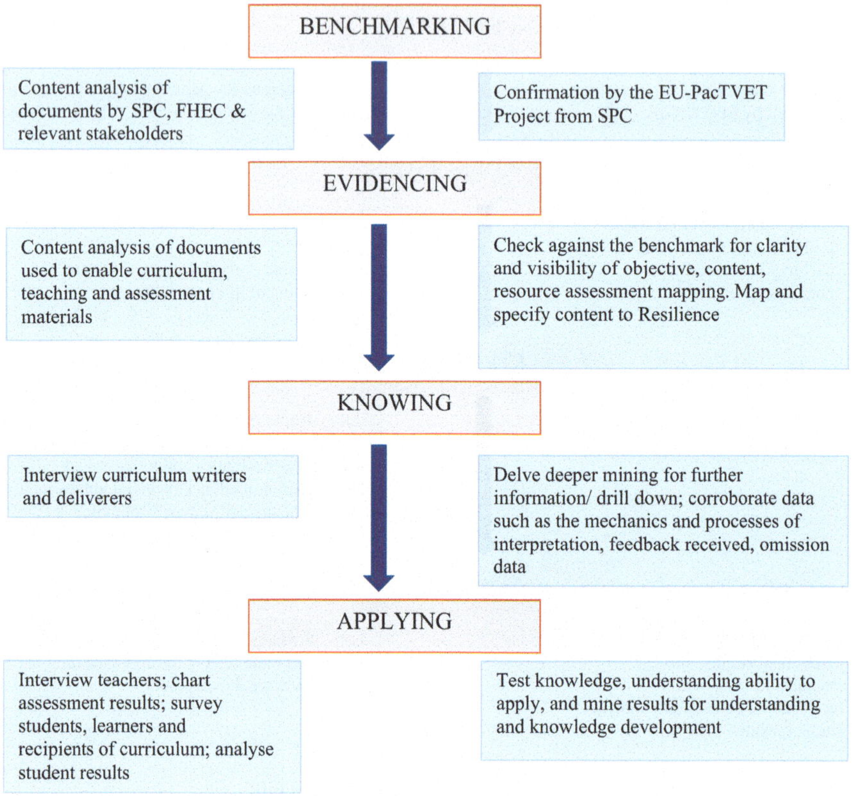

Fig. 2 Model for resilience in education for the region for the TVET

consultations and expert presentations, write-shops, and stock-take or collection of CCA/DRM education awareness materials for Fiji and from the regional climate change websites.

The study found the integration of climate change in education for primary and secondary schools is more appropriate than developing a standing-alone climate change subject for two reasons. First, it is more relevant to what teachers are doing to strengthening the learning outcome of the unit subject by linking climate change to these subjects (e.g. science and social science), so to prevent work overload.

Second, according to the Ministry of Education, integration of climate change is cheaper and by doing so helps the school system to overcome staff turnover (e.g. from teachers' relocation to other schools, teaching other subjects and retirement), now and in the future. More importantly, since the primary and the secondary education in Fiji is now in the second phase such as monitoring and evaluation of their integration programme, it is recommended that other Pacific Islanders who have not yet integrated climate change to their school curriculum can adapt Fiji's climate change education programme and perform a pilot study.

Addressing climate change to the children at a young age is one way of improving the understanding of the Pacific citizenry, the dynamics and processes that would help them build resilient Pacific Islanders now and in the future. As a Tongan concept to value the education of the Tongan children indicated that the "reef of today is an island of tomorrow" may mean to Tonga, but in the context of Fiji and the Pacific, this framework could also apply to ESD and CCE. As a result, not only has the Government of Fiji contributed significantly to its national development and adaptation plan, but it has appreciated all concerned from all parties and relevant stakeholders into account.

Resilience in Education—EU PacTVET Project

The other model to be used by TVET institution in the region is the Resilience in education that was developed by the EU PacTVET project. The EU PacTVET in Sustainable Energy (SE) and Climate Change Adaptation (CCA) project is the third component of a larger programme: Adapting to Climate Change and Sustainable Energy (ACCSE). The European Union funded PacTVET project is a €6.2 million project currently implemented by the Pacific Community (SPC) and the University of the South Pacific (USP) (EU PacTVET Project et al. 2016a, b).

One of the goals of the EU PacTVET project that had been achieved recently is the development of a model of resilience curriculum from certificate levels 1–4 to be used by the participating Pacific African Caribbean Pacific (P-ACP) countries: Fiji, Samoa, Tonga, Vanuatu, Papua New Guinea (PNG), Solomon Islands, Cook Islands, Kiribati, Tuvalu, Federated States of Micronesia (FSM), Marshall Islands, Niue, Nauru, and Timor Leste. This program of study is an accredited-competency based curriculum that is universal (e.g. open and flexible), which can be delivered and taught at eight streams: agriculture, coastal management, energy and infrastructure, fishery, forestry, health, tourism and water resource management—at any institution as long as they met the expected countries' national qualification criteria of deliverables.

As this study is the first to undertake a curriculum analysis on climate change, it is expected that this will be a benchmark to the milestone of developing a state-of-the-art system of support to build a world-class and renowned institution for resilience, owned and operated by PICs, as internationally recognised, vibrant, forward-thinking and innovative as USP. As a result, it is significant that the Pacific countries must continue with this sustainable model for capacity development to 2030 and beyond.

Using the Concept of Ako to Understand Resilience in Education

Using the concept of ako (e.g. to learn, study or educate) as a hallmark to understand the learning of climate change and/or resilience in education, students at primary and secondary schools and TVET can be able to learn climate change through a formal learning environment using a mix of students and/or teachers-based pedagogical approach. Because information and technology are very advanced in this era, this paper proposes using VARK (video, aural, reading/writing and kinesthetic) model of learning (Othman and Amiruddin 2010). In this approach, students are not only prepared academically but also with a placement component whether classroom-based or industrial-based assessment or visitation. All theoretical treatment and practical placement will be climate change and hazards related (e.g. sea level rise, temperature rise).

Policy Implication: Resilience in Education for the Pacific by Integration or Teaching

Based on the results of this study, this paper suggested the governments in the Pacific to use this evidence to propose a policy to their own Ministry of Education depending on their situation. First, is to integrate climate change into their school curricula. Second, is to teach climate change as a compulsory in schools. Nevertheless, other options are that some schools could mix these two methods and employ them concurrently.

This climate change policy is vital especially from a resilience point of view because it is also a part of the National Adaptation Plan for Fiji. At the regional level, this is where the EU-PacTVET came into support the 14 PICs. The Pacific schools should consider this alternative paradigm as a state-of-the-art ESD for the region.

Conclusions

In conclusion, there are three major lessons learnt from this paper. First, Resilience can be integrated into the current school curriculum using Fiji as a guide. Second, Resilience should be made as a leading course of study for schools and TVET. For primary and secondary education, they will be granted a certificate in resilience. For the TVET, the award is a certificate level 1–4. Third, the two models developed under this study should be used as a guide. Because of this, the model was designed to be universal and open, so it should be flexible enough to solve issues of other cultures.

For other cultures, since the model presented in this paper is universal and can go beyond Fiji, as a result, the TVET institutions in respective countries should architect their Resilience curriculum to their environment and context. For example, since different countries experience issues surrounding climate change differently, it is significant for the education sector (e.g. Curriculum Development Unit) to recognise these and address them accordingly in their schools' program of study.

The reason is that a five-year-old boy who has experienced climate change impacts and adaptation in New York City would differ from the same five-year-old boy whose livelihoods, health and well-being are affected in Suva, Fiji, or Nuku'alofa, Tonga. Therefore, it is imperative for children to be well-versed with what is affecting them (e.g. climate change impacts) and how to overcome adversity (e.g. climate change adaptation) if the plan is to build resilient and sustainable Pacific Islanders by 2030 and beyond.

For the future, further studies on evaluating resilience in education may include but are not limited to:

(1) longitudinal studies on cost versus benefit;
(2) cost-benefit analyses;
(3) studies on the impact of Resilience courses of study on community-based adaptation.

Acknowledgements We wanted to acknowledge the GIZ and the Ministry of Education in Fiji for allowing us to use their data. Most importantly, we would like also to thank the EU PacTVET project at SPC for sharing their data. Without these stakeholders, we would not be able to write this paper.

References

APTC (2018) APTC recognises students' achievements in Honiara. Australia Pacific Technical College (APTC). Suva, Fiji. Retrieved 15 Apr 2019. From http://www.aptc.edu.au/news/news/2018/11/22/aptc-recognises-students-achievements-in-honiara

Barrow Cadbury Trust (2012) Adapting to change: the role of community resilience. The Young Foundation, London, UK

Bell J, Johnson J, Ganachaud A, Gehrke P, Hobday A, Hoegh-Guldberg O, Le Borgne R, Lehodey P, Lough J, Pickering T, Pratchett M, Waycott M (2011) Vulnerability of tropical Pacific fisheries and aquaculture to climate change: summary for Pacific Island Countries and Territories. Secretariat of the Pacific Community, Noumea, New Caledonia

Boakye C (2015) Climate change education: the role of pre-tertiary science curricula in Ghana 5(4):1–10

Boniwell I, David SA, Ayers AC (2013) Oxford handbook of happiness. Oxford University Press, UK

Brady B, O'Regan C (2009) Meeting the challenge of doing an RCT evaluation of youth mentoring in Ireland: a journey in mixed methods. J Mixed Methods Res 3(3):265–280

Cefai C, Matsopoulos A, Bartolo P, Galea K, Gavogiannaki M, Zanetti A, Renati R, Cavioni V, Pavin Ivanec T, Šarić M, Kimber B, Eriksson C, Simões C, Lebre P (2014) A resilience curriculum for early years and primary schools in Europe: enhancing quality education. Croatian J Educ 16(Sp.Ed.No.2):11–32

Chirwa G, Naidoo D, Chirwa G (2014) Curriculum change and development in Malawi: a historical overview. Mediterr J SocSciences 5(16):336
Clark VLP, Ivankova NV (2016) Mixed method research: a guide to the field. Sage, London
Creswell JW, Plano Clark VL (2011) Designing and conducting mixed methods research, 2nd edn. Sage, USA
Curry L, Nunez-Smith M (2015) Mixed methods in health sciences research: a practical primer. Sage, London
EU PacTVET Project (2017) The Pacific Regional Federation of Resilience Professionals (PRFRP). A professional organization to build capacity in the climate change and disaster risk management sectors in 15 small island developing states in the Pacific—Submission to the Paris Committee on Capacity-Building. SPC, Fiji
EU PacTVET Project, SPC and USP (2016a) Regional certificate 3 in resilience (Climate Change Adaptation and Disaster Risk Reduction/CCA & DRR). EU PacTVET Project, Suva, Fiji
EU PacTVET Project, SPC and USP (2016b) Regional certificate 4 in resilience (Climate Change Adaptation and Disaster Risk Reduction/CCA & DRR). EU PacTVET Project, Suva, Fiji
Flinders University (2018) Concepts—origins of the term. Torrens Resilience Institute Australia. Retrieved 11 May 2018. From http://www.flinders.edu.au/torrens-resilience-institute/concepts/concepts_home.cfm
Hall CEA (2014) Toward a model of curriculum analysis and evaluation—Beka: a case study from Australia. Nurse Educ Today 34(3):343–348. https://doi.org/10.1016/j.nedt.2013.04.007
Havea PH, Hemstock SL, Combes HJD (2017) Preparing for better livelihoods, health and well-being—a key to climate change adaptation. In: Leal Filho W (ed) Climate change adaptation in Pacific Countries: fostering resilience and improving the quality of life. Springer International Publishing, Cham, pp 87–99
Havea PH, Jacot des Combes H, Hemstock SL, Gee K, Han D, Khan MGM, Moore T (2018) Climate change impact on livelihood, health and well-being: a mixed method approach, (Ph.D. thesis). The University of the South Pacific, Fiji
Hemstock SL, Jacot Des Combes H, Buliruarua L-A, Maitava K, Senikula R, Smith R, Martin T (2018) Professionalising the 'resilience' sector in the Pacific Islands Region: formal education for capacity building. In: Klepp S, Chavez-Rodriguez L (eds) A critical approach to climate change adaptation: discourses, policies and practices. Routledge Advances in Climate Change Research, Abigdon Oxon (In Press)
Heyvaert M, Hannes K, Onghena P (2017) Using mixed methods research synthesis for literature reviews. Sage, Washington DC
IPCC (2014a) Climate change 2014: impacts, adaptation, and vulnerability. Part A: global and sectoral aspects. Contribution of working group II to the fifth assessment report of the intergovernmental panel on climate change. Cambridge University Press, Cambridge, 1132 pp
IPCC (2014b) Climate change 2014: impacts, adaptation, and vulnerability. Part B: regional aspects. Contribution of working group II to the fifth assessment report of the intergovernmental panel on climate change. Cambridge University Press, Cambridge, 688 pp
Luetz J, Havea PH (2018) We're not refugees, and we'll stay here until we die!—climate change adaptation and migration experiences gathered from the Tulun and Nissan Atolls of Bougainville, Papua New Guinea. In: Leal Filho W (ed) Climate change impacts and adaptation strategies to coastal communities. Springer, Switzerland
Luetz JM, Sultana N (2019) Disaster risk reduction begins at school: research in Bangladesh highlights education as a key success factor for building disaster ready and resilient communities—a manifesto for mainstreaming disaster risk education. In: Leal Filho W, Lackner B, McGhie H (eds) Addressing the challenges in communicating climate change across various audiences. Climate change management. Springer Nature, Cham, Switzerland, pp 617–646
Maini R, Clarke L, Blanchard K, Murray V (2017) The Sendai framework for disaster risk reduction and its indicators—where does health fit in? Int J Disaster Risk Sci. https://doi.org/10.1007/s13753-017-0120-2

McIver L, Kim R, Woodward A, Hales S, Spickett J, Katscherian D, Hashizume M, Honda Y, Kim H, Iddings S, Naicker J, Bambrick H, McMichael AJ, Ebi KL (2016) Health impacts of climate change in Pacific island countries: a regional assessment of vulnerabilities and adaptation priorities. Environ Health Perspect 124(11):1707–1714

Mokaya D, Kariuki C, Mwangi D, Glara Sr JK (2016) Infusing climate change content into Kenya's secondary school taught curriculum: successes and challenges. IOSR 10(9 v1):23–29

Nastasi BK, Hitchcock J, Sarkar S, Burkholder G, Varjas K, Jayasena A (2007) Mixed methods in intervention research: theory to adaptation. J Mixed Methods Res 1(2):164–182

Nunn PD (2013) The end of the Pacific? Effects of sea level rise on Pacific Island livelihoods. Singap J Trop Geogr 34(2):143–171

Nunn PD, Aalbersberg W, Lata S, Gwilliam M (2014) Beyond the core: community governance for climate-change adaptation in peripheral parts of Pacific Island Countries. Reg Environ Change 14:221–235. https://doi.org/10.1007/s10113-013-0486-7

Othman N, Amiruddin MH (2010) Different perspectives of learning styles from VARK model. Proc Soc Behav Sci 7:652–660. https://doi.org/10.1016/j.sbspro.2010.10.088

Pacific Community, Secretariat of the Pacific Regional Environment Programme, Pacific Islands Forum Secretariat, United Nations Development Programme, United Nations Office for Disaster Risk Reduction, & University of the South Pacific (2016) Framework for resilience development in the Pacific: an integrated approach to address climate change and disaster risk management (FRDP) 2017–2030. The Pacific Community, Geoscience Division, Suva, Fiji

Pacific Community (SPC) (2011) Climate change education. SPC, Fiji. Retrieved 7 May 2018. From http://www.spc.int/coping-with-climate-change-in-the-pacific-islands-region-cccpir-programme/climate-change-education/

Pacific Community (SPC) (2014) Climate change adaptation in Kiribati—Solar Disinfection (SODIS) in Kiribati. SPC, Suva, Fiji

Pacific Community (SPC) (2015) Pacific community strategic plan 2016–2020. SPC, New Caledonia

Parvaiz G, Mufti O, Wahab M (2016) Pragmatism for mixed method research at higher education level. Bus Econ Rev 8(2):67–79

Regmi S (2015) Gender and health adaptation measures to climate change in the Pacific: a case study of Papua New Guinea. In: Leal Filho W (ed) Handbook of climate change adaptation. Springer, Heidelberg

Rochat P (2003) Five levels of self-awareness as they unfold early in life. Conscious Cogn 12(4):717–731

Singh RB, Hales S, de Wet N, Raj R, Hearnden M, Weinstein P (2001) The influence of climate variation and change on diarrheal disease in the Pacific Islands. Environ Health Perspect 109(2):155–159

SPC and GIZ (2016) Coping with Climate Change in the Pacific Island Region (CCCPIR) annual report—Vanuatu Programme 2016. SPC-GIZ, Vanuatu

SPREP (2017) Celebrating island biodiversity achievements in the Pacific: case studies from the GEF-PAS Integrated Island Biodiversity Project. SPREP, Apia, Samoa

Tanyanyiwa VI, Kanyepi T (2015) Mixed methods research strategies in climate change adaptation in Mt Darwin District, Zimbabwe. INIRD 4(9):213–221

Taylor M, McGregor A, Dawson B (2016) Vulnerability of Pacific Island agriculture and forestry to climate change. SPC, New Caledonia

UNDP (2015) Sustainable Development Goals (SDGs). UNDP, USA

UNESCO (2006) Pacific education for sustainable development framework. UNESCO, Fiji

United Nation Office for Disaster Risk Reduction (UNISDR) (2015) Sendai framework for disaster risk reduction 2015–2030. UNISDR, Geneva, Switzerland

United Nation University (2016) World risk report 2016: focus (logistic and infrastructure). UNU, Tokyo, Japan

United Nations University, Institute for Environment and Human Security (UNU-EHS) and Universität Bonn (2013) World risk report 2013—focus: health and healthcare. Bündnis Entwicklung Hilft, Berlin

USP (2018) Certificate III & IV in resilience (Climate change adaptation & disaster risk reduction). University of the South Pacific (USP), Suva, Fiji

Vernon RF (2004) A brief history of resilience. In: Clauss-Ehlers CS, Weist MD (eds) Community planning to foster resilience in children. Springer, Boston, MA

WHO (2015) Human health and climate change in Pacific Island countries. World Health Organisation, Geneva

Woods DD (2015) Four concepts for resilience and the implications for the future of resilience engineering. Reliab Eng Syst Saf 141:5–9. https://doi.org/10.1016/j.ress.2015.03.018

Wosnitza M, Peixoto F, Beltman S, Mansfield CF (2018) Resilience in education: concepts, contexts and connections. Springer, Switzerland

Yamamoto L, Esteban M (2017) Migration as an adaptation strategy for Atoll Island States. Int Migr 55(2):144–158

Engaging Communities and Government in Biodiversity Conservation and Climate Adaptation in Papua New Guinea

Brent Jacobs, Kylie McKenna, Louise Boronyak, Francesca Dem, Shen Sui, Kenneth Pomoh, Mavis Jimbudo and Heveakore Maraia

Abstract Papua New Guinea's (PNG) mainland consists of 33 million hectares of forests. The third largest intact rainforest in the world, it contains about 7% of the world's species, 2/3 of which are unique to PNG. PNG's ecosystems face multiple and interdependent threats associated with economic development, population growth and a changing climate. Academic and policy analysis on environmental change in PNG is extensive, particularly associated with the minerals and energy extraction sector. To counterbalance the negative impacts of this sector on affected communities, much of the focus has been on devising compensation packages and formal regulatory mechanisms to increase 'landowner' participation. Less attention has been afforded to the development activities undertaken by communities (e.g. development of new roads, expansion of settlements, land clearance from fires and logging), which also impact on ecosystem services. PNG's rural communities are eager for more support to identify existing threats to supplement their own processes for determining trade-offs of development particularly under a changing climate. This paper describes the use, in facilitated workshops, of participatory techniques to engage communities in managing ecosystem services and biodiversity conservation to inform the development of community-led adaptive strategies.

Keywords Climate change · Development · Ecosystem services · Biodiversity · Community engagement

---

B. Jacobs (✉) · K. McKenna · L. Boronyak
University of Technology Sydney, Institute for Sustainable Futures, PO Box 123, Broadway, NSW 2007, Australia
e-mail: brent.jacobs@uts.edu.au

F. Dem · S. Sui · K. Pomoh · M. Jimbudo · H. Maraia
New Guinea Binatang Research Centre, PO Box 604, Madang 511, Madang Province, Papua New Guinea

W. Leal Filho (ed.), *Managing Climate Change Adaptation in the Pacific Region*, Climate Change Management, https://doi.org/10.1007/978-3-030-40552-6_11

Introduction

Changes in ecosystems from human activities have accelerated over the past 50 years (IPBES 2019; Millennium Ecosystem Assessment 2005). Climate change, land modification, habitat loss, pollution and over-harvesting of natural resources are causing widespread degradation of ecosystems (IPBES 2018; Millennium Ecosystem Assessment 2005; Thomas et al. 2004; Foley et al. 2005). For Papua New Guinea (PNG) and other Pacific nations, a major challenge is how to generate economic development while preventing further environmental damage. With almost 40% of the PNG population living on less than US$1.25/day (UNDP PNG n.d.), opportunities to pursue economic development and supporting infrastructure are highly sought-after. Many communities across PNG however, have already experienced both positive and negative economic, social and environmental costs of large development projects. The positive benefits for communities in PNG include employment, business spin-offs, roads, hospitals and training, yet these benefits have largely been "off-set by a range of economic, social, political, and environmental risks and costs" (Filer and Macintyre 2006, p. 217).

Ecosystem change and loss in PNG will have global ramifications for biodiversity because the PNG mainland contains 33 million hectares of forests representing the largest rainforest in the Asia- Pacific region, and the 3rd largest intact rainforest in the world. PNG is estimated to contain at least 7% of the world's species; 2/3 of which are unique to PNG (UNDP n.d.). A current population of approximately 7.5 million could double by 2050 (UNDP n.d.), raising concerns about future land shortages and the sustainability of current livelihood practices. The PNG economy is dominated by two sectors: the agricultural, forestry and fishing sector and the minerals and energy extraction sector (World Bank 2018). Both sectors are important sources of economic development yet contribute to ecosystem degradation stemming from over-exploitation of natural resources, unsustainable land use, habitat destruction, pollution and poor environmental governance (UNDP n.d.). Additional to the impacts of population growth and economic development, PNG is expected to experience largely negative impacts of climate change.

A range of efforts to address the negative impacts of large-scale mineral and gas developments has been made by the Government of PNG and extractives companies. Examples include but are not limited to, the Development Forum established by the Government of PNG to formulate benefit sharing agreements between landowners and various levels of government (Filer 2008); landowner compensation (Filer et al. 2000) and corporate social responsibility policies and programs driven by individual companies (McKenna 2016). While these kinds of processes have resulted in safer technologies and better stakeholder engagement with PNG communities (Gilberthorpe and Banks 2012, p. 186), instances of environmental disaster (e.g. Ok Tedi), human rights violations (e.g. Porgera) and community actions that interfere with production (e.g. Ramu Nickel mine, Lihir gold mine) have continued to plague the minerals and energy sector. As these cases involve large multinational corporations, there is a tendency for commentary on environmental change in PNG to focus on relationships between 'affected communities' and 'external' actors. Less attention is afforded to the activities that communities themselves undertake, such as the extension of roads to enable access to town markets, the establishment of new buildings, and selective logging that can degrade local ecosystems on one hand, and local reforestation efforts, establishment of community conservation reserves and sustainable natural resource-based businesses on the other. These community actions can also affect the availability of ecosystem services (ES) that are essential to the maintenance of local livelihoods and wellbeing.

Climate Change in Papua New Guinea

Across the Asia–Pacific region climate change presents many challenges from the inundation of low lying islands to the risk that droughts pose to food security (IPBES 2018; IPCC 2014). Climate change alters critical climate variables, such as rainfall and temperature that can result in changes in natural climate variability and extreme events, such as floods and drought. These changes directly influence the health and function of ecosystems which in turn influence the supply of the goods and services they provide. Therefore, effectively conserving ecosystems will depend on the ability to predict the risk of extreme climatic effects and to facilitate the capacity of communities to cope or adapt (SEG 2007).

The Pacific Climate Change Science Program (PCCSP) identified 18 best fit Global Climate Change models (GCM's) to represent the climate of PNG (PCCSP 2011). These models were used to develop PNG climate projections to 2090 (Table 1 for a summary to 2055). Key features of projections for PNG are that: (1) temperatures will continue to increase; (2) incidence of very hot days will increase; (3) rainfall patterns will change; (4) incidence of extreme rainfall days will increase; and, (5) tropical cyclones will become less frequent but more intense (PCCSP 2011, p. 6).

In addition, storms are expected to be less frequent but more intense with an increase in average maximum wind speed and an increase in rainfall intensity. Increased rainfall during storms could also heighten risk of soil erosion, nutrient loss, landslides and floods. Extreme weather events have already affected communities from the highlands to the islands. In 2015, much of rural PNG suffered a severe drought and repeated frosts in the Highlands (Bourke et al. 2016) and in 2017 alone, 2000 households across 30–35 coastal communities in PNG saw people relocated to higher ground due to coastal erosion and flooding (Caritas 2017).

As carbon emissions continue to rise the risk of ocean warming and acidification increases, which poses a major threat to coral reef ecosystems and food security in PNG (IPBES 2018). Coral reefs are of critical cultural, ecological, and economic importance across the Asia–Pacific region. They form an important buffer against the impacts of storm surge for coastal communities and provide critical habitat for marine species. Projections indicated that coral reef fisheries could decline by a further 20% by 2050 (IPBES 2018; PCCSP 2011).

Table 1 Future climate projections[a] for PNG

	Current	2030	2055
Rainfall	Annual rainfall is highly variable throughout PNG depending on El Nino and La Nina events	Increase in average annual rainfall and seasonal rainfall	Increase in average annual rainfall and seasonal rainfall with more extreme rainfall events particularity in the traditionally wet seasons
Average annual temperature	Minimum recorded temperatures have increased by 0.3 °C per decade from 1940 to 2010, while maximum temperatures have increased by 0.13 °C per decade	Increase of between 0.4 and 1.2 °C Increase in number of hot days	Increase of between 1 and 2 °C Further increase in hot days
Sea level rise		Increase of between 5 and 14 cm	Increase by between 9 and 30 cm

[a]Based on the IPCC medium emissions scenario A1B
Source PCCSP (2011)

Ecosystem Services

Ecosystem Services (ES) are the benefits people obtain from ecological systems or ecosystems. Costanza and Daly (1992) presented the concept of natural capital to describe ecosystem goods and services, later refined in the 2005 Millennium Ecosystem Assessment (MA) that grouped ecosystem services into four broad categories: supporting; provisioning, regulating and cultural services (MA 2005). Supporting services, include aspects such as nutrient cycling, soil formation and primary production (MA 2005). Provisioning services involve the production of renewable resources for example, food, wood, and fresh water (MA 2005). Regulating services are those that lessen environmental change such as climate regulation provided by forests or water purification (MA 2005; Cardinale et al. 2012). Lastly, cultural services include the aesthetic, spiritual, educational and recreational benefits from ecosystems (MA 2005). The interactions between the biotic and abiotic components of ecosystems as well as ecological and evolutionary processes, create the stocks and flows that underpin the ES (Mace et al. 2012). ES are critical for supporting human wellbeing and livelihoods, and changes in these services affect human well-being.

The biological composition of ecosystems, measured as biodiversity, is a factor regulating the ecosystem processes that underpin ES (Mace et al. 2012). The diversity of genes, species, and ecological processes makes a vital contribution to ecosystem services (Cardinale et al. 2012; Mace et al. 2012). Biodiversity can be defined as the 'variability among living organisms from all sources including, *inter alia*, terrestrial, marine and other aquatic ecosystems and the ecological complexes of which they are part; this includes diversity within species, between species and of ecosystems' (UNEP 1992, p. 3). Due to the complexity of ecosystems, the interlinkages between ecosystem services and biodiversity are not well understood (Nicholson et al. 2009). Biodiversity contributes to ES in three key ways: firstly, through contributing to ecosystem function and processes in terms of regulating and supporting services; secondly, as a final consumable ecosystem service, e.g. wild medicines; and thirdly, as a 'good' that has a direct value (Cardinale et al. 2012; Mace et al. 2012). Biodiversity also has spiritual and cultural value, such as the appreciation of wildlife (Mace et al. 2012; MA 2005). Economic development and climate change create new challenges for biodiversity conservation (Adams et al. 2004; Heller and Zavaleta 2009) that require close engagement with communities to better understand place-based impacts.

Participatory mapping is a technique frequently used to engage communities in aspects of land use planning, including the location of ecosystem services using geographical information systems or GIS (Brown and Fagerholm 2014; Martínez-Harms and Balvanera 2012). However, reliance on technology for engagement with Indigenous communities in developing-world settings can be problematical (Dyson et al. 2007). The availability of satellite imagery through Google Earth presents an opportunity for 'lo-fi' approaches to participatory mapping of ecosystem services in difficult settings, such as the highlands of PNG, to allow exploration of changes in land use patterns, the trade-offs inherent in land use change and identification of areas for biodiversity conservation, among other uses (Liang et al. 2018).

This paper describes the use, in a series facilitated workshops, of participatory techniques to engage communities in managing ecosystem services and biodiversity conservation. The engagement process was designed to:

(1) identify locally important ecosystem services related to biodiversity;
(2) identify threats to biodiversity and to develop possible adaptation and preventative management strategies;
(3) consider the relative values of local ecosystem services and how these services may be affected by climate change and land use change such as, intensification of agriculture, mining or expansion of roads; and
(4) identify positive solutions to address biodiversity loss and other changes to ecosystems.

The workshops formed the major component of a broader USAID-funded project to (1) aid communities to identify and develop community-led, rather than negotiated or imposed, adaptive strategies, for a range of threats to ES at local scale; and, (2) engage provincial government decision makers in supporting communities in climate adaptation planning and action.

Methods

Over 190 people participated in the project through community and government workshops and qualitative interviews. Participants comprised residents, women and church leaders, current and former council members; teachers; and government representatives from the provincial, district and ward levels drawn from four PNG communities: Riwo (Madang Province); Ohu (Madang Province); Bundi District (Madang Province); and Hogave (Eastern Highlands Province). These communities were chosen because they effectively formed a transect from the coast (Riwo) through the hinterland (Ohu) to the highlands (Bundi and Hogave), spanned a broad range of climate and development issues faced by communities throughout PNG, and were accessible by road from the provincial capital of Madang. Figure 1 shows a map of the research sites in Madang and Eastern Highlands Provinces.

Project researchers conducted semi-structured interviews with 15 representatives of the Madang and Eastern Highlands Provincial Governments. The representatives were drawn from relevant divisions including: Forestry; Business; Economics; Agriculture and Livestock; Disaster; Planning and Works, and Community Development. The interviews provided a means to elicit information on plans for specific infrastructure developments such as roads, new mines or logging concessions. These data were used to inform development scenarios to engage community members in a

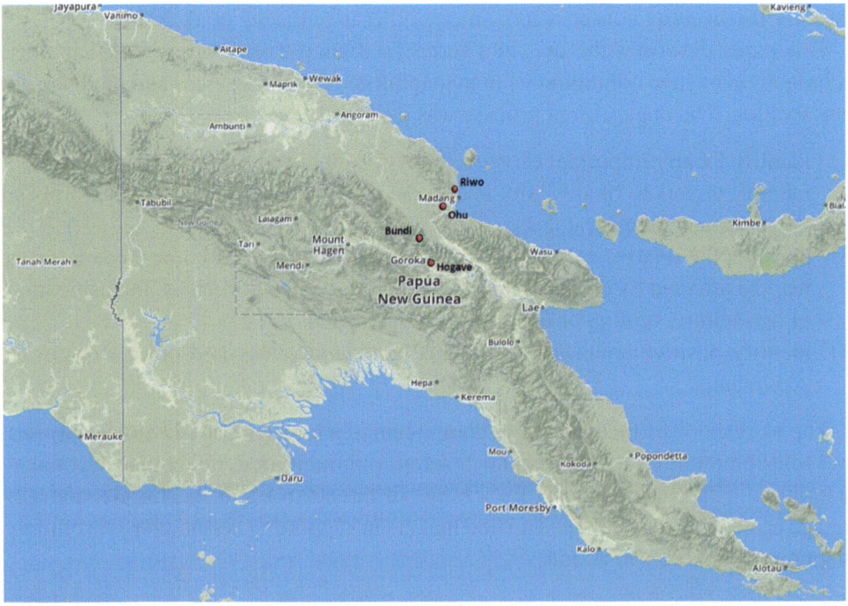

Fig. 1 Map of research sites in Papua New Guinea. Red dots denote each of the four workshop locations. *Source* Adapted from Mapbox

series of four *Climate Change Adaptation and Biodiversity Planning* workshops. The workshops consisted of six key activities:

(1) **Storytelling**: to create space for discussions around peoples' experience with change in their local environment (Iseke 2013). Storytelling activities, led by community leaders (e.g. women's leaders, religious leaders and village elders), generated discussion on the lived experiences of environmental change such loss of biodiversity and/or climate variability.
(2) **Identification of local ecosystem service values**: that are of importance both individually and collectively (e.g. fish derived from a local river or building materials from local forests). Participants situated, onto large (A1-sized) Google Earth images of their village and its surrounds, icons that represented a range of ecosystem services related to biodiversity to build a geo-spatial picture of local ecosystem values.
(3) **Discussion of climate impacts for the region**: The facilitators presented both the historical changes in temperature and rainfall for PNG as well as climate projections. The participants were asked to indicate with 'sticky dots' the changes in climate variables they were most concerned about (e.g. more intense storms, changing rainfall, drought, increasing temperatures etc.) to aid in prioritising subsequent discussion of climate impacts of concern.
(4) **Visualizing the impacts of environmental change (impact mapping)**: This activity involved mapping the impacts of a range of drivers of change (changes in climate, society, and economic development) selected by consensus among participants. The process used causal diagrams to depict the changes in local systems (Jonassen and Ionas 2008).
(5) **Identification of ecosystem service values at risk from climate change and development**: Participants used the local ES maps to identify which ecosystem services, with a focus on biodiversity, may be at risk from a changing climate and local economic development in the short and long-term.
(6) **Plans for collective action**: using a simple project planning template, participants identified a range of practical actions at local scale to manage impacts and capitalise on the benefits of social-ecological change for their communities. The planning template called for participants to decide on a project name and brief description, and identify steps to implementation, stakeholders involved and resources needed.

The activities were detailed in a facilitator's handbook for *Engagement on Biodiversity Conservation and Climate Change Adaptation in Papua New Guinea* (Boronyak et al. 2018). The handbook aimed to guide individuals and organisations through the process to extend the project's reach beyond the target communities to a wider audience in PNG and across the Pacific.

A significant amount of data was generated across the four community workshops. We will draw on selected workshop outputs (ecosystem services and impact mapping) to illustrate the most pertinent issues raised by the four focal communities.

Results

The aerial images were a useful boundary object (Star and Griesemer 1989) that focused the attention of participants on consideration of ecosystem services (ES). Up to eight people at a time were actively engaged in situating icons on the maps and included men, women and youth (Fig. 2). Facilitators were careful to ensure that no single participant dominated the exercise. The activity also generated much discussion in Tok Pisin and English, which led to revision of the map throughout the process as consensus was reached on where specific ecosystem service icons should be best located.

While the maps produced by the participants were 'messy' (e.g. Fig. 3), they nevertheless documented their tacit knowledge of the region. The focus was primarily on locating provisioning ecosystem services (i.e. agriculture, food and bush materials, and fresh water supply). However, participants also located other types of services, e.g. supporting services in areas of primary forest, regulating services along streams where potable water was collected, and cultural services in areas of high recreational amenity (e.g. for swimming) or of spiritual value (e.g. for collection of ceremonial feathers).

Fig. 2 Community workshop participants engaged in ecosystem services mapping. *Source* Authors

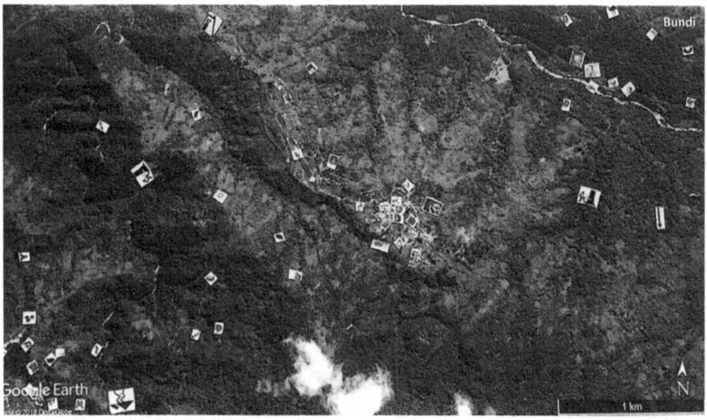

Fig. 3 A community ecosystem services map from Bundi workshop. *Source* Authors

From one of the workshop maps (the hinterland village of Ohu), we produced a simplified layout of the village (Fig. 4) to demonstrate the utility of the mapping process. The simplified map divides Ohu village into three zones of roughly equal area. Firstly, a zone of food production, where the vegetation consists primarily of early regrowth forest in varying stages of recovery interspersed with areas under crop. Secondly, a zone of primary forest, which made up the village's conservation area and was the site of hunting, gathering, and collection of bush materials. This zone encompassed the major water catchment in the area. Thirdly, an intermediate zone, made up of mixed secondary forest and agriculture, the community's two main settlements (school, meeting houses etc.) and the main road.

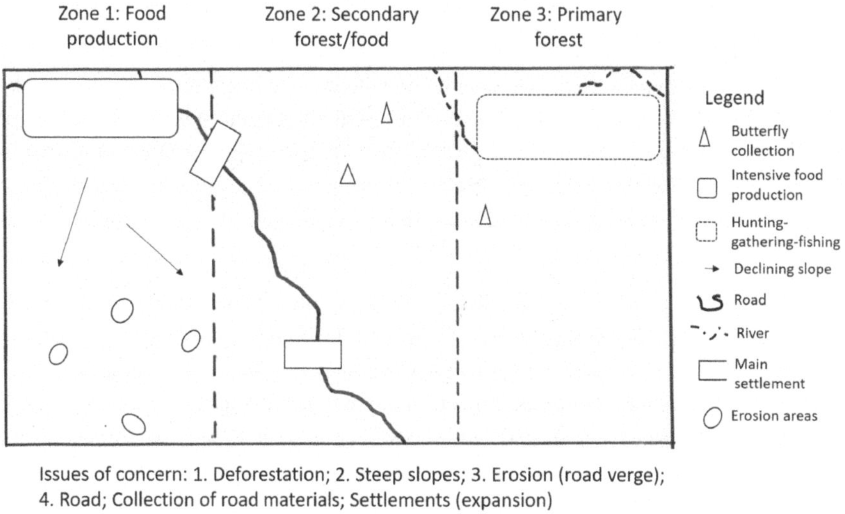

Fig. 4 Simplified ecosystem services map from Ohu Village. *Source* Authors

Discussion of the ES map with participants during the workshop identified concern in the community about the need to intensify agricultural production to provide food for a growing population. Intensification of agriculture led to shorter fallow periods and, declining crop yields (particularly of kumara). The combination of the removal of large trees and the sloping terrain has led to the erosion of top soil and land slippage, and subsequent pressure for expanded crop production in the secondary forest zone. Erosion along the road and its verge, which were subject to failure during extreme rainfall events cutting the village off from produce markets, and the collection of materials (bush rock and logs) to effect road repairs and maintenance, were seen as increasingly damaging to the local environment. Finally, population growth required construction of additional houses, which put pressure on land resources and supply of bush materials.

It was clear from the four community workshops, that climate change is already having significant effects on communities. Changes in rainfall and rising temperatures were considered to be the most important climate variables. For example, residents of Ohu village reported experiencing changes in rainfall (e.g. more frequent, high intensity events), which have damaged the road to Madang town. An increase in the number of hot days is also limiting the number of hours that women reported spending on working in the garden. Similarly, residents of Bundi District were concerned about the potential for more hot days and changes in rainfall patterns. In common with Ohu, heavy rain is damaging the already dangerous road into Madang town. Residents of Hogave (Eastern Highlands) were overwhelmingly concerned about the potential for wildfires resulting from changed rainfall patterns and higher temperatures, particularly in the community forest conservation area. These changes have flow on effects to food security, human health and ecosystem health and function (see Table 2).

In addition to the impacts posed by climate change, community members shared their views on the advantages and disadvantages of development projects on local social-ecological systems. During the workshops, the facilitators presented a series of development scenarios that were established through findings of the scoping workshop and government interviews. These developments ranged in scale and intensity from the impacts of mining in Bundi district (Fig. 5) to the maintenance of a road in Ohu.

In a facilitated workshop exercise, participants worked through the impacts of local developments drawing on their tacit knowledge to reveal the processes required for making complex, development decisions with multiple trade-offs. For example, Fig. 5 depicts both the direct and indirect impacts that nickel mining has on the communities in Bundi and Sinopas as identified by the community participants. The direct impacts are the initial effects that occur at the beginning of the chain as a result of the development, while indirect impacts occur at the end of the chain as the positive and negative affects escalate in scale and scope. The first set of negative impacts from the mine include logging to make space for the mine and its infrastructure as well as the water, air and noise pollution that come from the construction and operation of the mine. Logging has flow on effects to forest ecosystems including loss of forest dwelling animals from habitat decline, which has flow on effects to food security

Table 2 Examples of concerns raised by communities in relation to increasing temperatures

Ohu village (hinterland)	Riwo village (coastal)
Damage to food crops and low crop yields e.g. betel nut; coconut etc.	Gardens affected by heat resulting in less food and it's too hot to work in the gardens with low yields
Farmers need to work harder due to shortage of incomes from low yields	People are hot and may become sick
Impacts to human health from water shortages, poor water quality, increasing hot days, and dust on the road	Increased intensity of storms with large waves causing shoreline erosion and sea water to wash into low lying coastal areas impacting soils; well water may become salty, spiritual places on the coast could be destroyed
More sick people at hospitals puts strain on health system	Coral bleaching from sustained high temperatures meaning less fish and less shells
Trees are dry and may burn in a fire	Heat impacts livestock e.g. diseases
Fish and other aquatic species affected if river levels are low	Trees and plants become water stressed, dry out and more prone to bushfires
	Increase in mosquitos and other vectors

Source Authors

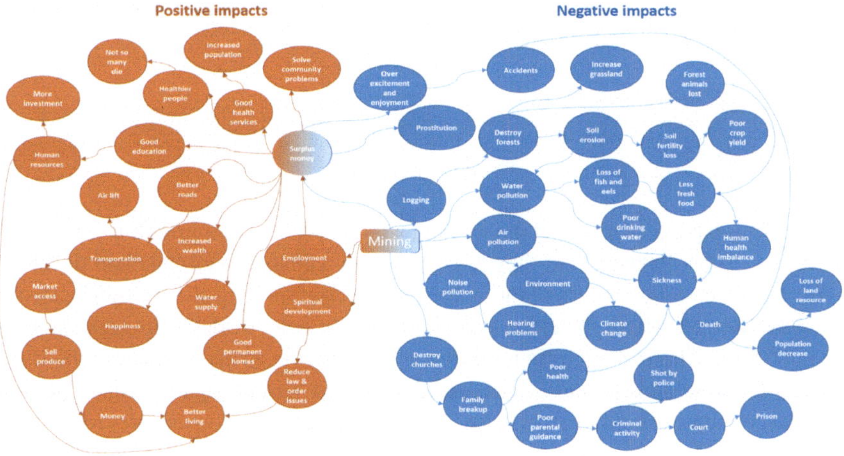

Fig. 5 Example impact map depicting the impacts of mining on communities in the Bundi district. *Source* Authors

(IPBES 2018). Furthermore, logging can lead to erosion of soils and the potential for loss of soil fertility if top soil were to wash away during a rain event, which also ultimately impacts food security. The operation of the mine has impacts on local water quality from dust and also the chemical runoff from the mine tailings into the local river, resulting in poor water quality, which has implications for ecosystem services and human health and mortality.

Participants also identified direct benefits of the mining development. These were primarily in the form of potential for economic development from increased employment. The community termed this 'surplus money', which could have benefits if used for improvements in infrastructure in the form of better road condition, and which aids transportation and increases access to markets to sell produce; thereby, increasing community wealth and improving local standards of living.

The final activity in the workshop process encouraged participants to develop project plans, using a simple template that called for them to consider collective action, at the scale of the village, to address impacts of concern from climate change and development (Table 3). These projects addressed typically place-based, direct environmental impacts but also, somewhat surprisingly, impacts of issues such as population growth and migration indirectly linked to ecosystem change through the need for economic development. For example, the coastal community of Riwo identified projects related to sea level rise (mangrove restoration and sea wall construction), the impacts of which were already being felt locally through coastal erosion. In Ohu village, participants developed projects related to the intensification of cropping (legume-tree mulching, composting) to address sweet potato yield decline. In Bundi and Hogave regions, soil erosion and landslides from land clearing for agriculture were addressed through reforestation projects, an issue of pressing concern to highlands communities. However, other projects sought to improve aspects of human health (e.g. reducing malaria incidence), family planning, climate change education and alternative livelihoods (commercial honey production).

Discussion

Our workshop process revealed the ways in which PNG communities are experiencing threats to biodiversity that they associate with a changing climate and development. The place-based nature of these threats was clear moving along a transect from the coast (Riwo), where the effects of sea level rise were evident, through the hinterland (Ohu), where changes in rainfall are causing erosion and damaging the local road, to the highlands (Bundi region and Hogave) where changed rainfall patterns increase erosion and land slip from clearing activities. Higher temperatures reportedly affected the ability of women to work in gardens and increased the risk of bushfires in the highlands. The ES mapping proved a useful mechanism to explore these threats to biodiversity in greater detail while also maximizing community participation by generating a rich dialogue among the participants about ecosystem service values, potential threats and trade-offs (Martinez-Harms and Balanerva 2012).

While workshop participants were aware of climate change impacts, their literacy about the science of climate change was relatively poor and is probably a reflection of the level of education in PNG, particularly in regional areas (Le Fanu 2013). Indigenous peoples have always adapted to environmental change and there are differences in the ways that climate change is interpreted locally (Jacka 2009, p. 198). Participants generally demonstrated a desire to acquire a greater understanding, often

Table 3 Community projects to address local ecosystem service change

Project name	Description
Riwo (coastal village)	
Community education—sea level rise	Educate the community about the past, present and future of sea level rise through a poster aimed at young people
Seawall construction	Build a seawall made of local materials (e.g. sandbags; river fill)
Maintaining the environment in Riwo	Local resource mapping to identify reforestation areas and educate clan leaders and youth about environmental issues
Bush restoration and connection	Replanting, protection and connection of bush areas
Mangrove restoration	Restoring mangroves (which were there in the past)
Family planning	Address the issue of over population, shortage of resources and associated social problems
Ohu (hinterland village)	
Mulching and making compost	Improve soil moisture in dry weather through mulching and composting
Legume trees	Planting legume trees to improve sweet potato nutrition
Agro Forestry Project	Tree planting with cash crops (e.g. cocoa, vanilla, and livestock) that integrates with forest conservation. It will be a sustainable project that links communities and will help the people
Bundi region (Sinopas—highland village)	
Healthy Highlands concept	Protect against malaria for a healthy community. Decrease the breeding of mosquitos through a community owned project
Sinopas community health projects	For healthy living, decrease health issues and death caused by climate change
Reforestation project	Protection from landslides and soil erosion
Hogave (Eastern Highlands village)	
Reforestation project	Replant in non-forested areas with native trees to address landslide problems
Bushfire prevention	Reduce and prevent bushfire- plant trees as a fire break—yar, pine and kumurere
Oloma honey project—alternative income project	Produce and supply honey to Goroka supermarkets

Source Authors

organising an impromptu question and answer session on climate science to close the workshop event. They also discussed the need for simple educational materials (fact sheets, posters) for schools and village decision makers on climate change science. At later workshops we added additional material to the workshop process to convey basic concepts of climate science to attendees.

Communities clearly understood the immediate affects, both positive and negative, of development as demonstrated in the impact mapping activities (Fig. 5), which emphasised the inter-relationships between environmental change and social problems (e.g. food security and human health implications). However, while they could describe the trade-offs, the desire for economic development from large scale natural resource extraction projects, with their promise of improved infrastructure (for health, communications, education and transport) and potential to generate community wealth, appeared to mostly outweigh the negative consequences for local environments (e.g. O'Faircheallaigh and Corbett 2005). For some resources (particularly coastal marine resources), there are frustrations about the scope of land ownership and resource rights and how large-scale development activities might impact upon local livelihoods and conservation efforts. Participants also expressed frustration at the lack of alternatives to economic development projects that do not over-exploit local natural resources or lead to ecosystem degradation (McShane et al. 2011).

It was clear from many of the impact diagrams and discussions throughout the workshops that women and children were most disadvantaged by local environmental change and often failed to benefit from resource-extraction developments (Sontheimer 1991). Female workshop participants were often most vocal about dimensions of vulnerability (such as population growth, internal migration) and their links to social issues (unemployment, alcoholism, domestic violence, petty crime and prostitution) that they viewed as likely to be heightened through ecosystem decline. Currently, the migration of highlanders to coastal centres seeking employment is viewed as contributing to the loss of local ecosystem services (Numbasa and Koczberski 2012). For example, in the coastal village of Riwo, participants attributed land use conflict and a lack of bush materials for construction of houses to local population change. When coupled to sea level rise, which erodes the community's land, a vicious cycle might result which heightens demand for unsustainable economic development and over-extraction of natural resources.

Adaptation requires community buy-in and participation (Chapin et al. 2006). To this end, conservation policies that foster learning and participation (Ramakrishnan 1998) and provide options that are culturally and economically appropriate, such as those that honour traditional management systems and do not rely on expensive technologies, are more likely to be embraced and implemented. However, to achieve these goals communities require support from outside actors such as government and NGOs (e.g. Van Aalst et al. 2008). In our workshops, communities identified a range of local scale initiatives to address ecosystem decline in the form of simple project plans (Table 3). While they expressed considerable enthusiasm to promote collective action, they nevertheless voiced concerns about a lack of resources (financial, knowledge and equipment) to implement these plans that, if not addressed, would likely result in little progress. As a final phase to our project, we brought together, in a workshop setting, members of each of the communities with representatives of their respective governments, at provincial (i.e. Madang and Eastern Highlands), district and ward levels. The aim of these workshops was to establish connections between government and communities for exchange of information on the community's need for assistance with adaptation and any resources that might be available through existing government programs. While we were able to match some projects to provincial government programs in agriculture and reforestation, an important limitation noted by government representatives was the absence of a functioning provincial office of climate change. For example, in Madang Province while the government has established a Climate Change Development Authority (EMTV 2016) it has no resources and currently is actively supported only by the agriculture agency. Biodiversity conservation and climate change adaptation are seen as cross-cutting policy issues that require cross-division policy and planning. Issues of governance, such as a lack of funding or funding delays, leadership and administrative instability, and communication weaknesses to other layers of government were identified as obstacles to effective action and must be resolved for successful adaptation to climate change and biodiversity conservation in PNG.

Our project was limited in scale and timing to four communities in two provinces, which reflects the difficulties inherent in attempting community engagement in the PNG context. How experience of climate change and development may vary in other locations throughout PNG is difficult to determine, although we tried to address this by selecting communities along a transect from the coast to highlands. In addition, participation in workshops was in part by self-selection and invitation from village 'elders'. Although we sought to ensure a diverse range of 'voices' was heard (especially women) by translating material into Tok Pisin, it was inevitable that those with a stronger command of English made the greatest contribution to the process. This may have inadvertently excluded some community members from participation. Finally, we attempted to extend the life and reach of the project through the production of a facilitator's handbook (Boronyak et al. 2018) that could be used as a guide for similar engagement processes throughout the Pacific. However, the continued use of the handbook depends on the capacity of institutions (government and NGOs) to use these techniques, which is beyond the scope of the project to influence.

Conclusion

Our engagement with four PNG communities on the effects of climate change and development on local environmental change indicates that communities:

1. are generally aware that climate change is already affecting ecosystem services and that changes will likely be exacerbated in the future;
2. understood many of the trade-offs implicit in development, but were often driven by short-term needs (for 'money' or to accommodate a growing population) at the expense of longer-term benefits of maintaining ecosystem services;

3. recognised the need to address other dimensions of social vulnerability (population growth, internal migration, employment, human health) and that these were intrinsically linked to climate change and development; and,
4. required assistance, particularly from provincial and district governments, to support local collective action on adaptation and ecosystem service conservation.

While recognising the difficulties, our experience is that focal PNG communities and provincial government representatives in Madang and Eastern Highlands Provinces were nonetheless eager for more information on ways to respond to maintain the health of their land and wellbeing of current and future generations. Our workshop process provided a way to rapidly gather information to assess community exposure and vulnerability to the impacts of climate change and development as well as the capacity to adapt to change. McClanahan et al. (2008) argue that climate-informed conservation planning requires a site-specific understanding of environmental susceptibility and societal capacity to cope and adapt.

This desire for local ecosystem conservation seems at odds with the side-lining of smaller-scale, potentially more sustainable, developments pursued by provincial governments in favour of large-scale resource extraction projects promoted nationally. Observations of the agreements resource extractive companies negotiate around access to land with the customary owners, suggest that land owners generally do not demand stricter environmental controls (Macintyre and Foale 2004). Rather, they demand more money. Based on this, Macintyre and Foale (2004) argue that the main concern of communities is not necessarily the preservation of the environment per se, but the lack of utilizable resources and amenity. While the information from these PNG communities is place-based, there are likely parallels with other Pacific island communities facing threats to their natural resource base from climate change and the promise of wealth through development. Adaptation initiatives that can simultaneously address carbon sequestration, biodiversity conservation and human livelihoods, at scale, remain the goal of sustainable development.

Acknowledgements The project *Engaging Communities and Government in Biodiversity Conservation and Climate Adaptation* was led by researchers from the New Guinea Binatang Research Centre (BRC) and the Institute for Sustainable Futures at the University of Technology, Sydney. The research project was funded by the United States Agency for International Development (USAID). We thank staff of BRC for their support in logistics and members of the four participating communities for sharing their knowledge with us.

References

Adams WM, Aveling R, Brockington D, Dickson B, Elliott J, Hutton J, Roe D, Vira B, Wolmer W (2004) Biodiversity conservation and the eradication of poverty. Science 306(5699):1146–1149

Boronyak L, Jacobs B, McKenna K, Dem F, Pomoh K, Sui S, Jimbudo M, Maraia H (2018) Engagement on biodiversity conservation and climate change adaptation in Papua New Guinea. Prepared for USAID by the Institute for Sustainable Futures, University of Technology Sydney, Australia and the New Guinea Binatang Research Centre. https://www.uts.edu.au/sites/default/files/article/

downloads/ENGAGEMENT%20ON%20BIODIVERSITY%20CONSERVATION%20AND%20CLIMATE%20CHANGE%20ADAPTATION%20IN%20PAPUA%20NEW%20GUINEA_1.pdf
Bourke RM, Allen B, Lowe M (2016) Estimated impact of drought and frost on food supply in rural PNG in 2015. Development Policy Centre Policy Brief 11
Brown G, Fagerholm N (2015) Empirical PPGIS/PGIS mapping of ecosystem services: a review and evaluation. Ecosyst Serv 13:119–133
Cardinale BJ, Duy JE, Gonzalez A, Hooper DU, Perrings C, Venail P, Narwani A, Mace GM, Tilman D, Wardle DA (2012) Biodiversity loss and its impact on humanity. Nature 486:59
Caritas (2017) Turning the tide: Caritas state of the environment for Oceania report. Wellington, New Zealand:Caritas ISSN 2463-3208
Chapin FS, Hoel M, Carpenter SR, Lubchenco J, Walker B, Callaghan TV, Folke C, Levin SA, Mäler KG, Nilsson C. Barrett S (2006) Building resilience and adaptation to manage Arctic change. AMBIO: J Hum Environ 35(4):198–203
Costanza R, Daly HE (1992) Natural capital and sustainable development. Conserv Biol 6:37–46
Dyson LE, Hendrix M, Grant S (eds) (2007) Information technology and indigenous people. Hershey, IGI Global
EMTV (2016) Madang: first province with provincial climate change committee. Available from: https://emtv.com.pg/madang-first-province-with-provincial-climate-change-committee/. Accessed 26 April 2019
Filer C (2008) Development forum in Papua New Guinea: upsides and downsides. J Energy Nat Resour Law 26(1):120–150
Filer C, Macintyre M (2006) Grass roots and deep holes: community responses to mining in Melanesia. Contemp Pacific 18(2). https://doi.org/10.1353/cp.2006.0012
Filer C, Jackson RT, Henton D (2000) Landowner compensation in Papua New Guinea's mining and petroleum sectors. PNG chamber of Mines and Petroleum, Port Moresby
Foley J, DeFries R, Asner GP, Barford C, Bonan G, Carpenter SR, Chapin FS, Coe MT, Daily GC, Gibbs HK, Helkowski JH (2005) Global consequences of land use. Science 309:570–574
Gilberthorpe E, Banks G (2012) Development on whose terms? CSR discourse and social realities in Papua New Guinea's extractive industries sector. Resour Policy 37(2). https://doi.org/10.1016/j.resourpol.2011.09.005
Heller N, Zavaleta E (2009) Biodiversity management in the face of climate change: a review of 22 years of recommendations. Biol Cons 142:14–32
Intergovernmental Science-Policy Platform on Biodiversity and Ecosystem Services (IPBES) (2018) Regional assessment report on biodiversity and ecosystem services for Asia and the Pacific. In: Karki M, Senaratna Sellamuttu S, Okayasu S, Suzuki W (eds) Secretariat of the Intergovernmental Science-Policy Platform on Biodiversity and Ecosystem Services, Bonn, Germany
Intergovernmental Science-Policy Platform on Biodiversity and Ecosystem Services (IPBES) (2019) Summary for policymakers of the global assessment report on biodiversity and ecosystem services of the Intergovernmental Science-Policy Platform on Biodiversity and Ecosystem Services. https://www.ipbes.net/system/tdf/spm_global_unedited_advance.pdf?file=1&type=node&id=35245. Accessed 17 May 2019
IPCC (2014) Climate change 2014: synthesis report. Contribution if working groups I, II and III to the fifth assessment report of the Intergovernmental Panel on Climate Change. In: Pachauri RK, Meyer LA (eds), 151 pp. IPCC Geneva, Switzerland
Iseke J (2013) Indigenous storytelling as research. Int Rev Qual Res 6(4):559–577
Jacka J (2009) Global averages, local extremes: the subtleties and complexities of climate change in Papua New Guinea. Left Coast Press, Walnut Creek, pp 197–208
Jonassen DH, Ionas IG (2008) Designing effective supports for causal reasoning. Educ Tech Res Dev 56(3):287–308
Le Fanu G (2013) The inclusion of inclusive education in international development: lessons from Papua New Guinea. Int J Educ Develop 33(2):139–148

Liang J, Gong J, Li W (2018) Applications and impacts of Google Earth: A decadal review (2006–2016). ISPRS J Photogramm Remote Sens 146:91–107

Mace G, Norris K, Fitter A (2012) Biodiversity and ecosystem services: a multilayered relationship. Trends Ecol Evol 27(1):18–25

Macintyre M, Foale S (2004) Politicized ecology: local responses to mining in Papua New Guinea. Oceania 74(3):231–251

Martínez-Harms MJ, Balvanera P (2012) Methods for mapping ecosystem service supply: a review. Int J Biodivers Sci Ecosyst Serv Manage 8(1–2):17–25. https://doi.org/10.1080/21513732.2012.663792

McClanahan TR, Cinner JE, Maina J, Graham NAJ, and 7 more authors (2008) Conservation action in a changing climate. Conserv Lett 1(2008):53–59

McKenna K (2016) Corporate social responsibility and natural resource conflict. Routledge, Abingdon

McShane TO, Hirsch PD, Trung TC, Songorwa AN, Kinzig A, Monteferri B, Mutekanga D, Van Thang H, Dammert JL, Pulgar-Vidal M, Welch-Devine M (2011) Hard choices: making trade-offs between biodiversity conservation and human well-being. Biol Cons 144(3):966–972

Millennium Ecosystem Assessment (MA) (2005) Ecosystems and human well-being, vol 5. Island Press, Washington, DC

Nicholson E, Mace GM, Armsworth PR, Atkinson G, Buckle S, Clements T, Ewers RM, Fa JE, Gardner TA, Gibbons J, Grenyer R (2009) Priority research areas for ecosystem services in a changing world. J Appl Ecol 46(6):139–1144

Numbasa G, Koczberski G (2012) Migration, informal urban settlements and non-market land transactions: a case study of Wewak, East Sepik Province, Papua New Guinea. Aust Geogr 43(2):143–161

O'Faircheallaigh C, Corbett T (2005) Indigenous participation in environmental management of mining projects: the role of negotiated agreements. Environ Politics 14(5):629–647

Pacific Climate Change Science Program (2011) Current and future climate of Papua New Guinea. Available at https://www.pacificclimatechangescience.org/wpcontent/uploads/2013/06/14_PCCSP_PNG_8pp.pdf

Ramakrishnan PS (1998) Ecology, economics and ethics: some key issues relevant to natural resource management in developing countries. Int J Social Econ 25(2/3/4):207–225

SEG (Scientific Expert Group on Climate Change) (2007) Confronting the climate challenge: avoiding the unmanageable and managing the unavoidable. Report prepared for the United Nations Commission on Sustainable Development. Sigma Xi, Research Triangle Park, NC, and the United Nations Foundation, Washington, DC, p 144

Sontheimer S (ed) (1991) Women and the environment: a reader: crisis and development in the Third World. Earthscan

Star SL, Griesemer JR (1989) Institutional ecology, translations' and boundary objects: amateurs and professionals in Berkeley's Museum of Vertebrate Zoology, 1907–39. Soc Stud Sci 19(3):387–420

Thomas CD, Cameron A, Green RE et al (2004) Extinction risk from climate change. Nature 427(6970):145–148

UNDP PNG (n.d.) About Papua New Guinea. Available at http://www.pg.undp.org/content/papua_new_guinea/en/home/countryinfo.html

UNEP (1992) The convention of biological diversity. Art. 2. Definitions. UNEP 1992 Convention on Biological Diversity (CBD). Available at http://www.cbd.int/convention/articles

Van Aalst MK, Cannon T, Burton I (2008) Community level adaptation to climate change: the potential role of participatory community risk assessment. Glob Environ Change 18(1):165–179

World Bank (2018) The World Bank in Papua New Guinea. Available at http://www.worldbank.org/en/country/png/overview. Accessed 26 April 2019

Multi-level Governance of Climate Change Adaptation: A Case Study of Country-Wide Adaptation Projects in Samoa

Anna McGinn and Anama Solofa

Abstract Countries across the Pacific region have experienced a surge in internationally funding for climate adaptation initiatives. In the Independent State of Samoa, two major projects—funded by the Adaptation Fund and the World Bank Pilot Program for Climate Resilience (PPCR)—have supported adaptation planning and activity implementation in most villages across the country. These country-wide initiatives range from conducting LiDAR studies and updating Community Integrated Management (CIM) plans to installing rainwater catchment and storage tanks and reforesting water catchment areas. These projects inherently present a multi-level governance challenge because they are developed at the national level, are funded and monitored at the international level, and ultimately implemented in communities. This chapter explores the extent to which interactions across governance levels and scales advance effective adaptation to climate change. Based on in-country interviews, site observations, and observations of the Adaptation Fund's terminal evaluation process, this chapter presents evidence from these major adaptation initiatives in Samoa to highlight where multi-level governance had been leveraged to enhance the governance of adaptation as well as areas of the projects where this has not occurred. It further examines the trade-offs inherent in efforts to work across governance scales and levels in conducting climate change adaptation.

Keywords Multi-level governance · Climate change adaptation · Samoa · Climate finance

Introduction

Pacific Island countries are on the front lines of climate change impacts. At a 1.5 °C temperature rise above pre-industrial levels, the Intergovernmental Panel on Climate

---

A. McGinn (✉)
Climate Change Institute and School of Policy and International Affairs, University of Maine, Orono, USA
e-mail: anna.mcginn@maine.edu

A. Solofa
School of Marine Sciences, University of Maine, Orono, USA

W. Leal Filho (ed.), *Managing Climate Change Adaptation in the Pacific Region*, Climate Change Management, https://doi.org/10.1007/978-3-030-40552-6_12

Change (IPCC) projects that small islands will experience freshwater stress, coastal flooding, negative impacts on marine ecosystems, and migration induced, at least partially, by sea level rise (Hoegh-Guldberg et al. 2018). In fact, island states such as Samoa are already experiencing some of these changes. In Samoa, since 1950, annual maximum temperatures have increased by about 0.22 °C each decade (MNRE 2011). Sea level rise in Samoa, at 4 mm per year since 1993 (MNRE 2011), is above the global average of 2.8–3.6 mm per year. Additional impacts include ocean acidification which is particularly detrimental to corals, shifting rainfall patterns (i.e., less rain in the dry season and more rain during the rainy season), and more intense tropical cyclones (MNRE 2011).

These external forces compel Pacific Island countries to place climate change at the core of their national priorities and to advocate for swift and robust international action to curb greenhouse gas emissions. Yet, many of the impacts listed above must be addressed regardless of future greenhouse gas mitigation efforts because livelihoods, healthy ecosystems, and assets are at stake. Adaptation efforts initiated around the Pacific Islands region range from incremental steps to more transformational change. Some countries in the region have already started to develop relocation strategies to prepare for worst case future climate scenarios. For example, Fiji has initiated discussions on building relocation into their National Climate Change Policy, and Kiribati has already developed a "migration with dignity" policy (Yamamoto and Esteban 2017, 146). Given that Pacific Island countries, and developing countries more broadly, are largely not responsible for the cause of these impacts, funding for adaptation is expected to come from high-emitting developed countries.

Significant international funding has been allocated to climate change projects, especially mitigation projects, in developing countries for over 25 years. Donor entities include the World Bank, developed countries and their international development agencies, the European Union, non-governmental organizations, UN programmes and offices, and funding mechanisms under the United Nations Framework Convention on Climate Change (UNFCCC). According to the UNFCCC Standing Committee on Finance's 2018 assessment of climate finance flows, from 2015–2016 multilateral climate funds—which include the Adaptation Fund and the Green Climate Fund (GCF)—provided an average of US$1.9 billion for climate mitigation and adaptation (UNFCCC 2018). By comparison, bilateral climate finance provided US$31.7 billion and multilateral development bank climate finance summed to US$24.4 billion. All three categories continue to primarily fund mitigation with between 21 and 29% of funds allocated to adaptation (UNFCCC 2018). This extensive funding apparatus, coupled with the latest reports from the IPCC detailing the wide-reaching extent of climate change impacts, suggest that international funding for climate mitigation and adaptation projects in developing countries will continue to be a central feature of international climate governance.

The projects stemming from these funding apparatuses inherently present a multi-level governance challenge. In the case of climate change adaptation projects, they are often developed by national level ministries or multilateral development organizations (i.e., UNDP, UNEP, World Bank, etc.). Project proposals are reviewed by international panels and funding is distributed based on internationally determined rubrics and standards. Funding moves back to countries, usually to the national government, where it can be administered in a multitude of different ways. Ultimately, project implementation happens at the community level where rainwater catchment and storage systems are installed, or riparian buffers are planted along rivers.

It is well understood that all these levels of governance exist and function within the overall system of climate adaptation (Hooghe and Marks 2001). This chapter explores the extent to which interactions across governance levels and scales advance effective climate change adaptation. Understanding these systems is critical, especially in the context of developing countries, because the involvement of so many actors can potentially cause more harm than good if not facilitated effectively (Chaudhari and Mishra 2016). To examine these systems, the chapter explores evidence from an Adaptation Fund project and a World Bank Pilot Program for Climate Resilience (PPCR) in Samoa.

The chapter first outlines a brief history of multi-level governance and describes its value as a framework to understand implementation of climate adaptation projects. Next, it suggests eight characteristics of multi-level governance that emerge from the literature. Through an analysis of the Adaptation Fund and PPCR projects in Samoa, the chapter then tests the extent to which these characteristics are apparent in a governance system which inherently involved many different levels and scales of governance. Finally, the chapter highlights the trade-offs embedded in this systems approach to adaptation governance.

Background: The Multi-level Governance Framework

Rationale

Histories of international development and sustainable development show that when projects are conducted exclusively by a top-down actor they usually fail (Adams 2009). As a result, funders now seek projects that emphasize collaborative, community-based, and stakeholder-driven approaches. But, writing this plan in a proposal is vastly different from carrying it out effectively. Further, community-based approaches face another set of challenges if the responsibility for these projects move to the community level, but the resources for implementation and long-term management do not (Elyachar 2005). Given that internationally-funded climate adaptation projects seem to be following in the footsteps of international and sustainable development initiatives (i.e., funded by many of the same sources, following similar monitoring and evaluation processes, and aligning proposals with the Sustainable Development Goals), avoiding these same pitfalls will be essential. This paper explores how multi-level governance (MLG) is a useful framework to understand the extent to which projects truly engage across the levels and scales of governance versus evoking the above buzzwords—community-based and locally-driven—in a more superficial form.

In order to see how MLG can be used to understand internationally funded climate adaptation projects, it is critical to understand the multiplicity of ways that MLG has been conceived since the term emerged in the 1990s. At its root, MLG describes the diffusion of power from a central government to other levels of governance. In this way, governance roles would be shared from the local to the international levels—an effort to correct for the failures of national, top-down approaches to development. Yet, today it is largely recognized that there is a need for balance across levels of governance after critiques of state interventions swung the pendulum away from state responsibility, as with the Washington Consensus and neo-liberalization, which moved power and responsibility away from the central government to the markets. As described by Steger and Roy, neo-liberalism is "a mode of governance that embraces the idea of the self-regulating free market, with its associated values of competition and self-interest, as a model for effective and efficient government" (2010, 12). These structures, embraced by the US and imposed on many developing countries, move responsibility to other levels of government, but do not provide any resources or support for those levels of government to manage the services previously administered by the central government (Elyachar 2005). This is chiefly because under neoliberalism, the state's goals are the "deregulation of the economy, the liberalization of trade and industry, and the privatization of state-owned enterprises," so a central government would assume that decentralized services would be supported via market mechanisms (Steger and Roy 2010, 14).

Today, scholars employ MLG as an instrument which recognizes the faults of both overly-centralized or overly market-based governance. Instead, contemporary MLG provides a frame to describe "political decentralization within states" which refers to distributing power to the supranational level, to the sub-national level, and out to private entities (Hooghe and Marks 2001, 3). This form of governance, originally conceived as a way to understand the governance of the European Union, was expected to be more efficient and inclusive of all stakeholders. There are many derivatives of MLG discussed in the literature as "multi-tiered governance; polycentric governance; multi-perspectival governance; functional, overlapping, competing jurisdictions (FOCJ); fragmegration (or SOAs); and consortio and condominio" (Hooghe and Marks 2001, 3). This idea of MLG as a frame to understand governance structures has been employed by scholars to study international environmental change (Janicke 2017). Drawing on Janicke (2017), this chapter uses MLG to study internationally funded climate change adaptation projects.

We aim to test what aspects of MLG are present in the Samoan adaptation projects as well as aspects of the framework that are less applicable to this specific context. Samoa is home to two country-wide adaptation projects, of which one is complete and the other is in its final phase. The Government of Samoa considers climate change a national priority. Thus, the Samoa case provided an ideal location for preliminary research to test this form of adaptation project analysis. Further, emerging trends from the Samoa case, while not generalizable to all Adaptation Fund or World Bank projects, may provide valuable takeaways to consider in future projects development and implementation.

We explore MLG within the context of the Cash et al. (2006) model of MLG interactions. Cash et al. define scales "as the spatial, temporal, quantitative or analytical dimensions used to measure and study any phenomenon" (Gibson et al. 2000 in Cash et al. 2006, 2). Levels are seen as the "unit of analysis that are located at different positions on a scale (Gibson et al. 2000 in Cash et al. 2006, 4). For example, the jurisdictional scale includes the local, regional, national, and international levels. Cash et al. (2006) provide a differentiation between cross- and multi- scale and level interactions. Cross-scale indicates that there is actual interaction between the scale, level, or both, while multi-scale simply implies that the different levels or scales exist. Throughout this chapter, the term MLG is intended to encompass the above definition. Using this understanding of existence and potential for interaction between scales and levels, we explore the characteristics of MLG as defined in the literature below.

Characteristics of MLG

Eight central characteristics of MLG are evident in the literature: (1) Devolution of power to the supra- and sub-national levels; (2) inclusion of state and non-state actors; (3) involvement of all scales; (4) nested levels of governance; (5) intentionality; (6) equal power across levels; (7) fluidity; and (8) new policy solutions.

Devolution of power to the supra- and sub-national levels: MLG includes actors at both the supra- and sub-national levels. However, it is not simply that these entities are named in a proposal, but rather that power (i.e., resources, decision-making ability, etc.) is devolved to them (Keskitalo 2010). This implies that the entities have access to participation in the governance process (Stephenson 2013).

Inclusion of state and non-state actors: While governments may have a central role in facilitating MLG processes, non-state actors are also integrated in this framework. Keskitalo (2010) adds that both private and public entities need to be engaged across all the levels meaning that inclusion of multilateral development organizations is not sufficient to consider non-state actors adequately integrated into the governance process. Gumeta-Gómez et al. (2016) cite government agencies, communities, non-governmental organizations, and the private sector as specific examples of stakeholders in a MLG process. This makes the system necessarily complex with the involvement of "a multitude of actors" (Brockhaus et al. 2012, 201).

Involvement of all scales: Cash et al. (2006) identifies seven scales that are necessary to consider when studying human–environment interactions (see Fig. 2). The scales are spatial (with levels being areas from global to a single landscape), temporal (dealing with rates, durations, and frequencies), jurisdictional (governance levels from international to national to provincial to localities), institutional (dealing with rules that exist from international law to local norms), management (from strategies to plans to tasks), networks (dealing with links between people from 'trans-society' to family), and knowledge (defining truth from universal to contextual) (Cash et al. 2006, 3). To understand how MLG functions, it is important to explore the interaction across all these scales and their levels which creates a complex matrix of actors, institutions, and structures (Bulkeley and Betsill 2005; Hooghe and Marks 2001; Keskitalo 2010).

Nested levels of governance: Drawing from Ostrom's (2012) polycentric governance, the idea of nested governance is that the levels of governance need simultaneous independence and interdependence (Cole 2011; Keskitalo 2010). While dependence creates linkages between entities, any level can also produce independent policy and decisions which are based on the outcome of interactions across the levels. Stephenson describes this as "a mutual dependency through the intertwining of policy-making activities" (2013, 817). This concept of the nature of interactions between levels and across scales is central to the functioning of a MLG system.

Intentionality: Cole (2011) emphasizes that these MLG systems and the interactions, nested or otherwise, that characterize the system do not happen by accident. Rather, they are an intentional strategy to "determine the appropriate division of responsibility and authority between governance institutions and organizations at global, national, state, and local levels" (Cole 2011, 2). Interesting, in the case of internationally funded adaptation projects, multiple levels of governance are always involved in an adaptation project. This makes intentionality paramount because otherwise the involvement of so many actors can be more of a burden than a support system (Chaudhari and Mishra 2016). So, the extent to which MLG is strategically embraced in order to facilitate adaptation may dictate certain elements of *success* in the project.

Equal power across levels: Stephenson (2013) suggests that a part of this intentional effort to foster effective MLG is creating an environment where the levels of governance interact on equal footing. He states, "MLG implies engagement and influence—no level of activity being superior to the other" (Stephenson 2013, 817). Similarly, Cash et al. (2006) highlights that knowledge should be co-produced in MLG systems. This may be one of the most challenging characteristics of MLG in the context of internationally funded climate adaptation because it envisions that a community has the same power as the national government or the funding entity. Understanding how power structures are reinforced, renegotiated, or redesigned as a result of a MLG process is central to this study.

Fluidity: All the characteristics listed above are evolving—not static. This fluidity in the MLG system is central to how Bulkeley and Betsill (2005) and Hooghe and

Marks (2001) describe MLG. It is supposed to be a system that is adaptive and formed by the sum of its parts.

New policy solutions: In terms of outcomes of a MLG system, Cash et al. (2006) argue that MLG should facilitate the generation of new policy solutions that were not accessible prior to the effective usage of MLG. These solutions incorporate, and build on, co-production of knowledge and continuous negotiation across levels and scales.

Applying MLG to Internationally Funded Climate Adaptation Projects in Developing Countries

From the broadest perspective, MLG translates well to the study of climate change governance because while climate change is a global phenomenon, the international scale is not always the most appropriate place to address the challenge (Adger 2001). Evaluating the characteristics of MLG can be particularly instructive when thinking about internationally funded climate adaptation projects taking place in developing countries because not only the concept of climate change, but the projects themselves are constantly maneuvering across and between governance levels and scales.

MLG in the developing country context can be complicated by a number of factors. First, effective institutions are needed for MLG to function, yet countries taking on climate adaption measures do not necessarily have those institutions in place. Second, scholars argue that MLG of climate change adaptation will only work if it is consistent with broader development objectives. Chaudhari and Mishra suggest that in the case of linking watershed development and climate change adaptation in India, "for multi-level governance, to be efficient for bringing in the climate adaptation, coordinating and integrating climate and non-climate strategies across jurisdictions and sectors would be essential. Without these, the multiplicity of actors, scales and levels might be more of a hindrance rather than of any assistance" (2016, 326). Adger et al. (2003) echo Chaudhari and Mishra (2016) by suggesting that climate change adaptation in the developing country context will not work at any level if it is not done in tandem with general sustainable development objectives. The third complicating factor, as Adger et al. (2003) describe, is that developing countries are often working with limited options often because of financial resource constraints. Finally, MLG of adaptation in developing countries often involves funding from international sources (e.g., UNFCCC, World Bank, and other development agencies) which changes the players involved with governance and who dictates the adaptation needs (Adger et al. 2003). Adaptation implementation faces a different set of challenges in the developing country context, and it critical to understand how MLG functions in this space given the magnitude and scope of funding mobilized at the international level to support adaptation in developing countries.

Research Methodology

This project uses a case study methodology to study climate change adaptation projects in Samoa (Creswell 2013; Yin 2014). The case study included a literature review, a review of project documents, and fieldwork in Samoa. We conducted the fieldwork in Samoa in July 2018.

During our 2.5 weeks of fieldwork, we conducted semi-structured interviews with seven key informants which ranged from 45 min to 3 h in length. The interviews are referred to in the chapter as interviews 1–6 and include the date conducted (one interview included two interviewees, referred to as A and B). We also spent two days observing the terminal evaluation of the Adaptation Fund project. This included traveling around the entire island of Savai'i visiting adaptation projects funded by the Adaptation Fund with three Ministry of Natural Resources and Environment (MNRE) officials, one United Nations Development Programme (UNDP) staff member, and the international independent consultant conducting the evaluation. During this process, we observed interviews between the consultant and community stakeholders, informal conversations between the consultant and the other staff on the trip, and we learned all of their observations of the Adaptation Fund projects we visited. Separate from the terminal evaluation, we also visited Adaptation Fund and PPCR funded activities on Upolu.

We analyzed the transcribed interviews, field notes, and project documents using discourse analysis (Gee 2004). Applied broadly, discourse analysis explores the use of language in a set of texts with the understanding that all language is both from a specific context and situated in a larger framework of socially constructed conventions (Abrams 1999). Gee suggests that critical discourse analysis views language as a way of communicating social practices which "always have implications for inherently political things like status, solidarity…and power" (Gee in Rogers 2004, 33). In this context, we were reading for descriptions of when, how, and to what extent the national government distributed power to other levels; the difference, if any, in perspective on the projects from stakeholder approaching the projects from different levels and scales; and the processes and outcomes that interviewees suggested were both effective and ineffective.

The discourse analysis was conducted using the qualitative data analysis software NVivo by QSR International. Through a process of inductive and then deductive coding, we categorized how elements of the projects interacted across scales and levels as well as their alignment with the characteristics of MLG to identify where MLG occurred, to some extent, as well as where it did not happen effectively.

The study's main limitation is the fieldwork duration. The chapter presents first-hand observations of a snapshot in time rather than observations starting at the beginning of the project process and following through to post-implementation project management. Thus, the chapter relies on interviewee's accounts of the entire process. Time constrains also meant that we did not have sufficient opportunity to build relationship with village leaders and community members to accurately capture their perceptions of the project process. Thus, this article draws on a small set of observed interviewees between community members and the international consultant. An additional limitation is that, while general lessons may be drawn to inform adaptation across the Pacific region, the case study focuses specifically on the Samoan context. Generalizations out to the regional level should be done with careful consideration of countries' specific history, culture, economy, and geography.

The Case Study

Case Study Introduction: Climate Change Adaptation Projects in Samoa

The Independent State of Samoa is a Small Island Developing State (SID) located in the Pacific Islands region (Fig. 1). Samoa is composed of ten islands, of which four are inhabited. The two main islands are Upolu, where the country's capital Apia is located, and Savai'i. As of 2017, the population of Samoa was just over 196,000 people. About the same number of Samoans that live in country, live abroad, and the economy is dominated by remittances (Meleisea et al. 2012). Given Samoa's long history and rich culture, the country is often referred to as the 'Cradle of Polynesia.'

From the 1830s to 1962, Germany, and then New Zealand held administrative authority in Samoa. During the last decade of this period, New Zealand initiated work with the United Nations and Samoan leaders to craft a path to independence which was achieved in 1962. Thus, Samoa became the first politically independent island state in the Pacific Island region (Meleisea et al. 2012). Since gaining independence, the Government of Samoa has faced significant challenges, including balancing traditional norms and rules with that of Western institutional structures that

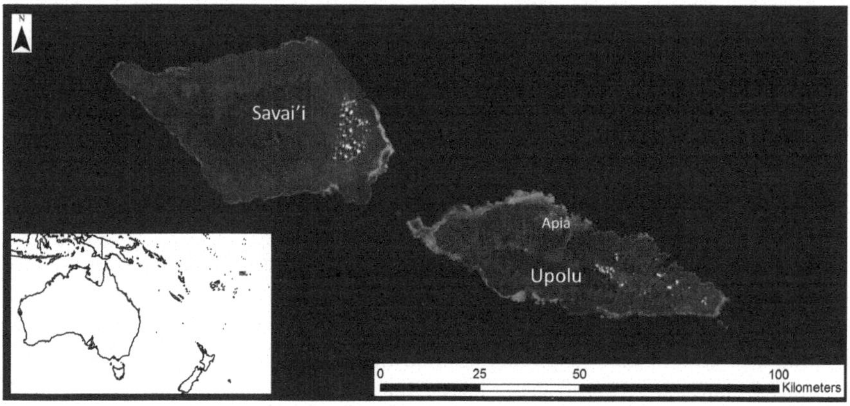

Fig. 1 Map of the independent state of Samoa

had been put in place prior to independence; broad pressures of globalization; and most recently the impacts of climate change (Macpherson and Macpherson 2009).

In the world of international climate change policy and planning, Samoa, and other Pacific island countries, are on the front lines—both in terms of facing climate change impacts, and in conveying the gravity of this challenge to the international community. The region is also at the forefront of climate change adaptation and are some of the main drivers behind obtaining climate change adaptation financing from the international community. Samoa has made climate change central to their overall development agenda and a broad national priority. This is seen through their focus on climate throughout the 2016–2020 Strategy for the Development of Samoa (SDS), their position as a founding member of the United Nations Group of Friends on Climate and Security, and their ambitious greenhouse gas mitigation goals laid out in their Nationally Determined Contribution (NDC) under the Paris Agreement. Because Samoa sees addressing climate change as central to their country's development and security, it seems fair to expect that the country would have the most motivation to effectively address the issue. Thus, a country like Samoa, that is ahead of the curve on thinking about climate change, represents a useful case to explore the extent to which stakeholders leverage MLG to facilitate adaptation. Further, the synergies and barriers that emerge from Samoa's experience implementing internationally funded adaptation projects could serve as a tool for monitoring adaptation project effectiveness more broadly.

This case study explores two of the most recent internationally-funded adaptation projects implemented in Samoa. The initiatives are considered sister projects designed to complement each other—carrying out the same set of activities in different parts of the country and sharing data throughout the duration of the projects. The first project falls under the Adaptation Fund which is a funding mechanism designed and overseen by parties to the UNFCCC. The Adaptation Fund project, *Enhancing Resilience of Samoa's Coastal Communities to Climate Change*, was proposed by the Samoan national government to the Adaptation Fund, and approved by the Adaptation Fund Board in December 2011 as a US$8.7 million project. It started in January 2013, and officially concluded mid-2018. The Samoa Ministry of Natural Resources and Environment (MNRE) served as the executing entity for the project, and the implementing entity was UNDP. The sister project is part of the World Bank's Pilot Program for Climate Resilience (PPCR) initiative which provided US$14.6 million for the project, *Enhancing the Climate Resilience of Coastal Resources and Communities*. The project was approved in 2013 and is set to close in June 2020. The Ministry of Finance (MOF) is responsible for the project.

Both projects have three main components: Adaptation planning, community grants, and major infrastructure projects. First, each project was responsible for working with a set of villages to update their Coastal Infrastructure Management (CIM) plans which were originally drafted between 2002 and 2005. During this process of planning and consulting with village leaders, the implementers updated the plan name to the Community Integrated Management (CIM) plans to reflect their new 'ridge-to-reef' approach to adaptation planning. The idea is that activities on the island have impact on the coastal ecosystem, and vice versa, so the focus of the management plans needs to broadly consider island-wide implications for the coast. The PPCR project additionally funded LiDAR mapping of Upolu and Savai'i for use in both projects CIM planning processes.

The second part of the projects were community grants. These grants, administered through the national government's Civil Society Support Program (CSSP), aimed to provide up to 50,000 Tala (~US$18,900) per grant to villages to implement adaptation projects in line with their CIM plans. Of the 45 Adaptation Fund supported CSSP projects, some of the major areas of work included 23 projects to install rainwater harvesting and storage systems; nine projects carried out some form of construction including retrofitting schools to serve as evacuation shelters and building revetment walls; and, three focused on mangrove rehabilitation and planting.

The final element of the projects was larger infrastructure activities including building bridges to replace river fords, and tar sealing (i.e., paving) inland access roads for storm evacuation and to encourage inland relocation of coastal villages.

Case Study Results

Building on the description by Cash et al. (2006) of the scales that are necessary to consider when studying human–environment interactions, we examine these adaptation projects in Samoa to identify examples where a series of scales and levels work effectively together and exhibit some of the eight characteristics of MLG (i.e., functioning MLG), and contrast that with where the levels and scales do not seem to interact despite evidence pointing to the value these interactions add. Figure 2 shows the Samoa case overlaid on the scales and levels identified by Cash et al. (2006) to display the multitude of factors, actors, institutions, and structures involved in this governance and implementation process.

The sections below describe the following Adaptation Fund and PPCR project elements: (1) programmatic and pragmatic approach; (2) management of safeguards; (3) resources available to project implementers; (4) project timelines; and, (5) stakeholder participation in decision-making and project implementation. While the sections outlined above may seem to provide disparate examples, they are selected to display the diversity of ways that MLG could be integrated into international donor-funded adaptation projects.

Programmatic and Pragmatic Approach

Technical and relevant government ministries in Samoa have substantial experience in seeking and accessing development funding assistance. Evidence of international development aid is visible on the two main islands, from schools' signs labeled 'built by the Chinese government' and park benches sponsored by 'The People of Japan,' to trash stands with signage from the Global Environment Facility.

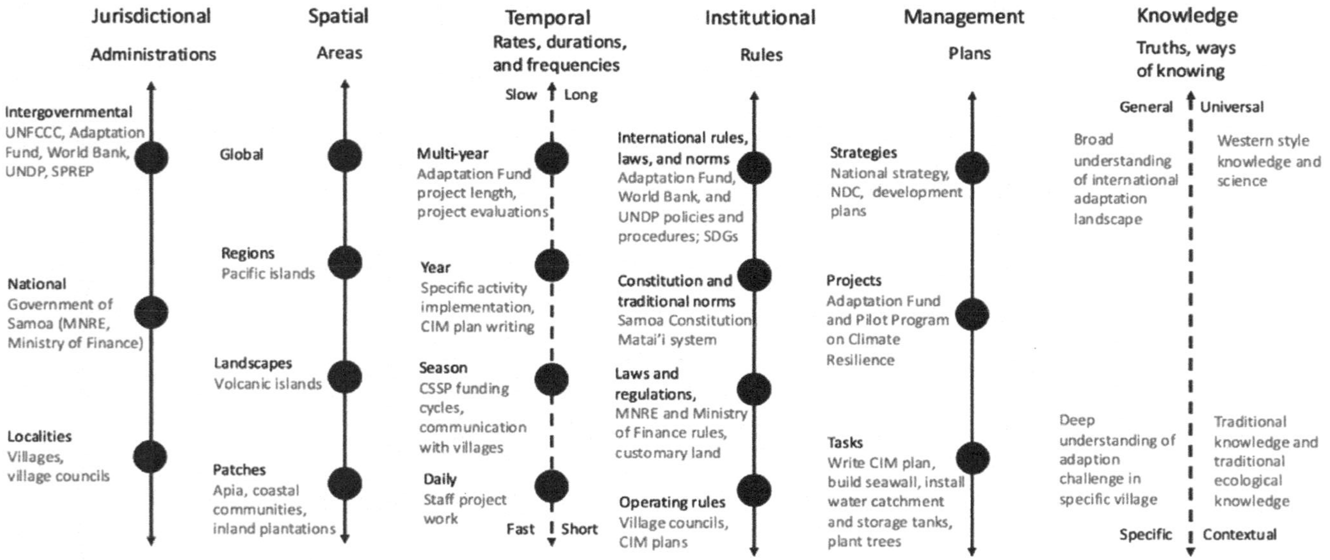

Fig. 2 Scales and levels involved in environment–human systems. Details list the factors, actors, institutions, and structures in some way connected to the internationally-funded adaptation projects in Samoa Adapted from Cash et al. (2006)

Interviewees remarked that the need to transition from a project-based approach—the dominant paradigm used by most of the above entities—to a programmatic and pragmatic approach to managing international funding became apparent during the development of the Community Integrated Management (CIM) plan process (Interview 6A and 6B, 9 July 2018). Adaptation Fund and PPCR implementing agencies saw the revised CIM plan process as an opportunity to increase the cohesiveness of the country's adaptation efforts by more effectively designing connections across scales and levels, demonstrating the intentionality characteristic of MLG. Interviewees also highlighted that a well-developed programmatic and pragmatic approach to adaptation could serve as a model for other Pacific island countries dealing with similar influxes of funds for climate adaptation and mitigation with convergent implementation objectives (Interview 4, 12 July 2018).

The PPCR project is housed within the Samoa Ministry of Finance (MOF). This placement has been key to developing the programmatic and pragmatic approach because MOF houses financial records for all government ministries and agencies within the Government of Samoa. Thus, they can identify areas of work within ministries and agencies that are unfunded and related to adaptation and can then direct PPCR funding to fill those gaps. Rather than develop new projects that may or may not align with agencies' long-term strategies, the funding supports on-going work that may not be implemented because of a lack of resources. For example, "SWA, [Samoa Water Authority], is already getting funding from the EU, but there are gaps in the funding that they need to source from other finance, so PPCR can fill those gaps. In that way, you have partner projects working together" (Interview 2 18 July 2018, also reflected in Interview 4, 12 July 2018). The same interviewee explained,

> A lot of what we are doing now with the large projects, it is aligned to government work plans, so what we did was collect the information on government priorities with districts, we cross checked with what districts identified in the community, so we come across, oh they need this road and it is identified here [in the government work plan], so it was already planned, but there was no funding, so this is one way to fund it. (Interview 2 18 July 2018, also reflected in Interview 4, 12 July 2018)

Going forward, another interviewee stated, "any money coming in, any proposal that goes out, we see it from this national planning level" (Interview 6B, 9 July 2018).

The implementing agencies of these adaptation projects see the programmatic and pragmatic approach as valuable because it may increase the likelihood that the activities carried out in these discrete projects will be sustained once the project funding ends. Interviewees noted heightened cross-ministry buy-in over the course of project implementation. For example, in the past, representatives from the Land Transportation Authority (LTA) and the Ministry of Works, Transportation, and Infrastructure (WTI) departments would not attend climate related meetings, but this has changed because they now feel ownership of projects that incorporate climate considerations (Interview 6B, 9 July 2018).

The approach is also beneficial in that it facilitates injection of additional funding resources into existing funding gaps to bolster implementation of activities that have already been approved, rather than creating new projects. One example of existing projects waiting for funding is improving inland roads. While these roads are in existing plans largely to facilitate improved economic conditions for inland plantations, they can also serve as evacuation routes. Thus, the project falls under the prevue of the adaptation funding. The expectation is that the ministries and agencies will maintain the activities over the long term because they had already been incorporated into their workplans before the Adaptation Fund and PPCR projects were developed. Here, we see that through the process of working across scales and levels, the implementing agencies generated new policy solutions that were not operationalized in this context before the initiation of this MLG process.

Interviewees indicated that this programmatic and pragmatic approach is already influencing change in that it has increased collaboration between government ministries on project implementation—one of the measures of effective MLG. There are interactions between the jurisdictional scale and institutional scale with efforts by the different government ministries influencing operational rules and norms around mainstreaming adaptation in their work areas. Finally, this example displays interactions between the jurisdictional scale and the management scale where the local-to-national level stakeholders move from only thinking in terms of ad hoc projects, to thinking about projects as a part of a higher-level strategy. A weaker aspect of the introduction of the programmatic and pragmatic approach is that, while local and international stakeholders did have some involvement in shaping this new approach, the national government played a central role. Thus, we see the involvement across the jurisdictional scale, but not at equal weight.

Embedded in this example of efforts to leverage MLG to support adaptation are trade-offs. Given that projects have finite resources and time, only certain areas of MLG are utilized, often at the expense of others. In this case, the national government heavily engaged its line ministries across the national level which created a sort of horizontal nested structure, but not a vertical one. This is particularly important in countries like Samoa where the local village and chiefly/social system structures play an important role in how initiatives are taken up and sustained.

Safeguards

Safeguards are a tool to identify and manage externalities of adaptation projects that could end up causing more harm than good to the people, ecosystem, or culture impacted by the project. Both the World Bank and the Adaptation Fund have safeguard policies which are supposed to guide project planning and implementation. Safeguards are an interesting challenge of MLG because the funding organizations provide broad expectations for the project, and the country implementers must interpret and apply the expectations to the projects. This challenge requires work across the jurisdictional scale, as well as interactions between the jurisdictional and institutional scales. It is also mediated by types of knowledge.

Inconsistencies between the World Bank and the Adaptation Fund project safeguards caused disagreement over how certain parts of projects were conducted. This suggests that, in practice, refinement of safeguards is necessary for them to function effectively across jurisdictional levels. This tension is captured by one interviewee working closely on the PPCR project,

> Safeguards and gender were put in [the project plan] and the environment and social criteria, those were developed under the PPCR, but were used by both projects. The PPCR also required the villages to go through the development consent process. That is like the government safeguards, so that is very important, but before not a lot of projects were going through this process. So, I think all the key players now understand the importance, because also, the Adaptation Fund through UNDP, they do not really emphasize the importance of safeguards unlike the World Bank, and that is a key difference because in Samoa, in a lot of projects, safeguards are the issues. And the reason why some partners insist on their procedures is because they see a gap in the national [procedure]. (Interview 3, 18 July 2018)

Interviewees put forward that Samoa is working to integrate the World Bank's safeguards into government procedures regardless of the funding source. This example of nested governance levels shows that the national level has observed and experienced a process developed at the supra-national level and sees a path to borrow that process to improve their governance procedures. Further, from the World Bank level, they are implementing a new approach to align safeguards with country priorities. An interviewee said that in a recent workshop, the World Bank presented a new framework which will prioritize "find[ing] a common approach with the government before, and agree on that approach, before [the project] gets implemented" (Interview 1, 20 July 2018). Again, a demonstration of how the levels have nested to improve overall governance.

However, this example again raises the question of power across the levels. While the national level appears to be reclaiming power over this process by adopting the World Bank safeguards on their own terms, the involvement of the sub-national level is variable. In some instances, the safeguard process did work to support the local level, from the perspective of the interviewees. For example, for all roads tar sealed by the government, the World Bank required the government to compensate people living on, or farming, land immediately adjacent to the roadwork, regardless of who actually owned the land. This certainly provided more protection for people than they would have received if the World Bank was not involved. However, the funds to pay these costs cannot come out of the project budget, they must come out of the government budget which causes the safeguards to be a strain on government resources. While interviewees reported that the government followed through on these payments, it is worth noting that there is less incentive to do so when the funds must be additional to the international funding allocated in the project. This is an instance where the different levels of rules and norms along the institutional scale can create barriers to following the safeguards.

In another case, a project under the Adaptation Fund built a revetment wall along a river which caused more flooding instead of preventing it. The wall was developed in its present location because of requests from the village leaders against the recommendation of engineers. This represents a lack of communication across the knowledge scale—perhaps with more discussion the village leaders' experienced and contextual knowledge could have been integrated with the analysis of the geography to come up with an output that would have worked. This points to a need to increase the involvement of all scales in this aspect of the projects. Additionally, the level to which these issues are reported back up to the international jurisdictional level is not clear. If safeguard issues are underreported, it gives a different perception on the effectiveness of safeguards as you move across the jurisdictional levels, which does not foster nested levels of governance.

Resources

Three types of interconnected resources were highlighted by interviewees: operational costs, staff, and institutional memory. We observe a juxtaposition between the influx of funding from the international level, and the expectation by the international level of additional resources at the national and local jurisdictional levels that may or may not exist. In the case of the PPCR project, one interviewee explained that the World Bank provides the funds for the project, but not the operational costs. The expectation is that the country shows its commitment to the project through supporting these operational costs which range from work computers to the payments for using land to improve roads as discussed in the safeguards section.

This raised a particular challenge in Samoa because the CIM plans required that ministry officials and consultants travel to every village in the country. But, to pay visits to villages means that they must practice "the cultural protocol which is very expensive. And those are not funded by the projects" (Interview 3, 18 July 2018). The interviewee suggested that the number of village consultations correlated with the funding that the government ministries could put forward to pay for the gifts and activities integral to the cultural protocol. This represents a blockage between the jurisdictional levels and the institutional levels where universal rules generated at the international level do not allow countries to use funding for what can be a critical part of the project process. This primary blockage also creates secondary issues like unequal power across levels because the national and the local had limited interactions given logical barriers.

One of the most popular lines among the consultants and ministry officials interviewed was that, in their one job, they wear about twelve different 'hats.' There are simply a limited number of qualified and/or experienced staff in the relevant agencies, and resources are also limited. Thus, staff working directly on climate adaptation are required to serve in a multitude of other capacities. This impacts the effectiveness of MLG because it is hard to be intentional about integrating new approaches and coordination when staff are already overworked. One interviewee explained,

> I can't think about replication. So, it is nothing to do with the capacity of our people to be able to do it. It has to do with the fact that it is overwhelming. You have a hat on and say you have twelve [hats] already, then you are asked to do this pragmatic approach because we are looking at replicating. (Interview 6B, 9 July 2018)

This same interviewee underscored the imperative of the pragmatic approach, but simply cautioned that implementation is challenging with limited personnel.

An interlinked challenge is that of institutional memory. Not only are the staff hard to find, but in the case of the Adaptation Fund project, they were transient. At the time of the project's terminal evaluation, only one ministry official who started with the project when the original Adaptation Fund proposal was drafted, was still on the team. An observer of the project noted that it is hard to effectively facilitate work across the jurisdictional levels—not to mention build higher-level strategy (management scale), work across types of knowledge, and align timescales—when staff are constantly changing (Interview 5, 18 July 2018). The observer continued by noting that this is a frequent challenge in internationally-funded climate projects across the board. Not only is institutional memory missing within projects, but it is often almost non-existent across large projects especially when they involve different sectoral partners. Thus, these resource barriers to effective MLG might be more broadly applicable across climate projects.

Aligning Timelines

The implementing agencies did not effectively use MGL to manage the different project timetables for the Adaptation Fund and the PPCR, respectively. This challenge traverses the management, temporal, jurisdictional and institutional scales. According to several interviewees, the ministries decided to link up the Adaptation Fund project and the PPCR project once they were both approved. The initial challenge this created was that the PPCR project was approved later and thus had more administrative matters to address before beginning implementation. Those implementing PPCR also noted that the World Bank requires more administrative reporting throughout the duration of the project. Conversely, the Adaptation Fund provided those implementing the Adaptation Fund project with the flexibility to change decisions during implementation without prior approval from the Fund. Given these different processes, the Adaptation Fund implementation was put on hold while the PPCR carried out its preparatory phase. This highlights that there was, from the start, intentionality to connect the projects linking up many different stakeholders and structures across scales. However, this connectivity became a problem when the Adaptation Fund Secretariat, according to interviewees, contacted the implementers to say that if they did not start implementation, they would lose the funding. So, while the in-country teams were working to align the workplan across the many different scales, this was prevented by international institutions' temporal expectations.

This announcement caused the Adaptation Fund project team to uncouple the sister projects, which removed the explicit alignment of the temporal and management scales. Hence, the project timelines were disconnected, and so, to some degree, were the project strategies. One observer of the Adaptation Fund project noted that this is, in their opinion, one of the most remarkable aspects of the project—that such rapid directional change was possible (Interview 5, 18 July 2018). While the separation process may have been impressive, it also resulted in some substantial challenges. Most importantly, since the Adaptation Fund still wanted to conduct the CIM planning in line with the PPCR, they started the implementation step with the infrastructure projects based off the 2002–2005 CIM plans, and then updated the plans after the projects were at least partially carried out. In the case of the CSSP projects for the Adaptation Fund, all the villages in Savai'i also ended up applying for projects without an updated CIM plan on which to base their project proposals. The Adaptation Fund implementers said that, while it was not ideal, they had no choice if they wanted the CIM planning to be effective countrywide and finish by the Adaptation Fund's deadline, even with a granted extension. Observations suggest that the implementation was also rushed which may connect closely with the safeguard challenges discussed above. Thus, the clash of the temporal scale with the institutional and jurisdictional scales significantly impacted the efficacy of the project process.

Stakeholder Participation in the Decision-Making Process

It is well documented in the literature that top-down development projects continuously fail to give critical stakeholders a seat at the table (Adams 2009). The Adaptation Fund was charged with shifting this paradigm by being a Fund that would be responsive to the voices of the 'vulnerable' people it was set up to serve. In terms of MLG, this means that the Adaptation Fund is structured for movement of power to the supra-national and sub-national levels. In Samoa's project proposal to the Adaptation Fund, the national government laid out their plan to engage local non-governmental organizations (NGOs) in the planning and implementation process. Namely, NGOs would be brought on board to work with communities to develop, apply for, and implement their CSSP projects. However, due to the shortened timeline and lack of resources for the NGO trainings, the NGOs were not involved in the project in favor of ministry staff putting on another hat to advice the villages. This exemplifies how limited time and resources snowball into a stakeholder's exclusion from the project which leads to forgoing a key characteristic of effective MLG which is the inclusion of both state and non-state actors. It should be noted that NGOs were engaged with the PPCR project in a limited capacity.

Observations suggest that the Adaptation Fund and PPCR planning and implementation was dominated by the national level ministries. The projects seem to have done an exemplary job engaging ministries across the government to support different elements of the project including housing and implementing many elements of both the Adaptation Fund project and the PPCR project. As one interviewee explained, "So apart from the community consultations, we also did site assessments with the technical experts that were on the team. So, we had an ecosystem specialist, a civil engineer, a spatial risk planner, a geomorphologist, and plus the representatives from implementing entities like LTA, SWA, MWTI, MNRE, and the Ministry of Women" (Interview 2, 18 July 2018). Village leaders had direct opportunities to run projects through the CSSP portion of the programs which proved to be a useful way to create ongoing conversation between the government ministries and the villages beyond the CIM plan consultations. However, the villages did not receive full autonomy over the project namely because they were not given the grant in money and the discretion on how to spend it; rather, they were provided the physical materials only. According to Stephenson (2013), in an effective MLG arrangement "no level of activity [is] superior to the other." The case in Samoa reflects a clear imbalance.

Discussion

Characteristics of MLG

This chapter sets out to test the extent to which eight characteristics of effective MLG, as described in the literature, are evident as stakeholders maneuver across levels and scales to implement internationally-funded climate change adaptation projects in Samoa. This is of particular interest because, in cases of international funding flowing to national governments who implement projects at the local level, multiple levels of governance are inherent in the project governance process. However, the extent to which stakeholders recognize and actively work to facilitate MLG is not predetermined and may play an important role in how effectively these forms of adaptation are managed.

The eight characteristics of MLG are evident in this case study to varied degrees. In a broad sense, there was an *intentional effort* by the Adaptation Fund and PPCR implementers to work within the MLG framework baked into the projects. Similarly, the implementers embraced the *fluid nature* of MLG especially in managing the different project timelines.

Despite the inherent multi-jurisdictional level interactions built into an internationally-funded adaptation project, the national government holds much of the control over both the Adaptation Fund and PPCR projects. The national government did work closely with *supra-national entities* such as the World Bank, the Adaptation Fund, and the UNDP. Especially in the case of the World Bank, the national government gave power to these organizations to vet each step of the project process including small logistical changes. Power was shared with the *local level* to a lesser extent. Villages were consulted as a part of the CIM planning process and had the opportunity to apply for and implement CSSP projects. However, both these aspects of the projects were closely overseen by the national government at every step. The interviewees summarize that the ministries saw the villages as dependent on them for support, but the ministries did not feel a mutual sense of dependency on the villages. This mutual dependency, which leads to more flattened power structures, is a tenant of Stephenson's (2013) understanding of MLG. This example highlights that the characteristic of *equal power across levels* did not manifest in these projects.

Returning to Keskitalo's (2010) understanding of MLG, both *private and public entities* are supposed to be involved at every level. At the international to national levels, we do see a mix of interactions between the World Bank, the UNDP, the Adaptation Fund, and the government ministries. However, as highlighted by the discussion of stakeholders engaged in the projects, there is less collaboration between state and non-state actors in-country. Further integrating Samoan civil society, in the form of NGOs and village leadership groups, into the project process could enhance their engagement with this characteristic of MLG.

In terms of engaging *across all scales*, the introduction of the programmatic and pragmatic approach is an example where the project does work across multiple scales (i.e., jurisdictional, temporal, institutional, and management). Yet, the approach does not necessarily integrate levels across the knowledge scale. Given village members and leaders' deep, contextual knowledge of the islands, this is a critical scale to include in the Samoan context. Thus, while we do observe continuous efforts to work across scales, it is apparent that resource and time constraints moderate the ability to do so.

Cash et al. (2006) suggests that, when effective, MLG should create opportunities for *new policy solutions* that were not possible before the levels and scales started working both interdependently and independently (Cole 2011). In Samoa, the introduction of the pragmatic and programmatic approach represents the most significant structural change inspired by these projects. Moving away from the approach that deals with all internationally-funded projects in a vacuum towards one that integrates the use of international funds with existing ministry projects and priorities will likely have a long-term impact on Samoa's approach to international development finance. Further, two of the implementers also explained that they were encouraging communities to use the new CIM plans to hold their elected officials accountable. Villages should request that their elected officials use the plan to form their policy positions in the Legislative Assembly. They explained that this could be one of the most effective paths to ensure long term usage of the CIM plans. If the CIM plans become a tool for villages to increase climate resilience through legislative actions this would represent another policy solution born out of the MLG process. Thus, while there were barriers to leveraging MLG for effective adaptation, we still see substantial policy impact coming out of these two sister projects.

Conclusion

MLG, polycentric governance, and nested governance have been discussed in the literature for over 25 years (Stephenson 2013). Yet, this case shows that despite the attention, implementation of MLG is challenging and there is no clear playbook for how to do it well. In Samoa, we see that the obstacles to carry out effective MLG are generated at various levels and scales with no particular actor or area serving as the central source of the issues. Interviewees suggested that, over the course of the projects, they were increasingly aware of the value of working across the levels and scales.

However, the analysis also suggests that even when project implementers embrace MLG, not all characteristics of MLG are apparent in each element of the project nor does each element of the project include cross-level and scale interactions. This illuminates the spectrum and quantities of trade-offs embedded in this systems approach to project management and governance. In fact, it brings into questions when, if ever, MLG aligned with all of the characteristics is possible. In the case of Samoa, decisions to work across all jurisdictional levels were constantly modified by access to resources and time. Thus, project implementers had to decide when to engage with the sub-national level entities rather than having a more nested governance approach. Since there are so many variables embedded in this systems approach, they cannot all be maximized at the same time. This chapter finds that effective MLG also requires an awareness of these diverse trade-offs throughout the governance process.

As Pacific island countries continue to apply for and receive international funding for climate adaptation, embracing MLG and understanding the trade-offs woven into the system can be informative in both the planning and project implementation phases. At the same time, it is critical that project funders place value in the characteristics of effective MLG and aim to support them through institutional policies and procedures. In particular, ministries responsible for writing project proposals ought to work with key stakeholders to build interactions between levels and scales into project proposals to set the project on a trajectory to be inclusive, strategic, and in-line with broader sustainable development goals. For example, providing funding for cultural protocols in the grants could enable more sustainable relationships between a national government and communities. At the implementation phase, project implementers need sufficient time to build relationships across scales and levels in order to see the programmatic and pragmatic approach to fruition. These are critical points for project funders to embrace.

Tracking future adaptation projects in Samoa to see how they build on or diverge from the project efforts described in this chapter as well as revisiting the Adaptation Fund and PPCR project elements in future years would be instructive to understand the extent to which MLG actually influenced the long-term sustainability of the projects. Further, an analysis of climate adaptation projects in other Pacific island countries, similar to the work throughout this book, will increase our understanding of the tools that are essential for effective climate adaptation in the Pacific context—essential knowledge for a region on the front lines of climate change impacts.

Acknowledgements This material is based upon work supported by the National Science Foundation Graduate Research Fellowship under Grant No. DGE-1144205. Any opinion, findings, and conclusions or recommendations expressed in this material are those of the authors and do not necessarily reflect the views of the National Science Foundation. This research is also supported by Dan and Betty Churchill Exploration Fund, the Richardson-Churchill SPIA Scholarship, and the University of Maine Graduate Student Government.

The authors are also grateful for the support of Dr. Cindy Isenhour for her guidance throughout the research process. The project would not have been possible without the participation and engagement from the interviewees and their colleagues who welcomed us into the climate adaptation project processes.

References

Abrams MH (1999) A glossary of literary terms, 7th edn. Heinle & Heinle, Thomson Learning, Boston

Adams WM (2009) Green development: environment and sustainability in a developing world. Routledge, New York

Adger WN (2001) Scales of governance and environmental justice for adaptation and mitigation of climate change. J Int Dev 13:921–931

Adger WN, Huq S, Brown K, Conway D, Hulme M (2003) Adaptation to climate change in the developing world. Progress Develop Stud 3(3):179–195

Brockhaus M, Djoudi H, Kambire H (2012) Multi-level governance and adaptive capacity in West Africa. Int J Commons 6(2):200–232

Bulkeley H, Betsill M (2005) Rethinking sustainable cities: multilevel governance and the 'urban' politics of climate change. Environ Politics 14(1):42–63

Cash DW, Adger WN, Berkes F, Garden P, Lebel L, Olsson P, Pritchard L, Young O (2006) Scale and cross-scale dynamics: governance and information in a multilevel world. Ecol Soc 11:2–8

Chaudhari VR, Mishra A (2016) Multilevel policy responses to mainstream climate adaptation through watershed development in rainfed farming systems in India. Climate Dev 8(4):324–335

Cole DH (2011) From global to polycentric climate governance. European University Institute: Robert Schuman Centre for Advanced Studies Global Governance Programme. Working paper

Creswell JW (2013) Qualitative inquiry and research design: choosing among five approaches. Sage, Los Angeles

Elyachar J (2005) Markets of dispossession: NGOs, economic development, and the state in Cairo. Duke University Press, Durham

Gee JP (2004) Discourse analysis: what makes it critical? In: Rogers R (ed) An introduction to critical discourse analysis in education. Lawrence Erlbaum Associates, Mahwah

Gumeta-Gómez F, Durán E, Bray DB (2016) Multilevel governance for local management of drinking water in Latin America: case studies from Costa Rica, Honduras, and Mexico. Acta Universitaria 26(3):3–13

Hoegh-Guldberg O, Jacob D, Taylor M, Bindi M, Brown S, Camilloni I, Diedhiou A, Djalante R, Ebi KL, Engelbrecht F, Guiot J, Hijioka Y, Mehrotra S, Payne A, Seneviratne SI, Thomas A, Warren R, Zhou G (2018) Impacts of 1.5 °C global warming on natural and human systems. In: Masson-Delmotte V, Zhai P, Pörtner H-O, Roberts D, Skea J, Shukla PR, Pirani A, Moufouma-Okia W, Péan C, Pidcock R, Connors S, Matthews JBR, Chen Y, Zhou X, Gomis MI, Lonnoy E, Maycock T, Tignor M, Waterfield T (eds) Global Warming of 1.5 °C. An IPCC Special report on the impacts of global warming of 1.5 °C above pre-industrial levels and related global greenhouse gas emission pathways, in the context of strengthening the global response to the threat of climate change, sustainable development, and efforts to eradicate poverty. World Meteorological Organization, Geneva, Switzerland

Hooghe L, Marks G (2001) Types of multi-level governance. Europ Integr Online Pap 5(11):1–32

Janicke M (2017) The multi-level system of global climate governance- the model and its current state. Environ Policy Govern 27:108–121

Keskitalo ECH (2010) Developing adaptation policy and practice in Europe: multi-level governance of climate change. Springer, Dordrecht

Macpherson C, Macpherson L (2009) The warm winds of change: globalization in contemporary Samoa. Auckland University Press, Auckland
Meleisea M, Meleisea SP, Meleisea E (2012) Samoa's journey 1962–2012: aspects of history. Victoria University Press, Wellington
MNRE, Australia Bureau of Meteorology, Commonwealth Scientific and Industrial Research Organization (2011) Pacific climate change science program: current and future climate of Samoa. International Climate Change Adaptation Initiative. Available online at https://tinyurl.com/y4kgocb2. Last access 23 June 2019
Ostrom E (2012) Nested externalities and polycentric institutions: must we wait for global solutions to climate change before taking actions at other scales? Econ Theor 49:353–369
Rogers R (ed) (2004) An introduction to critical discourse analysis in education. Lawrence Erlbaum Associates, New Jersey
Steger MB, Roy RK (2010) Neoliberalism: a very short introduction. Oxford University Press, Oxford
Stephenson P (2013) Twenty years of multi-level governance: 'where does it come from? What is it? Where is it going? J Europ Public Policy 20(6):817–837
UNFCCC (2018) UNFCCC Standing Committee on Finance 2018. Biennial assessment and overview of climate finance flows technical report. United Nations Framework Convention on Climate Change. Available online at https://tinyurl.com/y2h5yc8c. Last accessed 22 June 2019
Yamamoto L, Esteban M (2017) Migration as an adaptation strategy for Atoll Island States. Int Migr 55(2):144–158
Yin RK (2014) Case study research: design and methods. Sage, Los Angeles

The Impact of Connectivity on Information Channel Use in Tonga During Cyclone Gita: Challenges and Opportunities for Disaster Risk Reduction in Island Peripheries

Aideen Foley

Abstract In island contexts, geographic characteristics such as remoteness and boundedness can substantially impact the capacity for connection, and how people experience the state of being connected, i.e. connectivity. Varying degrees of connectivity may, in turn, affect how island communities prepare for and respond to the impacts of climate change, including extreme weather risks. Tropical cyclone warnings are a pertinent example; vital information must reach citizens in isolated peripheral locations, where both telecommunications infrastructure and cultural contexts may differ from the island core, leading to differences in how people access information. Drawing primarily on the case of Tonga, which was affected by Cyclone Gita in 2018, this paper explores these core-periphery patterns relating to how information channels are engaged with when facing extreme weather risks. Census data and cyclone impact data are used to assess spatial patterns in the extent of material and non-material connectivities based on communications, economic and linguistic variables, and to explore the impact of connectivity on the processes of risk reduction and natural hazard response, through dissemination of hazard information. Low-value clusters, i.e. coldspots, in receipt of warnings via 'modern' information channels (internet and SMS) are identified on 'Eua, while hotspots are identified in western Tongatapu. These hotspots and coldspots do not appear to be linked to the accessibility of internet and mobile phones, but do overlap with hotspots and coldspots in linguistic and economic variables, illustrating the potential for non-material differences in socio-cultural context to influence how information channels are engaged with. The implications of these results, including challenges and opportunities for disaster risk dissemination in island peripheries, are also discussed.

Keywords Disaster risk reduction · Communication · Cyclone warning · Islands · Connectivity

---

A. Foley (✉)
Department of Geography, Birkbeck University of London,
32 Tavistock Square, London WC1H 9EZ, UK
e-mail: a.foley@bbk.ac.uk

W. Leal Filho (ed.), *Managing Climate Change Adaptation in the Pacific Region*,
Climate Change Management, https://doi.org/10.1007/978-3-030-40552-6_13

Introduction

Small Island Developing States (SIDS) face profound physical impacts from climate change (e.g. sea level rise, land loss, extreme weather), which in turn have social and cultural impacts (Adger et al. 2011, 2013; Graham et al. 2013) on island peoples. Of the climate-related hazards facing SIDS, tropical cyclones are a key concern, with considerable risk of both human and economic cost; For example, in 2015, Cyclone Pam left 11 dead and 75,000 people homeless (Nishijima et al. 2015), with the cost of recovery over a 4-year period estimated as 43.3% of GDP (Dornan and Newton Cain 2015).

There has been an increase in recent decades in the number and proportion of category 4 and 5 tropical cyclones, particularly in the North Pacific, Indian, and Southwest Pacific Oceans (Webster et al. 2005), alongside an increasing trend in tropical cyclone destructiveness (Emanuel 2005). Given the limitations of the historical record, it is uncertain whether these observed changes are beyond the range of natural variability (Knutson et al. 2010). Modelling studies, however, have shown that observed poleward shifts in tropical cyclone exposure are likely to continue under representative concentration pathway (RCP) 8.5 (Kossin et al. 2016). There is also the prospect of so-called 'grey swan' tropical cyclones, which are storms that, based on historical data alone, would be considered to be of unusual magnitude, or occurring in unexpected geographic regions, but which can be anticipated based on an understanding of storm physics and paleotempestology (Lin and Emanuel 2016).

The geography of many SIDS factors in their vulnerability and responses to this and other hazards. Their land boundedness places significant tracts of the population at risk from cyclone-induced storm surges, which can have lasting impacts on communities (Martin et al. 2018). The scope for evacuation away from the coast is often limited, given the smallness of most SIDS and the logistical challenges of mass off-island evacuation (Shultz et al. 2018). Research suggests that the burden of disaster on Pacific islands, especially, is underestimated due to deficiencies in the quantitative data available (Noy 2016; Edmonds and Noy 2018).

Yet, vulnerability is increasingly recognised as a social construct (Jackson et al. 2017). Recent scholarship in disaster risk reduction has highlighted the need for holistic approaches to understanding vulnerability and managing risk, recognising the role of context and nuance (Kelman 2018). Within, as well as between, islands, diverse environmental and social contexts contribute to vulnerability and resilience. In Dominica, Barclay et al. (2019) highlight the role of historic social, economic, and political processes in determining where people and assets are located today, and thus, who and what is at most risk during a natural disaster. Warrick et al. (2017) finds that that in addition to access to resources, values or belief systems are of particular importance in shaping adaptive capacity in the Pacific region. For example, Brown et al. (2018) report that being affected by Cyclone Evan increases future expectations of disaster risk for Indo-Fijians but not *iTaukei*, surmising that the "collectivist" social structure of *iTaukei* may help to absorb such shocks. Disaster risk reduction is therefore an arena in which environmental and social factors dynamically interact to shape outcomes, with characteristics such as land boundedness, comparative smallness and isolation presenting further challenges but also opportunities in small island settings.

However, relational approaches are increasingly recognising islands as spaces of interconnection (Chandler and Pugh 2018; Grydehøj and Casagrande 2019), and the implications of this for climate change action and disaster risk reduction demand consideration. The characteristics of islandness such as smallness, isolation and fragmentation can be conceptualised as advantages, with potential to be harnessed as assets with appropriate policy interventions (Deidda 2016). However, it is vital to understand the intersections between connectivity and the processes of island resilience, to support sustainable adaptation and reduce risks of maladaptation.

Connectivity, defined here as the capacity for connection, and how people experience the state of being connected, manifests at different spatial scales and in different forms, both material (e.g. roads, ferries, power and telecommunications infrastructure, etc.) and non-material (e.g. familial and cultural links, governance structures, etc.). Climate change has the potential to disrupt these connectivities, e.g. coastal airports at risk from sea level rise (Monioudi et al. 2018). Other key contemporary trends, such as migration, technological change and exposure to global knowledge, also have the potential to both enhance and break down such connectivities. For example, migration from peripheral to core communities can improve access to financial and social capital, which can foster greater adaptive capacity (Birk and Rasmussen 2014). However, migration can also challenge a community's sense of identity and lead to a loss of culture (Connell 2016), which may include traditional/local adaptation that has sustained island communities for centuries.

Varying degrees of connectivity may, in turn, affect how island communities prepare for and respond to extreme weather risks. Pre-cyclone warnings provide a salient example. Vital information must reach citizens in isolated peripheral locations, but here, geography presents more than simply logistical challenges. The core–periphery relation, positioning rural/outer island communities as geographically, economically and politically marginal (e.g. Nunn and Kumar 2017; Overton 1994; Sofer 1988) may manifest in not only variable infrastructure and service provision but also, persistence of traditional livelihoods and practices in remote areas (Connell 2010; Nunn et al. 2014; Sofer 2018). Thus, material differences in the extent of communication networks alongside non-material differences in socio-cultural context, may both engender different preferences around how information channels are used and how warnings are acted upon.

There has been some focus on associations between demographic or socio-economic factors and information channel preferences, but little has focused specifically on islands and core-periphery communities. In the context of hurricane warnings in the USA, research indicates that demographic characteristics are associated with different preferences around information channels, with minority residents favouring community or local government information sources (DeYoung et al. 2016). In an Australian study, older people tended to prefer 'traditional' communicative technologies like radio and landline telephone to newer technologies such as apps, while culturally and linguistically diverse participants favoured in-person communication (Howard et al. 2017).

There has been some study of the information channel preferences of urban Pacific islanders in relation to tropical cyclones (Magee et al. 2016), and the scholarship on island experiences of tsunamis also highlights the relative effectiveness of different information channels (Perry 2007) alongside the vital role that traditional knowledge capital continues to play in mediating natural hazards in remote island locations (McAdoo et al. 2006, 2009; Hall et al. 2017). Yet, within-island state patterns in information channel penetration have received little attention.

Using Tonga as a case study, this research interrogates the assumptions inherent in the core-periphery paradigm by assessing the extent to which rural/outer island communities exhibit different behaviours around information channels, and the relative importance of different aspects of connectivity in explaining those patterns. It discusses the potential for interaction between 'islandness', characterised by smallness, boundedness, isolation and fragmentation (Fernandes and Pinho 2017), and disaster risk communication behaviours, which is of increasing importance in light of climate change. Key questions to be investigated are:

(1) Are there differences in the patterns of information channel usage (radio, television, SMS, internet, word of mouth, etc.) between core and peripheral areas?
(2) Can any differences be explained by either patterns in the use of communication technology, or patterns in indicators acting as proxies for traditional practices?

Case Study Context: Tonga and Cyclone Gita

This paper focuses on the islands of Tongatapu and 'Eua in Tonga. Tongatapu is the main island of the Kingdom of Tonga. It is the country's most populous island with just over 74,600 people in 2016. In contrast, 'Eua reported a population of 4945 in the 2016 census. Yet, even in Tongatapu, there is considerable variation in population density between villages, with the greatest concentration of people in the capital, Nuku'alofa, which comprises the districts of Kolomotu'a and Kolofo'ou (Fig. 1).

Traditionally, agriculture and fishing are the main sources of livelihood (Nunn and Waddell 1992; Brown Pulu 2013). Traditional practices are of importance, e.g. in the healthcare system (McGrath 1999), where social relationships are deeply connected to cultural conceptualisations of health (Capstick et al. 2009). While English and Tongan are both official languages, secondary and post-secondary education is largely in English, and radio and television programmes are mainly in English (Otsuka 2007). Despite this, there are many villages where a large proportion of households speak only Tongan at home, according to census data (Fig. 5).

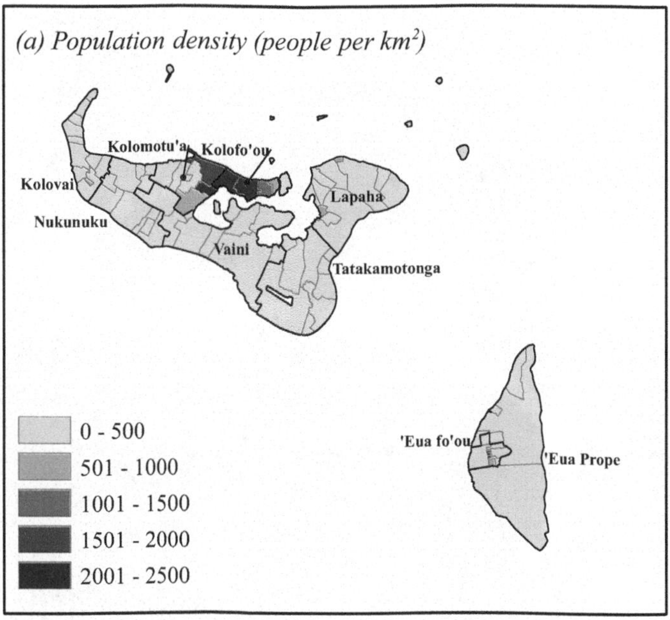

Fig. 1 Population density of Tongatapu and 'Eua islands

Tongatapu and 'Eua were the two Tongan islands most severely impacted by Cyclone Gita in 2018. Gita was near peak intensity (Category 4 on the five-point Saffir-Simpson scale) as the storm approached Tongatapu (Fig. 2).

Almost 80% of the population of Tonga were directly impacted by the cyclone, with total damage estimated at T$356.1 million (US$164.1 million) (Global Facility for Disaster Reduction and Recovery 2018)

Data and Methods

Data used in this study is obtained from the 2016 Population and Housing Census[1] and the 2018 Gita Impact Assessment Survey,[2] both produced by the Tonga Department of Statistics.

The mode by which households were warned is not an exclusive variable. I.e. Some households received warnings via multiple information channels, but data relating to the use of single versus multiple channels was not available.

[1]TONGA CENSUS 2016 DATA: Tonga Department of Statistics, 2016 Population and Housing Census. Data available at: http://tonga.popgis.spc.int/#c=home.

[2]GITA IMPACT ASSESSMENT: Tonga Department of Statistics, 2018 Gita Impact Assessment Survey. Data available at: http://tonga.popgis.spc.int/#c=home.

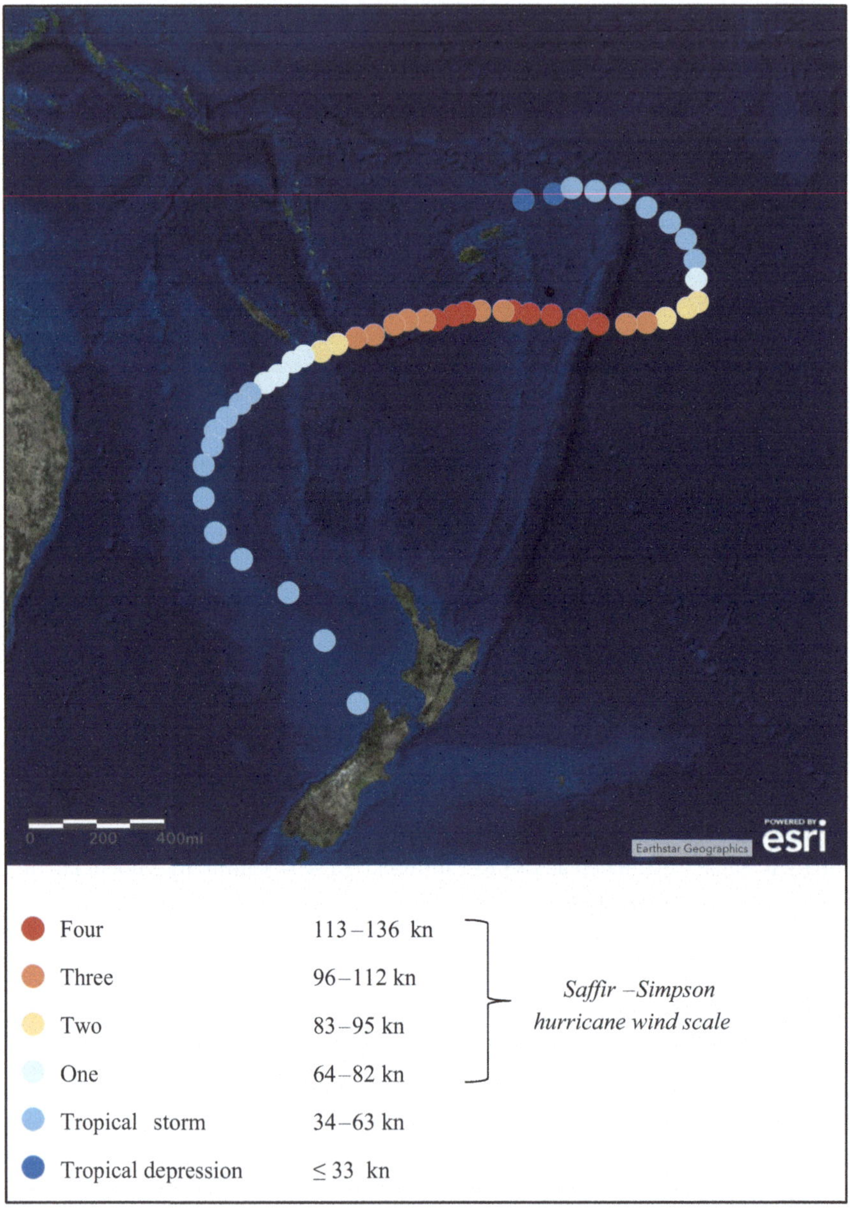

Fig. 2 Track and intensity of Gita, according to the Saffir–Simpson scale (Data: NOAA, Data available at: https://ftp.emc.ncep.noaa.gov/wd20vxt/hwrf-init/decks/bsh092018.dat)

A selection of variables corresponding to communications, economic activity and linguistic preferences are also utilised. Proportions of the population with access to mobile phones and internet are analysed to assess whether greater access to such technologies results in higher proportions of households receiving warnings by these means. The proportion of the population employed in traditional livelihoods (agriculture, fishing and forestry) is used as a proxy to identify areas which may be operating in more traditional contexts. The possible impact of linguistic characteristics on interaction with different warning information channels is explored also (Table 1).

The Getis-Ord Gi* statistic is used to identify high or low-value clusters in the spatial data (Getis and Ord 1992). A statistically significant positive/negative z-score is returned when the local sum for a feature and its neighbours (defined as features sharing a boundary) is very different to the expected local sum, based on all features. Statistically significant positive/negative z-scores indicate clustering of high/low values (hot/cold spots). As spatial dependency can artificially inflate statistical significant in local pattern analysis, a false discovery rate correction is applied. Statistically significant p-values are ranked from smallest to largest, and the weakest significant p-values are omitted, with the number of values to be discounted based on the false positive estimate.

Table 1 Data utilised in this study

Country	Dataset	Variables	Level
Tonga	Cyclone Gita Impact Assessment	*Households warned* – Radio—Proportion of total yes – Television—Proportion of total yes – SMS—Proportion of total yes – Internet—Proportion of total yes – Word of mouth—Proportion of total yes – Other- Proportion of total yes	Village
	Tonga Census 2016	*Communications* – Own a working mobile phone—Proportion of total yes—(10+) – Access internet—Proportion of total yes—(10+) *Economic* – Main industry—Proportion of total agriculture, forestry and fishing—(15+ act. pop.) *Linguistic* – Speaking Tongan at home—Proportion of total yes, only this language—(5+) – Difficulty reading English—Proportion of total no difficulty—(5+)	Village

For each variable, the analysis is performed on the value as a proportion of the population, rather than on the total number of positive responses for that variable. It would naturally be expected that the number of households warned via each mode of communication, for example, would be higher in villages where there are more households; by analysing a rate, this relationship is controlled for.

The method can be replicated and applied in other island contexts where similar data is available, to explore similar questions of core-periphery relationships.

Results

Several notable hotspots and coldspots in information channels via which warnings were received at household level. Low-value clusters, i.e. coldspots, in 'modern' information channels (internet and SMS) are observed on 'Eua. Conversely, hotspots in these information channels are identified in western Tongatapu. A hotspot for television warnings is also observed here. Several more isolated, but highly significant, word-of-mouth warning hotspots are observable in western Tongatapu. Radio is by far the most successful information channel in terms of reaching households. As values are almost universally high, no hotspots are identified, although a single village was flagged as a radio warning coldspot.

While access to mobile phones is not universal, access is evenly distributed across both islands (i.e. no hotspots or coldspots). SMS warning hotspots have considerable overlaps with the 'Only Tongan spoken at home' coldspot. To express this another way, there appear to be higher proportions of households receiving these types of warnings in villages where higher proportions of people speak an additional language in the home. Internet warning hotspots also have considerable overlaps with the 'Only Tongan spoken at home' coldspot. There is also an internet warning hotspot corresponding to the only internet access hotspot, Kolofu'ou.

'Eua represents a coldspot of SMS warning activity and, to a lesser extent, internet warnings. It is not a coldspot in terms of mobile phone or internet access, suggesting that access to information and communication technology is not a barrier. It is also a hotspot for the variable 'no difficulty reading in English', suggesting that capacity to engage with web-based English language information sources is high. The island is also a hotspot in terms of traditional livelihoods (agriculture, fishing, forestry).

Hotspot analysis was also carried out to determine if there was a greater percentage of tourists and visitors present in villages corresponding to the 'Only Tongan spoken at home' coldspot, as this would have implications for the nature of tropical cyclone warnings, but no hotspots were identified (not shown) (Figs. 3, 4 and 5).

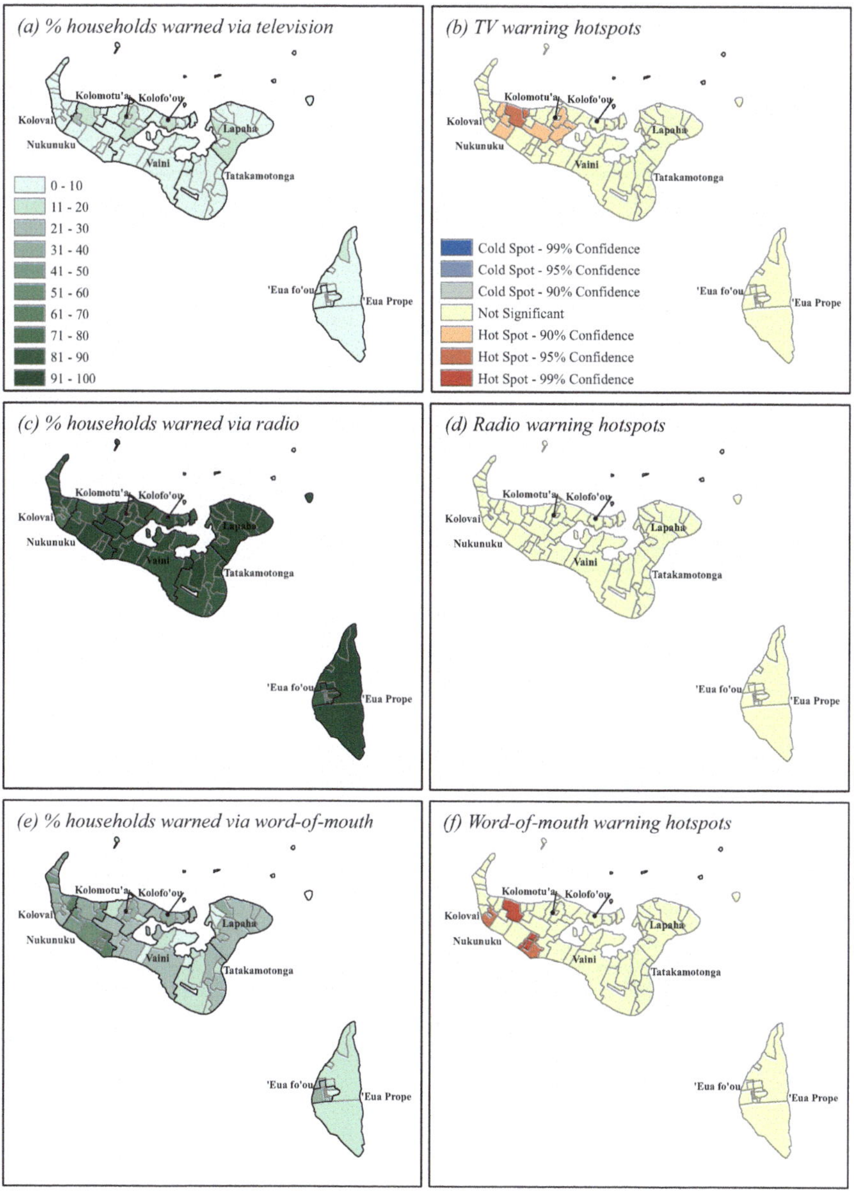

Fig. 3 Percentage of households warned via each information channel (left) and hotspot analysis results for that information channel (right)

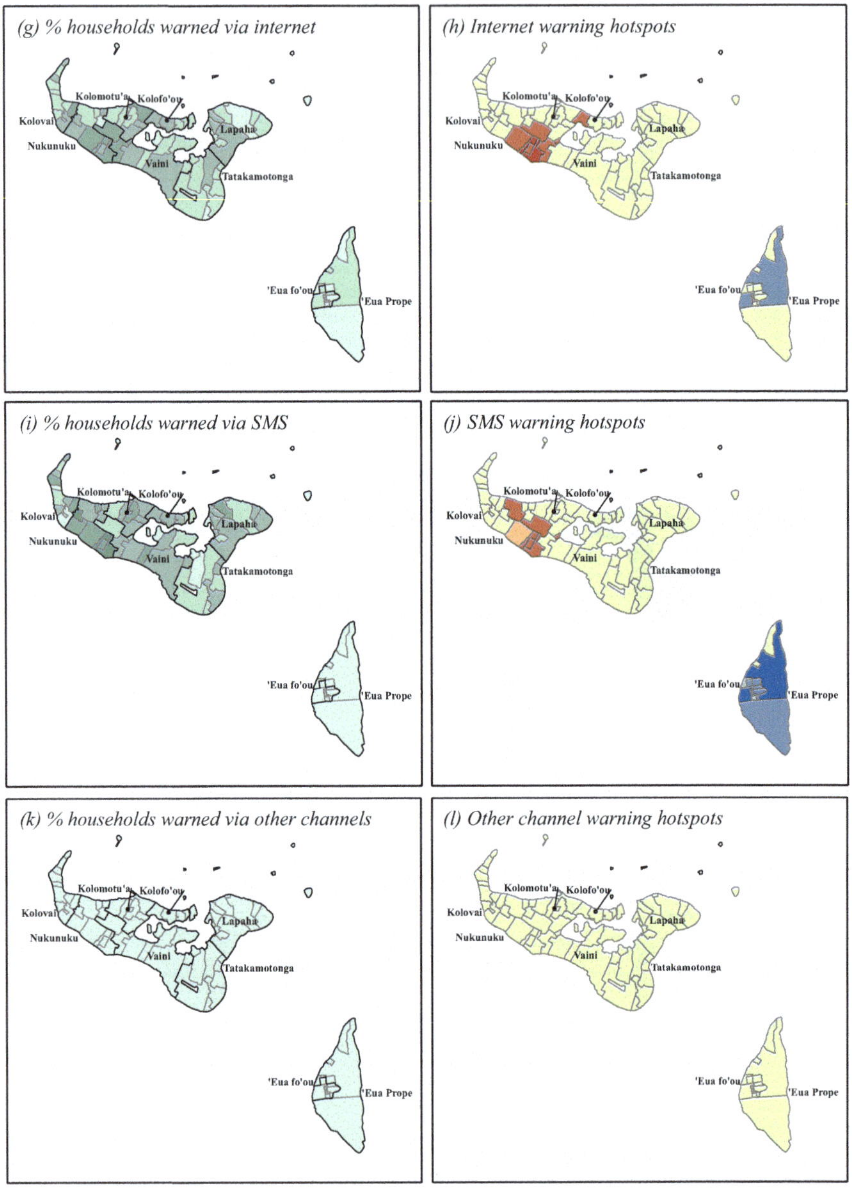

Fig. 3 (continued)

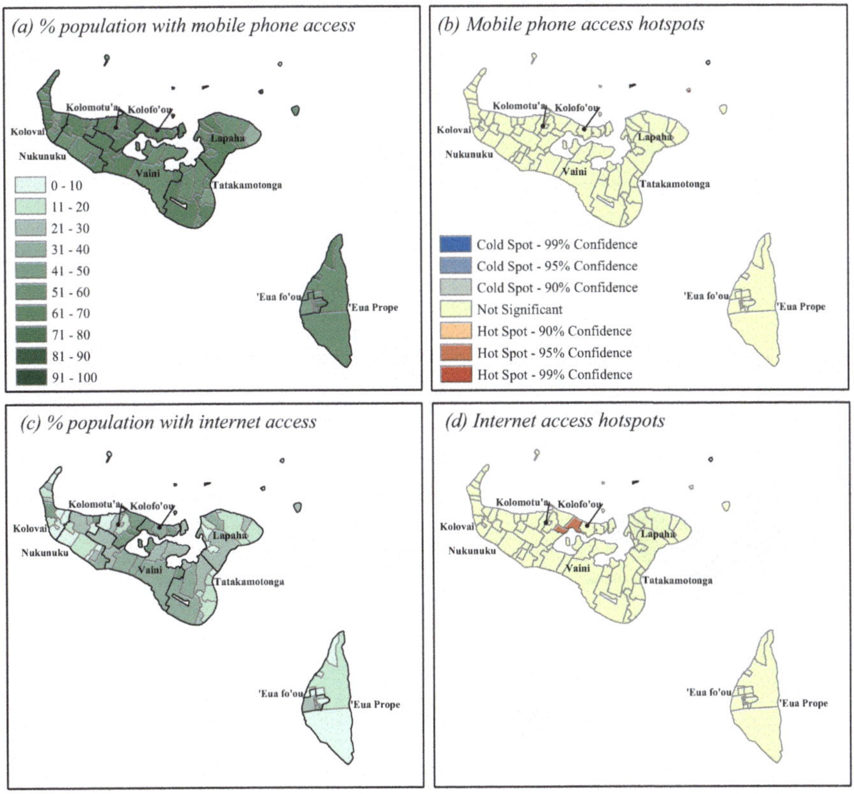

Fig. 4 Percentage of population with access to mobile phones and internet (left) and hotspot analysis results for these variables (right)

Discussion

Radio is by far the most successful information channel in terms of reach. Radio Tonga One is a key partner in Tonga's tropical cyclone warning system, due to the maximum coverage it provides, although on 'Eua the reach by radio is limited (I. Tora, Pacific Environmental Journalists Network, personal communication, 26 March 2019). However, there are limits to the kinds of information that radio, and particularly mainstream media outlets can provide. Information may not always meet the needs of rural and peripheral communities (Singh and Naidu 2018). Community radio can fill this gap, but given that half of all community radio stations in the South Pacific islands are faith-based (Austin 2014), it is also necessary to consider how this cultural context might influence messaging about natural hazards, in order to implement culturally-grounded adaptive practice (Chester 2005; Nunn et al. 2016).

Radio offers no means of integrating visual data such as maps, which may improve understanding of disaster messaging (Liu et al. 2017). Furthermore, lessons from the

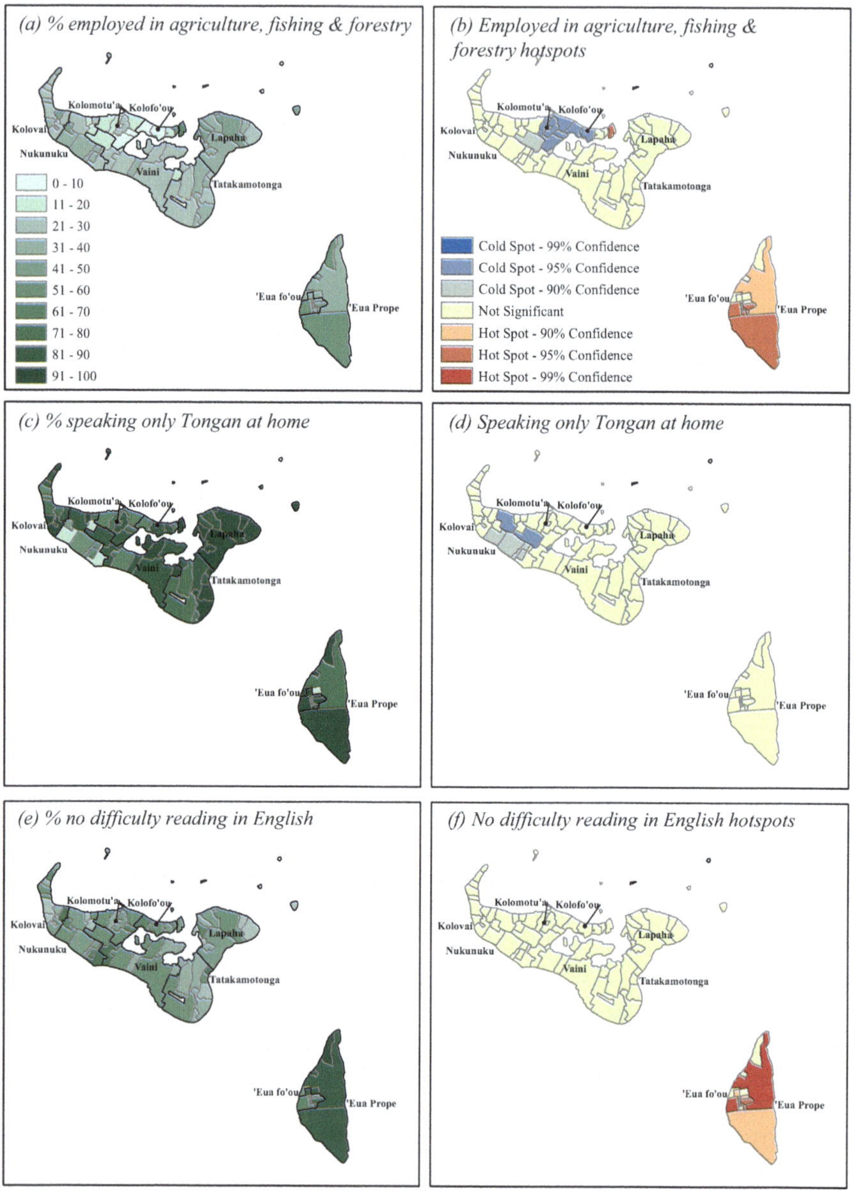

Fig. 5 Economic and linguistic variables (left) and hotspot analysis results for these variables (right)

fields of volcanic hazard suggest that peoples can struggle to orientate themselves using traditional plan view maps, but that this improves when using aerial photographs (Haynes et al. 2007). Mobile phones have considerable potential, then, to provide localised evacuation guidance, which would greatly benefit visitors to an island as well as locals (Rahman et al. 2012).

In Pacific countries, use of mobile phones is on the rise, and consequently, they are increasingly incorporated into disaster risk communication (Noske-Turner et al. 2014). The content of these brief messages matters, as there is evidence from the US that people find mobile alerts and tweets uninformative and impersonal (Bean et al. 2016). Analysis of experiences in Vanuata during Cyclone Pam demonstrated the effectiveness of SMS warning, with this information channel largely replacing radio (Handmer and Iveson 2017). Given the emerging role of mobile and internet technology, the limited role these technologies appear to play in the dissemination of cyclone warning information in Tonga, particularly in rural areas, is a trend worthy of further investigation.

Information and communication technologies also allow people to take a more active, participatory role in sharing information (Oxendine and Waters 2014; Harris et al. 2016); before and during Cyclone Winston in Fiji, social media played a role in both disseminating information and documenting the unfolding situation (Finau et al. 2018). Understanding community preferences for accessing information may help to foster contextually appropriate, community-centric warning systems that optimally use media channels (Baudoin et al. 2016).

The overlap of the 'internet warning' hotspot and the 'Only Tongan spoken at home' coldspot suggest that linguistic preferences may play a role in engagement with different information channels. TV, radio and SMS messages are transmitted in both Tongan and English (I. Tora, personal communication, 26 March 2019). However, several relevant pages on the Tonga Meteorological Service website, such as the Tropical Cyclone Outlook, and the Natural Disaster Information, are available only in English. Tropical Cyclone Advisories are made available in both English and Tongan, but must be navigate to via the main webpage, which is in English. Yet, the 'internet warning' coldspot on 'Eua overlaps with a 'No difficulty reading English' hotspot, further suggesting that if there is a link between engagement with certain information channels and linguistic characteristics, it is driven by linguistic preferences, rather than abilities.

Conclusions

This paper has explored the impact of connectivity on dissemination of information about disaster risk in Tonga, with two key conclusions. Firstly, this research identifies notable core-periphery patterns in how communities were warned about Cyclone Gita. Coldspots in 'modern' information channels (internet and SMS) are observed on 'Eua, in contrast with hotspots in internet, SMS and television warnings in western

Tongatapu. Secondly, linguistic preferences and traditional context, rather than access to ICT, appear to play a role in explaining these patterns.

The empirical-analytical approach taken in this paper identifies compelling patterns in the available data. However, there are several limitations that must be kept in mind. Firstly, the unit of analysis is villages and conclusions must be interpreted within this context. Inferences cannot be made about the relationship between information channel use, linguistic preferences and traditional context at the individual level. Secondly, as this study uses secondary data, there are certain dimensions of connectivity that cannot be examined as the datasets do not contain relevant variables (e.g. relating to access to transport).

Additional community-based participatory research is needed in order to understand the drivers underlying the spatial patterns identified in a comprehensive and holistic way. Furthermore, while the data analysed here focused on how households were warned about Cyclone Gita, the question of how households acted upon that information is also of vital importance. Here, too, cultural context may contribute to different behaviours in communities operating within traditional contexts. As disaster response is a social process (Drabek 1999), further research is needed into how 'islandness' impacts the understanding people make of warnings, and the actions they take as a result.

This paper illustrates the value of data-driven, statistical approaches in generating insights into island practices around DRR and climate resilience, which can then be used in formulating additional, context-specific research questions and hypotheses. The approach used in this paper can be applied in other island contexts where similar data is available, which may help to identify common issues across island communities, as well as transferable practices for fostering climate resilience.

Acknowledgements I gratefully acknowledge the Tonga Department of Statistics, which makes available the data used in this research (TONGA CENSUS 2016 DATA: Tonga Department of Statistics, 2016 Population and Housing Census. GITA IMPACT ASSESSMENT: Tonga Department of Statistics, 2018 Gita Impact Assessment Survey). I also acknowledge the assistance of Mr Iliesa Tora (Pacific Environmental Journalists Network). Maps throughout this paper were created using ArcGIS® software by Esri. ArcGIS® and ArcMap™ are the intellectual property of Esri and are used herein under license. Copyright © Esri. All rights reserved. For more information about Esri® software, please visit www.esri.com.

References

Adger WN, Barnett J, Brown K et al (2013) Cultural dimensions of climate change impacts and adaptation. Nat Clim Change 3:112–117. https://doi.org/10.1038/nclimate1666

Adger WN, Barnett J, Chapin FS, Ellemor H (2011) This must be the place: underrepresentation of identity and meaning in climate change decision-making. Glob Environ Politics 11:1–25. https://doi.org/10.1162/GLEP_a_00051

Austin L (2014) Faith-based community radio and development in the South Pacific Islands. Media Int Aust 150:114–121. https://doi.org/10.1177/1329878X1415000122

Barclay J, Wilkinson E, White CS (2019) Historical trajectories of disaster risk in Dominica. Int J Disaster Risk Sci. https://doi.org/10.1007/s13753-019-0215-z

Baudoin M-A, Henly-Shepard S, Fernando N et al (2016) From top-down to "Community-Centric" approaches to early warning systems: exploring pathways to improve disaster risk reduction through community participation. Int J Disaster Risk Sci 7:163–174. https://doi.org/10.1007/s13753-016-0085-6

Bean H, Liu BF, Madden S et al (2016) Disaster warnings in your pocket: how audiences interpret mobile alerts for an unfamiliar hazard. J Contingencies Crisis Manage 24:136–147. https://doi.org/10.1111/1468-5973.12108

Birk T, Rasmussen K (2014) Migration from atolls as climate change adaptation: current practices, barriers and options in Solomon Islands. Nat Resour Forum 38:1–13. https://doi.org/10.1111/1477-8947.12038

Brown P, Daigneault AJ, Tjernström E, Zou W (2018) Natural disasters, social protection, and risk perceptions. World Dev 104:310–325. https://doi.org/10.1016/j.worlddev.2017.12.002

Brown Pulu TJ (2013) Climate change blues: sustaining village life in Tonga. 6:260–305

Capstick S, Norris P, Sopoaga F, Tobata W (2009) Relationships between health and culture in Polynesia—a review. Soc Sci Med 68:1341–1348. https://doi.org/10.1016/j.socscimed.2009.01.002

Chandler D, Pugh J (2018) Islands of relationality and resilience: the shifting stakes of the anthropocene. Area. https://doi.org/10.1111/area.12459

Chester DK (2005) Theology and disaster studies: the need for dialogue. J Volcanol Geoth Res 146:319–328. https://doi.org/10.1016/j.jvolgeores.2005.03.004

Connell J (2010) Pacific islands in the global economy: paradoxes of migration and culture. Singap J Tropic Geogr 31:115–129. https://doi.org/10.1111/j.1467-9493.2010.00387.x

Connell J (2016) Last days in the Carteret Islands? Climate change, livelihoods and migration on coral atolls. Asia Pac Viewpoint 57:3–15. https://doi.org/10.1111/apv.12118

Deidda M (2016) Insularity and economic development: a survey. Int Rev Econ 63:107–128. https://doi.org/10.1007/s12232-015-0238-8

DeYoung SE, Wachtendorf T, Farmer AK, Penta SC (2016) NOAA radios and neighbourhood networks: demographic factors for channel preference for Hurricane evacuation information. J Contingencies Crisis Manage 24:275–285. https://doi.org/10.1111/1468-5973.12123

Dornan M, Newton Cain T (2015) Vanuatu and cyclone Pam: an update on economic, fiscal and development impacts. http://www.adb.org/publications/pacific-economic-monitor-july-2015. Accessed 8 Mar 2019

Drabek TE (1999) Understanding disaster warning responses. Soc Sci J 36:515–523. https://doi.org/10.1016/S0362-3319(99)00021-X

Edmonds C, Noy I (2018) The economics of disaster risks and impacts in the Pacific. Disaster Prev Manage 27:478–494. https://doi.org/10.1108/DPM-02-2018-0057

Emanuel K (2005) Increasing destructiveness of tropical cyclones over the past 30 years. Nature 436:686–688. https://doi.org/10.1038/nature03906

Fernandes R, Pinho P (2017) The distinctive nature of spatial development on small islands. Prog Plann 112:1–18. https://doi.org/10.1016/j.progress.2015.08.001

Finau G, Cox J, Tarai J et al (2018) Social media and disaster communication: a case study of Cyclone Winston. Pac Journalism Rev Te Koakoa 24:123–137. https://doi.org/10.24135/pjr.v24i1.400

Getis A, Ord J (1992) The analysis of spatial association by use of distance statistics. Geogr Anal 24:189–206

Global Facility for Disaster Reduction and Recovery (2018) Tonga: rapid damage assessment and recovery framework for tropical Cyclone Gita. https://www.gfdrr.org/en/tonga-rapid-damage-assessment-and-recovery-framework-tropical-cyclone-gita. Accessed 13 Mar 2019

Graham S, Barnett J, Fincher R et al (2013) The social values at risk from sea-level rise. Environ Impact Assess Rev 41:45–52. https://doi.org/10.1016/j.eiar.2013.02.002

Grydehøj A, Casagrande M (2019) Islands of connectivity: archipelago relationality and transport infrastructure in Venice Lagoon. Area. https://doi.org/10.1111/area.12529

Hall S, Pettersson J, Meservy W et al (2017) Awareness of Tsunami natural warning signs and intended evacuation behaviors in Java, Indonesia. Nat Hazards 89:473–496. https://doi.org/10.1007/s11069-017-2975-3

Handmer J, Iveson H (2017) Cyclone Pam in Vanuatu: learning from the low death toll. Aust J Emerg Manage 32:60–65

Harris US, Papoutsaki E, Kailahi S (2016) ICTs in climate change communication in the Pacific Islands. Inf Technol Int Dev 12:47–58

Haynes K, Barclay J, Pidgeon N (2007) Volcanic hazard communication using maps: an evaluation of their effectiveness. Bull Volcanol 70:123–138. https://doi.org/10.1007/s00445-007-0124-7

Howard A, Agllias K, Bevis M, Blakemore T (2017) "They'll tell us when to evacuate": the experiences and expectations of disaster-related communication in vulnerable groups. Int J Disaster Risk Reduction 22:139–146. https://doi.org/10.1016/j.ijdrr.2017.03.002

Jackson G, McNamara K, Witt B (2017) A framework for disaster vulnerability in a Small Island in the Southwest Pacific: a case study of Emae Island, Vanuatu. Int J Disaster Risk Sci 8:358–373. https://doi.org/10.1007/s13753-017-0145-6

Kelman I (2018) Islands of vulnerability and resilience: manufactured stereotypes? Area. https://doi.org/10.1111/area.12457

Knutson TR, McBride JL, Chan J et al (2010) Tropical cyclones and climate change. Nat Geosci 3:157–163. https://doi.org/10.1038/ngeo779

Kossin JP, Emanuel KA, Camargo SJ (2016) Past and projected changes in Western North Pacific Tropical Cyclone exposure. J Clim 29:5725–5739. https://doi.org/10.1175/JCLI-D-16-0076.1

Lin N, Emanuel K (2016) Grey swan tropical cyclones. Nat Clim Change 6:106–111. https://doi.org/10.1038/nclimate2777

Liu BF, Wood MM, Egnoto M et al (2017) Is a picture worth a thousand words? The effects of maps and warning messages on how publics respond to disaster information. Public Relat Rev 43:493–506. https://doi.org/10.1016/j.pubrev.2017.04.004

Magee AD, Verdon-Kidd DC, Kiem AS, Royle SA (2016) Tropical cyclone perceptions, impacts and adaptation in the Southwest Pacific: an urban perspective from Fiji, Vanuatu and Tonga. Nat Hazards Earth Syst Sci 16:1091–1105. https://doi.org/10.5194/nhess-16-1091-2016

Martin PCM, Nunn P, Leon J, Tindale N (2018) Responding to multiple climate-linked stressors in a remote island context: the example of Yadua Island, Fiji. Clim Risk Manage 21:7–15. https://doi.org/10.1016/j.crm.2018.04.003

McAdoo BG, Dengler L, Prasetya G, Titov V (2006) Smong: how an oral history saved thousands on Indonesia's Simeulue Island during the December 2004 and March 2005 Tsunamis. Earthquake Spectra 22:661–669. https://doi.org/10.1193/1.2204966

McAdoo BG, Moore A, Baumwoll J (2009) Indigenous knowledge and the near field population response during the 2007 Solomon Islands tsunami. Nat Hazards 48:73–82. https://doi.org/10.1007/s11069-008-9249-z

McGrath BB (1999) Swimming from Island to Island: healing practice in Tonga. Med Anthropol Q 13:483–505. https://doi.org/10.1525/maq.1999.13.4.483

Monioudi IN, Asariotis R, Becker A et al (2018) Climate change impacts on critical international transportation assets of Caribbean Small Island Developing States (SIDS): the case of Jamaica and Saint Lucia. Reg Environ Change 18:2211–2225. https://doi.org/10.1007/s10113-018-1360-4

Nishijima K, Mori N, Yasuda T et al (2015) DPRI-VMGD joint survey for Cyclone Pam damages. http://www.taifu.dpri.kyoto-u.ac.jp/wp-content/uploads/2015/05/DPRI-VMGD-survey-first-report-Final.pdf. Accessed 16 May 2019

Noske-Turner J, Tacchi J, Horst H, Papoutsaki E (2014) Locating disaster communication in changing communicative ecologies across Pacfiic. http://doi.org/10.7790/ajtde.v2n4.66

Noy I (2016) Natural disasters in the Pacific Island Countries: new measurements of impacts. Nat Hazards 84:7–18. https://doi.org/10.1007/s11069-015-1957-6

Nunn P, Kumar R (2017) Understanding climate-human interactions in Small Island Developing States (SIDS): implications for future livelihood sustainability. Int J Clim Change Strat Manag 10:245–271. https://doi.org/10.1108/IJCCSM-01-2017-0012

Nunn PD, Aalbersberg W, Lata S, Gwilliam M (2014) Beyond the core: community governance for climate-change adaptation in peripheral parts of Pacific Island Countries. Reg Environ Change 14:221–235. https://doi.org/10.1007/s10113-013-0486-7
Nunn PD, Mulgrew K, Scott-Parker B et al (2016) Spirituality and attitudes towards Nature in the Pacific Islands: insights for enabling climate-change adaptation. Clim Change 136:477–493. https://doi.org/10.1007/s10584-016-1646-9
Nunn PD, Waddell E (1992) Implications of climate change and sea level rise for the Kingdom of Tonga. South Pacific Regional Environment Programme, Apia
Otsuka Y (2007) Making a case for Tongan as an endangered language. Contemp Pac 19:446–473
Overton J (1994) Small states, big issues? Human geography in the Pacific Islands. Singapore J Trop Geo 14:265–276. https://doi.org/10.1111/j.1467-9493.1993.tb00052.x
Oxendine CE, Waters N (2014) No-notice urban evacuations: using crowdsourced mobile data to minimize risk. Geogr Compass 8:49–62. https://doi.org/10.1111/gec3.12104
Perry SD (2007) Tsunami warning dissemination in Mauritius. J Appl Commun Res 35:399–417. https://doi.org/10.1080/00909880701611060
Rahman KM, Alam T, Chowdhury M (2012) Location based early disaster warning and evacuation system on mobile phones using OpenStreetMap. In: 2012 IEEE conference on open systems, pp 1–6
Shultz JM, Kossin JP, Ettman C et al (2018) The 2017 perfect storm season, climate change, and environmental injustice. Lancet Planet Health 2:e370–e371. https://doi.org/10.1016/S2542-5196(18)30168-2
Singh S, Naidu V (2018) Coverage of extreme weather events and natural hazards in Pacific island countries: the need for media capacity-building. Pac Journalism Rev 24:87–101
Sofer M (1988) Core–periphery structure in Fiji. Environ Plan D 6:55–74. https://doi.org/10.1068/d060055
Sofer M (2018) 11 revisiting the Fijian periphery. In: John Connell HL (ed) Change and continuity in the Pacific: revisiting the region. Routledge, Abingdon, New York
Warrick O, Aalbersberg W, Dumaru P et al (2017) The ‘Pacific Adaptive Capacity Analysis Framework’: guiding the assessment of adaptive capacity in Pacific island communities. Reg Environ Change 17:1039–1051. https://doi.org/10.1007/s10113-016-1036-x
Webster PJ, Holland GJ, Curry JA, Chang H-R (2005) Changes in tropical cyclone number, duration, and intensity in a warming environment. Science 309:1844–1846. https://doi.org/10.1126/science.1116448

Climate Change in Tonga: Risk Perception and Behavioral Adaptation

David N. Sattler, Uili Lousi, James M. Graham, Viliami Latu, James Johnson and Siosaia Langitoto Helu

Abstract This chapter develops a model of climate change risk perception and behavioral intention to take action to prevent climate change in the Kingdom of Tonga. The project was conducted in the wake of Cyclone Gita, one of the strongest cyclones to strike Tonga in recorded history. Climate change may have increased the strength of Cyclone Gita. We examined how resource loss as a result of the storm, coping, community norms, value orientation, and posttraumatic stress influence climate change risk perception and behavioral intention to prevent climate change. The project was guided by climate change models we developed after Cyclone Winston made landfall in Fiji. The participants were 230 people (49% men, 51% women) in communities in Tonga (age: $M = 42$, $SD = 15.5$). They completed assessment instruments measuring resource loss as a result of the cyclone, coping, community norms regarding climate change, personal values, posttraumatic stress, climate change risk perceptions, behavioral intention to prevent climate change, and demographics. Two models with similar paths predicated climate change risk perceptions and behavioral intention to prevent climate change. In one path, loss of resources as a result of Cyclone Gita was associated with coping, and coping was associated

D. N. Sattler (✉) · J. M. Graham
Department of Psychology, Western Washington University, Bellingham, WA 98225-9172, USA
e-mail: David.Sattler@wwu.edu

J. M. Graham
e-mail: Jim.Graham@wwu.edu

U. Lousi
OHAI Foundation, Nuku'Alofa, Tonga
e-mail: uililousiart@yahoo.co.nz

V. Latu
Edwards Law, Nuku'Alofa, Tonga
e-mail: viliuasike@gmail.com

J. Johnson
School of Social Sciences, University of the South Pacific, Suva, Fiji
e-mail: james.johnson@usp.ac.fj

S. L. Helu
'Atenisi Institute and Pacific Resilience Program Tonga, Nuku'Alofa, Tonga
e-mail: helulangitoto@gmail.com

W. Leal Filho (ed.), *Managing Climate Change Adaptation in the Pacific Region*, Climate Change Management, https://doi.org/10.1007/978-3-030-40552-6_14

with climate change risk perceptions. In a second path, loss of resources as a result of Cyclone Gita was associated with posttraumatic stress, and posttraumatic stress was associated with posttraumatic growth. An additional analysis examined personal values and community norms. In one path, personal values were directly associated with behavioral intention to take action to prevent climate change. In a second path, coping mediated resource loss as a result of Cyclone Gita, community norms, and personal values. The findings extend our climate change model developed in Fiji and support and extend van der Linden's (J Environ Psychol 41:112–124, 2015) climate change risk perception model. We discuss implications of the model for climate change education campaigns.

Keywords Climate change · Risk perception · Behavioral adaptation · Posttraumatic growth · Tonga · Cyclone Gita · South Pacific · Climate change model

Introduction

Climate change is altering environmental conditions in the Kingdom of Tonga and other small island developing states in the Pacific Ocean. Oceans retain much of the heat reflected back to Earth, and warming ocean temperatures cause water to expand, which raises sea level. Some small island states, especially those on atoll islands, may need to relocate as a result of land loss due to the increase in sea level (Nurse et al. 2014). Ocean acidification increases with warming waters and threatens marine ecosystems, including coral reefs and marine fisheries (Laffoley and Baxter 2016). Warm ocean water also provides the fuel for cyclones, and in recent years, warmer waters have increased the magnitude of cyclones. As the strength of these storms surpasses the current classification system, the need for a new category 6 cyclone is being considered (Masters 2019). Stronger storms have significantly greater potential to threaten lives, create extensive property damage, and damage food crops (National Oceanic and Atmospheric Administration 2013; Pacific Climate Change Science Program 2016). Air temperature is increasing and precipitation patterns are becoming more unpredictable. Small island developing states contribute less than 1% of greenhouse gas emissions into the atmosphere, yet they experience a disproportionate degree of negative consequences as a result of climate change (United Nations Development Programme 2017).

In Tonga, air temperature has increased 0.10 °C in the capital of Nuku'Alofa during each of the past seven decades, and rainfall has decreased. Sea level has risen by 6 mm in Tonga, or nearly double the global average, in the past 25 years. At the September 2018 meeting of the United Nations, King Tupou VI of the Kingdom of Tonga underscored the challenges: "Climate change continues to pose significant security threats to us as island States," and he stressed "the devastating impacts of climate change on our marine environment" (United Nations News 2018).

Tonga is an archipelago located 500 miles southeast of the Fiji Islands and has a population of approximately 108,000 persons. Nearly three-quarters live on the main island of Tongatapu, and about one-quarter live on 35 other islands. The World Risk Report ranks Tonga as the second most vulnerable country in the world to natural disasters (Alliance Development Works 2013). Stronger cyclones as a result of climate change and increasing sea level pose real risk to life and property, agriculture, and individual and societal functioning. A review of the PsycINFO database shows no published studies examining any aspect of disaster preparedness and response or climate change risk perceptions and adaptation conducted in Tonga. There is great need for research examining climate change adaptation and response in Tonga and other small island developing states (Petzold and Magnan 2019).

This chapter presents a study that is among the first in Tonga and the Pacific to assess climate change risk perceptions and intention to modify behaviors that contribute to climate change. We were especially interested in examining climate change risk perceptions and behavioral intention in the wake Cyclone Gita, a powerful cyclone whose strength was likely associated with climate change. On February 12, 2018, Cyclone Gita made landfall in Tonga with sustained wind speeds of 230 and 280 km/h gusts. It ranks as one of the strongest cyclones to strike the country in recorded history (George and Sidhu 2018; Tapaleao and Dunlop 2018). Cyclone Gita affected nearly 70% of all persons in Tonga, destroyed more than 800 homes, and damaged approximately 4000 homes and infrastructure such as electric power lines and crops. Damage estimates were US$164 million. There have been significant discussions concerning the role of climate change in strengthening Cyclone Gita and other storms in the Pacific Ocean (Australia Department of Foreign Affairs and Trade 2018; Pearlman and Krol 2018).

A Climate Change Risk Perception and Behavioral Adaptation Model

This study was guided by a climate change risk perception and behavioral adaption model we developed in the wake of Cyclone Winston in the Fiji Islands (Sattler et al. 2018a, b). Cyclone Winston was the strongest storm in recorded history to make landfall in Fiji, and climate change may have contributed to increasing its size and strength (Earth Observatory 2016). The model considers how experiences during the cyclone as well as knowledge of and social norms about climate change influence risk perceptions and adaption. In particular, the model examines loss of resources; psychological reactions such as posttraumatic stress, coping, social support, and posttraumatic growth; knowledge about climate change; social norms concerning climate change; values; and demographic variables (e.g., age, gender, and education level). A brief discussion of each of these variables is presented below. For a more thorough discussion of these variables, see Sattler (2017) and Sattler et al. (2018a).

Resource loss and posttraumatic stress. According to conservation of resources stress theory, the threat of resource loss or loss of resources as a result of a cyclone or natural disaster can create psychological distress (e.g., depression, anxiety) and posttraumatic stress (Hobfoll 2012). The theory identifies four types of resources that may be at risk: object (e.g., home, possessions), condition (e.g., employment), energy (e.g., time for adequate sleep), and personal characteristic (e.g., sense of optimism; Hobfoll 2012). Posttraumatic stress can occur in response to life threat, and symptoms include intrusive thoughts about the event, difficult sleeping as a result of nightmares, severe anxiety, and flashbacks about the incident (National Center for PTSD 2017). Threat of loss or actual loss of resources is associated with psychological distress (Sattler 2017).

Coping and posttraumatic growth. People may cope with stress experienced as a result of a disaster by using a variety of methods, including confronting the situation directly to resolve it, disengaging from the situation, focusing on emotions, and receiving assistance from others. As people cope with the situation and rebuild their lives, they may reflect on their life priorities and what gives their life meaning. They may become more aware of the importance of developing a disaster plan with actions to take to prepare for a disaster, the importance of neighbors helping neighbors, and ways of coping with adversity. As they rebuild their lives and experience resource gains, they may have increased appreciation for support provided by others, feel a stronger sense of community, and reflect on the value of community and assistance provided to others. As a result, they may experience posttraumatic growth (Calhoun and Tedeschi 2001; Sattler and Smith, 2020).

Norms and values. According to van der Linden's (2015) climate change risk perception model, personal experience with climate change, community norms, and values influence the degree of risk people may perceive as a result of climate change. Personal experience concerns how experience with climate change has affected the individual and his or her emotional reactions. Community norms involve expectations of others behavior, including actions toward the environment that are associated with climate change. Biospheric values involve an awareness of the environment, egoistic values center on individual outcomes, and socio-altruistic values consider the needs of other people (van der Linden 2015).

Developing the model after Cyclone Winston in Fiji. After Cyclone Winston created extensive damage in Fiji, people living in coastal villages completed a questionnaire that assessed each variable in the model. They were without basic resources (food, water) for many days and were living in tents or damaged homes for many weeks after the storm. We used structural equation modeling to develop the model. The results showed three pathways to behavioral intention to take action to prevent climate change. In one path, resource loss as a result of the cyclone and education level were each independently associated with posttraumatic growth, and posttraumatic growth mediated their relationship with behavioral intention. In a second path, community norms concerning climate change mediated the relationship between resource loss and behavioral intention. In a third path, risk perceptions mediated the relationship between community norms, knowledge about climate change, and affect (Sattler et al. 2018a).

Present Study

The present study extends Sattler et al. (2018a) by developing a climate change risk perception and behavioral adaptation model in Tonga in the wake of Cyclone Gita. It includes the same variables as our Cyclone Winston study in Fiji. Our first research question was "Will the climate change behavioral model developed in Fiji replicate in Tonga?" The second question was "If the model does not replicate, what variables will contribute to climate change risk perception and to behavioral adaption for climate change in Tonga?" Because the study was exploratory, we did not have specific hypotheses.

Method

Participants and Procedure

Eight weeks after Cyclone Gita made landfall, we administered an anonymous questionnaire to persons living in coastal communities in Tonga. The participants were 230 persons (49% men, 51% women; age: $M = 42$, $SD = 15.5$, range: 18–89 years). More than one-third (36%) had less than a secondary school education, about one-third (34%) completed secondary school, and about one-third (30%) had some college or a college degree. The median distance they lived from the coast was less than 1 km.

The study was approved by the Human Participants Research Committee at Western Washington University and followed the American Psychological Association ethical guidelines. We recruited two college students in Tonga to serve as research assistants, and trained them in research ethics and questionnaire administration. We identified neighborhoods that experienced damage as a result of Cyclone Gita, and individuals in these neighborhoods completed the questionnaires in their homes. Participation was voluntary and no inducements were offered. Almost all individuals asked to participate did so; the response rate was 94%. It took about 25 min to complete the survey.

Assessment Instruments

We translated the questionnaire from English to Tongan. To do so, we used a version of the committee approach as discussed by Matsumoto and van de Vijver (2011) and van de Vijver and Leung (1997). We have successfully used this approach in numerous countries, including Thailand, Indonesia, and El Salvador (e.g., Sattler et al. 2014, 2018b). The fourth author translated the items and the research assistants reviewed the translation and provided feedback. The participants were asked about the readability of the survey, and they reported it was easy to understand.

A cover letter on the questionnaire packet contained informed consent information. The assessment instruments were presented in the following order on the questionnaire (see Table 1 for the means, standard deviations, and Cronbach's alpha reliabilities for each assessment instrument).

Climate change risk perceptions and affect (5 items). We adapted van der Linden's (2015) items to assess level of concern and expectations about experiencing negative consequences as a result of climate change. An example of an item is "In your judgment, how likely is it that climate change will have very harmful, long-term impacts on our society?" We summed the items to create a risk perception score. Higher scores reflect stronger perceived risk as a result of climate change.

We also included two items from van der Linden (2015) to assess affect concerning climate change. An example is "Overall, do you feel that climate change is not favorable?" Higher scores indicate higher negative affect toward climate change. Participants used a 5-point scale (1 = not at all to 5 = very much) to indicate their answers.

Community norms concerning climate change (7 items). We used van der Linden's (2015) items to assess community norms concerning climate change. An example is "Most people I care about are doing their bit to help slow climate change." Participants used a 5-point scale (1 = not at all to 5 = very much) to indicate their answers. Higher scores indicate more support for actions to reduce climate change. We summed the items.

Personal value orientations (12 items). We used van der Linden's (2015) items to assess biospheric, socio-altruistic, and egoistic values. Participants used a 6-point scale (1 = very much opposed to 6 = very important) to indicate how well each item matched their personal values. An example is "Protecting the environment and preserving nature." Higher scores indicate more personal importance.

Posttraumatic growth (21 items). We used Calhoun and Tedeschi's (2001) Posttraumatic Growth Inventory to assess degree of growth. An example is "Greater appreciation for the value of my own life." Participants used a 7-point scale (1 = great decrease to 4 = no change to 7 = great increase). Higher scores indicate greater degree of posttraumatic growth. We summed the items.

Behavioral intention to prevent climate change (12 items). We used items developed by the first and fifth authors to assess willingness to take action regarding climate change (Sattler et al. 2018a). The items include actions that may reduce or prevent climate change and were derived, in part, from actions discussed by Stern (2010), such as household behaviors and sustainable practices. An example is "Use less energy at home to help stop climate change." Participants used a 5-point scale (1 = not at all to 5 = very much). Higher scores indicate greater willingness to take action. We summed the items.

Table 1 Descriptive statistics and correlations (N = 230)

Variable	1	2	3	4	5	6	7	8	
(1) Education	–								
(2) Resource loss	0.07								
(3) Posttraumatic stress	0.02	0.61							
(4) Coping	0.04	0.41	0.33						
(5) Posttraumatic growth	−0.02	0.41	0.57	0.41					
(6) Personal value orientation	0.16	0.06	−0.08	0.28	0.04				
(7) Community norms	0.15	0.40	0.26	0.56	0.33	0.33			
(8) Risk perception	0.09	0.21	0.18	0.47	0.23	0.36	0.46		
(9) Behavioral intent	0.08	0.32	0.11	0.52	0.13	0.37	0.45	0.57	
Mean	2.97	2.54	43.32	3.16	2.95	5.03	3.06	3.81	3.63
Standard deviation	1.04	1.04	15.13	0.79	1.29	1.21	0.94	1.10	1.22
Cronbach's alpha	–	0.93	0.94	0.89	0.97	0.95	0.89	0.89	0.96

Note Correlations greater than 0.23 are statistically significant at the $p < 0.001$ level, greater than 0.18 at the $p < 0.01$ level, and greater than 0.13 at the $p < 0.05$ level

Demographics. Participants reported their gender, age, and level of education. Participants checked their choices or wrote in a number to indicate their answers.

Results

Model Building Data Analytic Plan and Procedures

Table 1 presents descriptive statistics and correlations among the variables.

To develop a path model to examine predictors of climate change risk perception and behavioral intention, we split the entire sample in half to create an exploratory sample ($N = 116$) and a confirmatory sample ($N = 114$). We then created a path model with the exploratory sample and assessed its generalizability with the confirmatory sample. This approach allowed us to confirm that the initial results with the exploratory sample were not solely the result of sampling error. This approach replicated our model building method to examine climate change risk perceptions following Cyclone Winston in Fiji (Sattler et al. 2018a). We performed the path analyses with maximum likelihood estimates using Amos version 24 (Arbuckle 2016).

We used object loss and personal characteristic/condition loss as congeneric indicators of loss, a latent variable. We averaged responses to the 20 posttraumatic stress items to create a single variable. We used the coping and social support variables as congeneric indicators of a latent coping variable. We averaged responses to the 21 posttraumatic growth items to create a single variable. Finally, we averaged responses to the seven risk assessment items to create a single climate change risk assessment variable. We allowed these five primary variables to correlate with one another and tested model fit using the exploratory sample. Table 2 shows that all fit indices of the measurement model indicated excellent fit. Because the measurement model allows all variables to correlate with one another, it represents the best possible fit for any subsequent structural models.

Table 2 Model fit statistics for structural regression models predicting climate change risk perception with loss, posttraumatic stress, coping, and posttraumatic growth

Sample	Model	X^2	df	*p*	SRMR	CFI	RMSEA
Exploratory (N = 116)	Measurement	15.54	13	0.28	0.038	0.993	0.041
	Structural (Fig. 1)	20.27	18	0.32	0.044	0.994	0.033
Confirmatory (N = 114)	Structural (Fig. 1)	36.24	19	0.01	0.060	0.035	0.095

Note SRMR—Standardized Root Mean Residual, CFI—Comparative Fit Index, RMSEA—Root Mean Square Error of Approximation

Climate Change Risk Perceptions

To examine climate change risk perceptions, we re-specified the measurement model into a structural regression model and converted correlations between variables into regression paths and added error terms to endogenous variables. We used loss to predict posttraumatic stress; loss and posttraumatic stress to predict coping; loss, posttraumatic stress and coping to predict posttraumatic growth; and all four other variables to predict climate change risk perception. Because this model allowed all variables to relate to one another, its fit indices were identical to those for the exploratory sample measurement model shown in Table 2.

To simplify the model, we examined each of the estimates from the structural regression model and removed the path with the lowest critical ratio. We then tested the fit of the resulting model and compared it to the fit of the measurement model. We continued this process until the change in fit indicated an ill-fitting model. We removed the following paths in order: posttraumatic stress to coping, posttraumatic growth to climate change risk perception, loss to posttraumatic growth, loss to climate change risk perception, and posttraumatic stress to climate change risk perception. At that point, the removal of the next path (coping to posttraumatic growth) would have resulted in a deterioration of fit. Figure 1 shows the resulting path diagram for the exploratory sample structural regression model, and Table 2 shows the fit indices. As shown in Table 2, the fit of the final structural model was excellent, and it was not statistically different from the measurement model, ΔX^2 (5) = 4.73. $p = 0.449$.

We then tested the fit of the model shown in Fig. 1 with the data from the confirmatory sample. Figure 1 shows the path coefficients and Table 2 shows the resulting fit indices. As seen here, the fit was generally acceptable, but not as strong as in the initial exploratory sample. The Chi-Square was statistically significant, but the other fit indices indicated a good, but not excellent fit.

The resulting model shown in Fig. 1 indicates that the effect of loss on posttraumatic growth was mediated by two variables. First, the effect of loss on posttraumatic growth was mediated by posttraumatic stress symptoms. Individuals experiencing greater levels of loss were more likely to experience posttraumatic stress symptoms, which in turn was associated with higher levels of posttraumatic growth. Second, the effect of loss on posttraumatic growth was mediated by coping. Individuals experiencing greater levels of loss were more likely to engage in coping behaviors, which in turn was associated with higher levels of posttraumatic growth. It is noteworthy that the experience of posttraumatic stress was not related to coping behaviors; this stands in contrast to the model from Sattler et al. (2018a), where the effect of loss on posttraumatic growth was both direct and mediated through posttraumatic stress and coping.

The results shown in Fig. 1 also indicate that posttraumatic growth was not associated with climate change risk perception. Thus, in the current sample, re-evaluation of life priorities associated with posttraumatic growth did not result in increased evaluation of climate change risk.

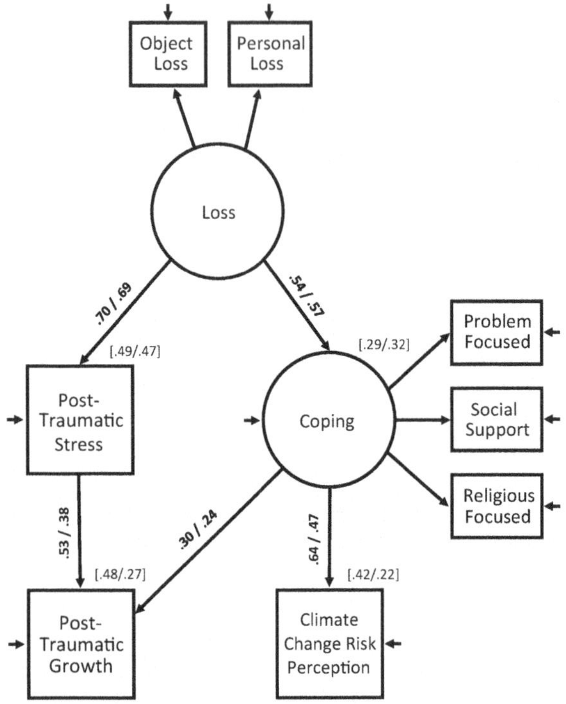

Fig. 1 Path diagram of the structural regression model predicting climate change risk perception. Path coefficients are from the exploratory/confirmatory samples. The squared multiple correlations in brackets are from the exploratory/confirmatory samples

Figure 1 shows that the effect of loss on climate change risk perceptions was mediated by coping. The more loss participants experienced, the more coping behaviors they displayed, which is in turn related to increased risk perceptions. Thus, it seems that the more coping resources an individual was required to use in the face of loss, the more risky they judged the impact of climate change.

Climate Change Behavioral Intention

The perception that climate change creates substantial risks does not always result in intention to take action to prevent climate change. To test how the variables predict climate change behavioral intention, we tested the same model shown in Fig. 1 and replaced the climate change risk perception variable with the climate change behavioral intention variable. We first tested the model using the exploratory sample. The resulting fit indices, shown in Table 3, indicated marginal fit. While the Chi-Square was statistically significant (indicating poor fit), it was only just so, and the

Table 3 Model fit statistics for structural regression models predicting climate change behavioral intention with loss, posttraumatic stress, coping, and posttraumatic growth

Sample	Model	X^2	df	p	SRMR	CFI	RMSEA
Exploratory (N = 116)	Without PTG to BI	31.20	18	0.03	0.055	0.966	0.080
	With PTG to BI	19.99	17	0.28	0.04	0.992	0.039
Confirmatory (N = 114)	With PTG to BI	28.91	17	0.05	0.056	0.957	0.079
	Without PTG to BI	30.51	18	0.03	0.058	0.955	0.078

Note SRMR—Standardized Root Mean Residual, CFI—Comparative Fit Index, RMSEA—Root Mean Square Error of Approximation, PTG—Posttraumatic Growth, BI—Behavioral Intention

CFI indicated excellent fit and the SRMR and RMSEA both indicated acceptable fit. The modification indices suggested that adding a path from posttraumatic growth to behavioral intention would improve the model fit. Doing so resulted in the fit indices shown in Table 3, all of which indicate good to excellent fit.

Next, we tested the model with the confirmatory sample. The resulting fit indices, shown in Table 3, were generally good; though the Chi-Square was statistically significant (indicating poor fit), it was only just so, the CFI indicated excellent fit and the SRMR and RMSEA both indicated acceptable fit. In the confirmatory sample, the path from posttraumatic growth to behavioral intention was not statistically significantly different from zero; as such, we tried removing that path and re-running the model with the confirmatory sample. The fit, shown in Table 3, was not meaningfully impacted by the removal of the path (ΔX^2 (1) = 1.60. $p = 0.205$), suggesting weak support for the effect of posttraumatic growth on behavioral intention.

Figure 2 shows the resulting model, with path coefficients from the exploratory and confirmatory samples. As seen here, the results are virtually identical to those using climate change risk perception. The experience of loss was associated with increased use of coping behaviors, which in turn was associated with increased behavioral intention to take action to prevent climate change. The effect of loss on posttraumatic growth was mediated by posttraumatic stress and coping.

Support for the path from posttraumatic growth to behavioral intention was less consistent. While its inclusion in the exploratory sample changed the fit of the model from good to excellent, its exclusion from the confirmatory sample did not have an appreciable effect on model fit. When present, it was a negative path, suggesting that higher levels of posttraumatic growth was associated with fewer behavioral intention to act against climate change. Because of the lack of clear support for this path, we did not include it in further analyses.

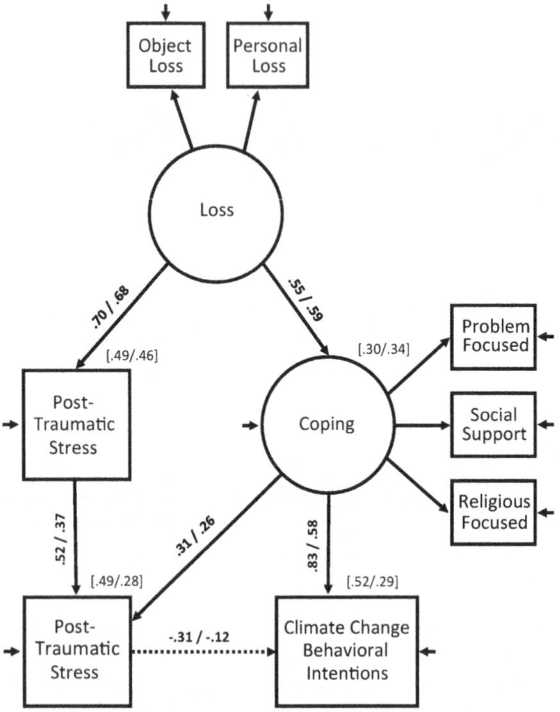

Fig. 2 Path diagram of the structural regression model predicting climate change behavioral intentions. Path coefficients are from the exploratory/confirmatory samples. The squared multiple correlations in brackets are from the exploratory/confirmatory samples. The path from post-traumatic growth to climate change behavioral intentions could be omitted in the confirmatory sample without affecting the overall fit of the model

Community Norms and Personal Values

The previous analyses indicated that coping fully mediated the effect of loss on behavioral intention. In the final set of analyses, we assessed if the addition of the community norms and personal values variables from van der Linden's (2015) climate change risk perception model might act as predictors of behavioral intention. We created a structural regression model, using the loss latent variable, the community norms measured variable, and personal values variable as correlated exogenous predictors. We allowed these three variables to predict the latent coping variable. All four of the preceding variables were then allowed to predict behavioral intention.

We tested this model using the exploratory sample. Table 4 shows the resulting fit indices. The results were mixed. Though the Chi-Square was statistically significant (indicating poor fit), it was only just so; the CFI and SRMR indicated excellent fit, and the RMSEA indicated poor fit. The modification indices suggested that fit might be improved by allowing the error terms for object loss and social support to correlate.

Table 4 Model fit statistics for structural regression models predicting climate change behavioral intention with norms, values, loss, and coping

Sample	Model	X^2	df	*p*	SRMR	CFI	RMSEA
Exploratory (N = 116)	Initial	27.719	14	0.037	0.049	0.967	0.082
	(Fig. 3)	14.728	9	0.099	0.042	0.979	0.074
Confirmatory (N = 114)	(Fig. 3)	16.036	9	0.066	0.043	0.972	0.083

Note SRMR—Standardized Root Mean Residual, CFI—Comparative Fit Index, RMSEA—Root Mean Square Error of Approximation

This would suggest that those who experienced the most loss of property and personal possessions were the most likely to make use of social support, presumably indicating a use of community relationships to cope with property loss. The critical ratios also suggested that both the path from community norms to behavioral intention and the correlation between loss and personal values could be removed. Table 4 shows the fit indices for this model with the exploratory sample, which were all good to excellent.

We tested the final model with the confirmatory sample. Table 4 shows the fit indices, which were all good to excellent, with the exception of the RMSEA. Because the RMSEA was only just outside the cut-off for acceptable fit, we deemed this final model acceptable.

Figure 3 presents the path coefficients for the final model for the exploratory and confirmatory models. Personal values had a direct effect on behavioral intention, with personal values more in line with biospheric and altruistic beliefs, and less in line with egoistic beliefs being related to more behavioral intention to act against climate change. Community norms did not directly predict behavioral intention; rather, the effect of community norms on behavioral intention was mediated by coping. Those reporting being from communities with norms supporting climate change action were more likely to use social support and problem focused coping, which was in turn related to a reported increased likelihood of taking action against climate change. Thus, while personal values both directly and indirectly (through coping) affect behavioral intention, community norms relate to behavioral intention through coping.

Discussion

This study is among the first, if not the first, to develop a model of climate change risk perception and behavioral adaptation in Tonga and the Pacific. The findings show the relationships among resource loss, distress, coping, posttraumatic growth, community norms, and personal values with climate change risk perceptions and climate change behavioral adaptation. Two models with similar paths predicated climate change risk perceptions (Fig. 1) and behavioral intention to prevent climate change (Fig. 2). In one path, loss of resources as a result of Cyclone Gita was associated with

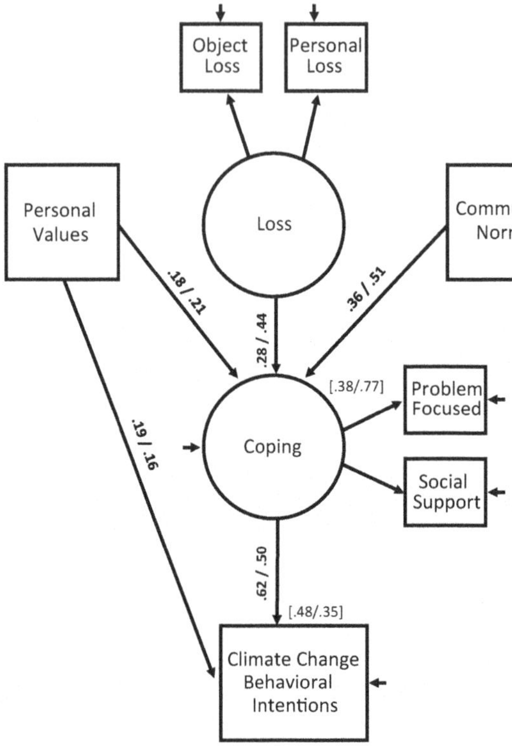

Fig. 3 Path diagram of the structural regression model predicting climate change behavioral intentions with community norms and personal values. Path coefficients are from the exploratory/confirmatory samples. The squared multiple correlations in brackets are from the exploratory/confirmatory samples. For the sake of clarity, we omitted the correlations between personal values and community norms, community norms and loss, and the error terms for object loss and social support

coping, and coping was associated with climate change risk perceptions. In a second path, loss of resources as a result of Cyclone Gita was associated with posttraumatic stress, and posttraumatic stress was associated with posttraumatic growth.

An additional analysis examined the role of personal values and community norms in predicting behavioral intention to take action to prevent climate change. In one path, personal values were directly associated with behavioral intention to take action to prevent climate change. In a second path, coping mediated resource loss as a result of Cyclone Gita, community norms, and personal values (Fig. 3). Participants reporting stronger community norms supporting climate change action and stronger personal value orientation were more likely to report higher levels of coping, which then was associated with behavioral intention to take action to prevent climate change. These findings support, in part, Roeser et al.'s (2012) model identifying factors that influence climate change behavior. These factors include beliefs and concern for climate change, self-efficacy, distress, perceived risk, feelings of responsibility, and adaptive behavior. The findings also support research showing the connection between disaster experience with climate change related events and climate change risk perceptions (Sattler et al. 2018a; Viscusi and Zeckhauser 2015).

The finding that personal value orientation was directly associated with behavioral intention to take action to prevent climate change supports and extends van der Linden (2015), and supports and extends our climate change models developed in Fiji. The personal value orientation variable in the present study included both biospheric values (concern about Earth's natural environment) and social-altruistic values (concern about the well-being of others). Unlike van der Linden, we included both values in one variable given the strong correlation between them. van der Linden found that biospheric values were directly associated with climate change risk perceptions, and the present study further found that these values are also directly associated with behavioral intention to take action to prevent climate change.

van der Linden suggests that broad value orientations may reflect the culture in which one lives and one's view the world. In Tonga, these personal value orientations might be based, in part, on a strong sense of community, shared identity, sense of shared history and future, traditional environmental knowledge, reliance on locally produced food items, and understanding of vulnerability to natural disasters and impending environmental threats as a result of climate change (cf. Barnett and Waters 2016; Thaman et al. 2014). Future research could examine in detail experiences that contribute to and support the development of these personal value orientations.

We also found that coping mediated the relationship between personal value orientation, community norms, and resource loss with behavioral intention to take action to prevent climate change. The coping variable includes problem focused coping and social support. Because Cyclone Gita threatened lives, livelihoods, and property, this finding may reflect the experience of living through Cyclone Gita and heightened awareness of the need to take action to reduce threats to life and property brought on by climate change. It also may reflect increased awareness as a result of steadfast educational efforts about climate change before and after Cyclone Gita made landfall. The Tonga Ministry of Meteorology, Energy, Information, Disaster Management, Environment, Climate Change and Communications and schools in Tonga provided educational information and lesson plans to raise awareness about factors that contribute to climate change, threats to Tonga and other Pacific island nations, and actions to take to minimize or prevent climate change.

The Kingdom of Tonga has developed strategies to reduce emissions that contribute to climate change and to increase resilience to threats associated with climate change. Tonga ratified the Paris Agreement and actively participates in United

Nations efforts to address climate change (Department of Climate Change 2019). These campaigns emphasize the importance of the community working together and the contributions each individual can make. Media reports in newspapers and television programming regularly inform citizens about these efforts, and conversations among community members may reinforce this information. The connection between personal values and community norms with behavioral intention to take action to prevent climate change may clarify, in part, the lack of association with posttraumatic growth with behavioral intention (cf. Lindell and Perry 2012). Future research could examine how educational activities about climate change can activate coping behaviors to confront climate change threats and reduce contributions to climate change.

The finding that resource loss predicted posttraumatic stress supports prior research, conservation of resources stress theory, and our climate change model (Hobfoll 2012; Sattler 2017; Sattler et al. 2018a). The finding that posttraumatic stress is associated with posttraumatic growth is consistent with prior research. As a result of life threat and resource loss, it is common for people to reflect on what gives their life meaning. This process may be a form a coping with the traumatic event (Sattler and Smith, 2020).

Regional characteristics may explain, in part, similarities and differences among climate change risk perception and behavioral adaption models we developed in Tonga and Fiji. For example, Fiji has a larger economy and a population almost nine times as large as Tonga. The highest point on Tongatapu, the main island in Tonga, is 213', whereas the highest point on Vita Levu, the main island in Fiji, is 3645'. People living in coastal villages on Tongatapu and the other islands may have fewer options to evacuate to higher areas in order to escape storm surge or to relocate as a result of sea level rise. The degree of damage caused by a disaster associated with climate change and aid provided by disaster agencies may vary by location (Sattler et al. 2002). Future research might examine how geography, economy, population size, and the threats associated with climate change influence behavioral intention to take action to prevent climate change.

The study has a few limitations. The sample is not a pure random sample. The findings may not generalize to all persons in Tonga who experienced the cyclone. Interrater reliability is not known.

Conclusion

This study is among the first to develop a model of climate change risk perception and behavioral adaptation in Tonga, a small island developing state. The findings underscore the multidimensional nature of climate change risk perception and behavioral adaption, and the role of experiential, cognitive, affective, and socio-cultural factors, and community norms. The findings also illustrate how exposure to a disaster whose strength is associated with climate change can influence climate change risk perceptions and behavioral adaptation. Future research exploring other variables that contribute to climate change risk perception is warranted. It may be particularly useful to examine how cultural variables influence risk perception and willingness to modify behaviors to mitigate climate change.

Given the role of community norms and personal values in the climate change risk perception and behavioral adaptation model, an implication of the findings is that educational campaigns should consider developing ways to highlight them. In particular, campaigns might emphasize descriptive and injunctive community norms. Descriptive norms indicate what others in the community are currently doing (e.g., to mitigate climate change), and injunctive norms indicate what community members should do (e.g., to mitigate climate change; Cialdini 2003). People are more likely to comply with information when they see others are concerned and when there is an expectation that people in the community are engaged in behaviors that mitigate climate change. In addition, campaigns should develop ways to present information in vivid ways that appeal to emotions (Roeser 2012; Sattler et al. 2014, 2018a).

A person's understanding about prior experiences, or mental representation, influences how the individual perceives, understands, and interprets new information and events. A mental representation of an event is vivid when the event is easy to remember and has emotional impact. Vivid information is likely to have more influence over judgments and decision-making, in part, because it is easier to remember than abstract or statistical information (Fiske and Taylor 1991). An information campaign might vividly describe the threats climate change poses to small island developing states, present personal stories of people who experience loss and the actions they are taking to minimize climate change, and present captivating imagery. The finding that resource loss played a critical role underscores the importance of clearly presenting the adverse consequences of climate change in an education campaign. This is especially important for people who have little to no experience with, or who are not aware of, the way in which climate change is negatively impacting ecosystems that they and others depend on for survival.

In order to increase their effectiveness, information campaigns could also be designed to present information in a manner that matches socio-cultural value orientations. For example, prescriptive messages suggesting actions to mitigate climate change might present one way for individuals with biospheric values, another for those with social-altruistic values, and a third for those with egoistic values. Whereas the recommended behaviors in each message might be the same, the arguments would differ depending on the value orientation that is targeted (Sattler and Kerr 1991). These campaigns also should consider barriers that might prevent people from taking action, and design specific remedies to overcome the barriers. Future research should develop and assess the effectiveness of various approaches to educate the public about climate change.

Acknowledgements This project was supported by funding from the Department of Psychology and the Office of Research and Sponsored Programs at Western Washington University. We thank Sione Fulivai, Folauhola Helu, Lau'aitu Lavulo, Saia Tahaafe, and Mehnaaz Sattler for their invaluable assistance, and extend our gratitude to the participants and communities in Tonga for participating in this project.

References

Alliance Development Works (2013) World risk report 2013. Retrieved from http://collections.unu.edu/eserv/UNU:2018/WorldRiskReport_2013_online_01.pdf

Arbuckle JL (2016) Amos (Version 24.0) [Computer Program]. IBM SPSS, Chicago

Australia Department of Foreign Affairs and Trade (2018) Tropical Cyclone Gita, 19 June 2018. Retrieved from https://dfat.gov.au/aid/topics/investment-priorities/building-resilience/humanitarian-preparedness-and-response/Pages/cyclone-gita.aspx

Barnett J, Waters E (2016) Rethinking the vulnerability of small island states: climate change and development in the Pacific Islands. In: Grugel J, Hammett D (eds) The Palgrave handbook of international development. Palgrave Macmillan, pp 731–748. https://doi.org/10.1057/978-1-137-42724-3_40

Calhoun LG, Tedeschi RG (2001) Posttraumatic growth: the positive lessons of loss. In: Neimeyer RA (ed) Meaning reconstruction and the experience of loss. American Psychological Association, Washington, DC, pp 157–172. https://doi.org/10.1037/10397-008

Cialdini R (2003) Crafting normative messages to protect the environment. Curr Dir Psychol Sci 12:105–109. https://doi.org/10.1111/1467-8721.01242

Department of Climate Change (2019) Pacific countries convene in Apia to prepare for the climate change conference in Madrid. Retrieved from https://www.climatechange.gov.to/news/pacific-countries-convene-apia-prepare-climate-change-conference-madrid

Earth Observatory (2016) Tropical cyclone Winston Slams Fiji. Retrieved from https://earthobservatory.nasa.gov/NaturalHazards/view.php?id=87562

Fiske ST, Taylor SE (1991) Social cognition, 2nd edn. McGraw-Hill, New York

George S, Sidhu S (2018) Worst storm in 60 years ravages Tonga. CNN, 13 Feb 2018. Retrieved from https://www.cnn.com/2018/02/13/asia/cyclone-gita-worst-tonga-intl/index.html

Hobfoll SE (2012) Conservation of resources and disaster in cultural context: the caravans and passageways for resources. Psychiatry 73:227–233. https://doi.org/10.1521/psyc.2012.75.3.227

Laffoley D, Baxter JM (2016) Explaining ocean warming: causes, scale, effects and consequences. Full report. Gland, Switzerland: IUCN

Lindell M, Perry RW (2012) The protection action decision model: theoretical modifications and additional evidence. Risk Anal 32:616–632

Masters J (2019) Hurricane Dorian was worthy of a category 6 rating: the category-1-to-5 Saffir-Simpson scale for rating hurricanes is inadequate. Scientific American, 3 Oct 2019. Retrieved from https://blogs.scientificamerican.com/eye-of-the-storm/hurricane-dorian-was-worthy-of-a-category-6-rating/

Matsumoto D, van de Vijver FJR (2011) Cross-cultural research methods in psychology. Cambridge University Press, New York

National Center for PTSD (2017) What is PTSD? Retrieved from https://www.ptsd.va.gov/public/ptsd-overview/basics/what-is-ptsd.asp

National Oceanic and Atmospheric Administration (2013) Hurricanes form over tropical oceans, where warm water and air interact to create these storms. Retrieved from http://oceanexplorer.noaa.gov/facts/hurricanes.html

Nurse LA et al (2014) Small islands. In: Barros VR et al (eds) Climate change 2014: impacts, adaptation, and vulnerability. Cambridge University, pp 1613–1654. https://doi.org/10.2134/jeq2008.0015br

Pacific Climate Change Science Program (2016) Current and future climate of Tonga. Retrieved from https://www.pacificclimatechange.net/document/current-and-future-climate-tonga-0

Pearlman J, Krol C (2018) Fiji PM warns of 'frightening new era' as he blames deadly cyclone on climate change. The Telegraph, 3 April 2018. Retrieved from https://www.telegraph.co.uk/news/2018/04/03/fiji-pm-warns-frightening-new-era-blames-deadly-cyclone-climate/

Petzold J, Magnan AK (2019) Climate change: thinking small islands beyond small island developing states (SIDS). Climactic Change 152:145–165. https://doi.org/10.1007/s10584-018-2363-3

Roeser S (2012) Risk communication, public engagement, and climate change: a role for emotions. Risk Anal 32:1033–1040. https://doi.org/10.1111/j.1539-6924.2012.01812.x

Sattler DN (2017) Climate change and extreme weather events: the mental health impact. In: Leal Filho W (ed) Climate change adaptation in Pacific countries: fostering resilience and up-keeping life quality. Springer, Berlin, pp 73–87. https://doi.org/10.1007/978-3-319-50094-2_4

Sattler DN, Kerr NL (1991) Might versus morality explored: motivational and cognitive bases for social motives. J Personal Social Psychol 60:756–765

Sattler DN, Smith A (2020) Facilitating posttraumatic growth in the wake of natural disasters: considerations for crisis response. In Schulenberg SE (ed) Disaster mental health and positive psychology. Springer, Berlin

Sattler DN, Preston A, Kaiser CF, Olivera VE, Valdez J, Schlueter S (2002) Hurricane georges: a cross-national study examining preparedness, resource loss, and psychological distress in the U. S. Virgin Islands, Puerto Rico, Dominican Republic, and the United States. J Trauma Stress 15:339–350

Sattler DN, Assanangkornchai S, Moller A, Kesavatana-Dohrs W, Graham JM (2014) Indian Ocean tsunami: relationships among posttraumatic stress, posttraumatic growth, resource loss, and social support at three and fifteen months. J Trauma and Dissoc Syst Trauma Special Issue 15:219–239. https://doi.org/10.1080/15299732.2014.869144

Sattler DN, Whippy A, Graham JM, Johnson J (2018a) A psychological model of climate change adaptation: Influence of resource loss, posttraumatic growth, norms, and risk perception following Cyclone Winston in Fiji. In: Leal Filho W (ed) Climate change impacts and adaptation strategies in coastal communities. Springer, Berlin

Sattler DN, Claramita M, Muskavage B (2018b) Natural disasters in Indonesia: relationships among posttraumatic stress, resource loss, depression, social support, and posttraumatic growth. J Loss Trauma 23. https://doi.org/10.1080/15325024.2017.1415740

Stern PC (2010) Environmentally significant behaviour and how to change it. Keynote address to the National Congress on Behaviour Change for Sustainability, Sydney, Australia, Oct 2010

Tapaleao V, Dunlop R (2018). Gita's fury: tropical cyclone hits Tonga and there are now fears for parts of Fiji. New Zealand Herald, 13 Feb 2018. Retrieved from https://www.nzherald.co.nz/world/news/article.cfm?c_id=2&objectid=11993109

Thaman R, Balawa A, Fong T (2014) Putting ancient winds into new sails: Indigenous knowledge as a basis for education for sustainable development. In: Otunuku M, Nabobo-Baba U, Johansson Fua S (eds) Of waves, winds and wonderful things: a decade-of rethinking pacific education. University of the South Pacific Press, Fiji

United Nations Development Programme (2017) Small Island nations at the frontline of climate action, 18 Sept 2017. Retrieved from https://www.undp.org/content/undp/en/home/presscenter/pressreleases/2017/09/18/small-island-nations-at-the-frontline-of-climate-action-.html

United Nations News (2018) At UN, Pacific Island leaders warn climate change poses dire security threat to their fragile countries and marine resources, 26 Sept 2018. Retrieved from https://news.un.org/en/story/2018/09/1020811

van de Vijver FJR, Leung K (1997) Methods and data analysis for cross-cultural research. Sage, London

van der Linden S (2015). The social-psychological determinants of climate change risk perceptions: towards a comprehensive model. J Environ Psychol 41:112–124. https://doi.org/10.1016/j.jenvp.2014.11.012

Viscusi WK, Zeckhauser RJ (2015) The relative weights of direct and indirect experiences in the formation of environmental risk beliefs. Risk Anal 35:318–331. https://doi.org/10.1111/risa.12271

David N. Sattler is a social psychologist and Professor of Psychology at Western Washington University. He conducts projects around the world that examine climate change risk perceptions and behavioral adaptation, and preparedness for and recovery following natural disasters. He develops interventions and produces educational videos to educate citizens about climate change and disaster threats. In Thailand, he established the International Tsunami Museum following the Indian Ocean Tsunami. He has been a delegate for the Kingdom of Tonga to multiple United Nations Climate Change Conferences. He is a Behavioral and Social Science Editor for Natural Hazards Review journal.

Uili Lousi is President of Ohai Foundation Tonga, a nonprofit non-governmental organization devoted to climate change work. He has been a delegate for the Kingdom of Tonga to multiple United Nations Climate Change Conferences and a panel member with U.S. Vice President Al Gore at the Climate Reality Leadership Training in Australia. He also is an internationally recognized artist.

James M. Graham is a Professor of Psychology at Western Washington University. He received his Ph.D. in Counseling Psychology from Texas A&M University. His substantive research focuses primarily on adaptive processes in romantic relationships. His research also focuses on modeling adaptation to stress in diverse populations, including statistically modeling climate change risk perceptions and reactions to natural disasters in Fiji, the Philippines, Tonga, and Mongolia, and examining the applicability of the social buffering model in same-sex couples.

Viliami Uasike Latu is a Legal Practitioner and a Political and Business Consultant at Edwards Law in the Kingdom of Tonga. He received an LL.B. and PDLP from the University of the South Pacific in Fiji, M.A. in Political Science from Auckland University in New Zealand, and a Ph.D. in Asia Pacific Studies from Ritsumeikan Asia Pacific University in Japan. He is a former Member of Parliament, a former Cabinet Minister in the Kingdom of Tonga, and also a former Dean of the Tonga International Academy.

James Johnson is a social psychologist and Head of the School of Social Sciences at the University of the South Pacific. His research interests are in interpersonal conflict and all forms of intergroup and intragroup stereotypes. He has secured over 4 million dollars in grant funding for his work, which has been featured in international news outlets such as CNN, Associated Press, and the Washington Post.

Siosaia Langitoto Helu is a Statistician-Mathematician who went to school at 'Atenisi Institute in Tonga, Victoria University of Wellington, University of Utah, and Oregon State University. He taught Statistics and Mathematics in the United States (Utah, Washington), New Zealand (Auckland), and Tonga. He is also a Climate Resilience Consultant in Tonga.

Climate Change Adaptation in the Pacific Islands: A Review of Faith-Engaged Approaches and Opportunities

Johannes M. Luetz and Patrick D. Nunn

Science without religion is lame, religion without science is blind.
(Attributed to Albert Einstein)

Abstract The Pacific Islands region is highlighted in the literature as one of the most vulnerable geographic areas in the world, with a high priority for adaptation to climate change. In consequence, many interventions have been proposed and implemented over the years that approach environmental sustainability and adaptation to climate change in the Pacific from a predominantly scientific and technocratic worldview perspective, in which climate change is seen as a science-informed issue, rather than a faith-informed issue. Overwhelmingly, adaptation initiatives are scientifically justified and externally conceived, funded and implemented. Regrettably, most interventions intended to reduce exposure to environmental risk and to enable effective and sustainable adaptation to climate change in the Pacific Islands region have failed to acknowledge influences on decision-making of spirituality and connectedness to Nature. In the light of the almost total Christianization of Pacific Islands within the past century, such intervention failures are surprising. The situation cannot continue because every day the need for adaptation to climate change that is effective and sustainable is growing. Given that in the Pacific Islands region decision makers are likely to be influenced more by tradition and local precedent than by science, makes

J. M. Luetz (✉)
School of Social Sciences, Adjunct Academic, University of New South Wales (UNSW), Sydney 2052, NSW, Australia
e-mail: jluetz@chc.edu.au; j.luetz@unsw.edu.au

School of Social Sciences, Christian Heritage College, Carindale/Brisbane, QLD 4152, Australia

P. D. Nunn
School of Social Sciences, University of the Sunshine Coast (USC), Locked Bag 4, Maroochydore, QLD 4558, Australia
e-mail: pnunn@usc.edu.au

W. Leal Filho (ed.), *Managing Climate Change Adaptation in the Pacific Region*, Climate Change Management, https://doi.org/10.1007/978-3-030-40552-6_15

the purposive exploration of faith-engaged approaches to climate change adaptation a fertile and promising undertaking. This paper extends previous research by means of a literary review of pertinent challenges and opportunities. The synthesised lessons are useful for both policy and practice serving the cause of climate change adaptation in Pacific island communities. A better understanding of the science-spirituality nexus in the Pacific will also improve the effectiveness and sustainability of adaptation responses to climate-driven environmental change.

Keywords Pacific islands · Christianisation · Science-spirituality nexus · Climate change · Adaptation · Consilience · Literary review

Science and Faith in the Pacific

The nexus between climate science and religion has garnered attention over the years as a fertile area of scholarly interest (Bergmann and Gerten 2010; Gerten and Bergmann 2012; Jenkins et al. 2018). There are good reasons for this. Roberts (2012) notes the link between Evangelicals in the United States and their influence on the politics of the Republican Party (pp. 107–131). His treatise has been synthesised by Veldman (2013) as follows: "evangelical climate skepticism drives skepticism in the United States' Republican party, rather than the influence working in the opposite direction, from conservative political spheres to religious ones" (p. 178). Following this argument, faith persuasion exerts significant influence on environmental concern, which works its way from belief to action, rather than from action to belief (Roberts 2012). A similar point has been repeatedly highlighted in correlational analyses (e.g., Morrison et al. 2013; Hagevi 2014; Arbuckle and Konisky 2015). While the findings of the studies are nuanced, there seems to be a common denominator that "attitudes to climate change and climate change policy differ across religious groups" (Morrison et al. 2013, p. 13).

The findings offer insights into denominational and geographical influences in respect of the science-faith nexus. Analysing the United States context, Arbuckle and Konisky (2015) conclude that "individuals in religious traditions that are more prone to teach biblical literalism are less likely to express high degrees of concern about the environment … Thus, for many, religiosity tends to move people even further away from stronger environmental attitudes" (p. 1260). For the Australian context, regression, factor and latent class analyses conducted by Morrison et al. (2013) suggest that "Christian Literalists (30.2%) are … more likely to have a much higher percentage of people located in the Doubtful and Dismissive segments than the sample overall (20.6%)" (p. 7). Finally, Hagevi (2014) investigated "environmental opinion in 22 countries in and near Europe" (p. 91) and concluded that "[t]he public opinion in countries dominated by Catholic and Eastern Orthodox culture tends to show a high level of environmental concern. The public opinion in predominantly Protestant countries shows lower levels of environmental concern" (Hagevi 2014, p. 105).

Notwithstanding some rather striking similarities in the findings of the three studies, all three investigations distance themselves to different degrees from the absolute human 'dominionist' hypothesis propagated by White (1967) who cautioned that "Christianity is the most anthropocentric religion the world has seen" (p. 1205) and that humans "shall continue to have a worsening ecologic crisis until we reject the Christian axiom that nature has no reason for existence save to serve man" (p. 1207).

In summary, there is a clear sense in correlational analyses that religiosity drives environmental action (or inaction), rather than vice versa. This synthesis holds important implications, given that religious 'worldview' effectively underpins how perspectives on nature, the world and the universe are formed, framed and propagated as being factually sound. One needs only to remember the virulent opposition Nicolaus Copernicus faced from the religious establishment of his day when he formulated and 'framed' a view of the world that placed the sun rather than the earth at the centre of the universe. The example of the famous Renaissance-era astronomer aptly reflects the fact that science can face formidable opposition when it challenges the established status quo and runs counter to entrenched religious worldviews (Clarke and Wilson 2011).

The example of Copernicus resonates with Nisbet's (2009) work on the ubiquity of 'framing'—whether done implicitly or explicitly, intentionally, tacitly or intuitively. He posits: "There is no such thing as unframed information … Frames are interpretive storylines that set a specific train of thought in motion, communicating why an issue might be a problem, who or what might be responsible for it, and what should be done about it" (p. 15). Furthermore, according to Kearns (2012), "frames provide a shorthand for believers, citizens, journalists, policymakers and others who count on the reader or listener making associations that can be left unsaid" (p. 141). According to Nisbet (2009),

> Framing … is not synonymous with placing a false spin on an issue, although some experts, advocates, journalists, and policymakers certainly spin evidence and facts. Rather, in an attempt to remain true to what is conventionally known about an issue, as a communication necessity, framing can be used to pare down information, giving greater weight to certain considerations and elements over others. (p. 16)

Tracing the beginning of research on 'framing' to the anthropologist Goffman (1974), Nisbet (2009) summarises early work in this area as "helping individuals negotiate meaning through the lens of existing cultural beliefs and worldviews" (p. 16). Goffman's work is reminiscent of the 'Cultural Iceberg' model, popularised around the same time by Hall (1976). It shows that visible human behaviour (surface culture) tends to be forcefully underpinned by invisible core values, worldviews and religious beliefs that may be concurrently imperceptible and intransigent (deep culture). Expressed in simple language, the small and visible part of the iceberg (above the waterline) reflects the 'What', while the larger invisible part (below the surface) conceals the 'Why' (Hall 1976).

Nisbet (2009) also refers to the Nobel Prize-winning research conducted by Kahneman and Tversky in the 1970s to reiterate the important point that "perception is reference dependent" (Kahneman 2003, p. 1454). In other words, it is impossible to

Natural Sciences-Informed Paradigm (Information impinges on human behaviour)	**Social Sciences-Informed Paradigm (Context + meanings motivate human behaviour)**
Advocated and supported by natural/physical sciences (contested by "soft sciences")	Advocated and supported by social sciences (contested by "hard sciences")
Supported by positivist paradigm (truth/knowledge objectively knowable)	Supported by interpretivist/constructivist paradigms (truth/knowledge subjectively constructed)
Positivism – "the belief that objective accounts of the world can be given, and that the function of science is to develop descriptions and explanations in the form of universal laws – that is, to develop nomothetic knowledge." (Punch 2014, p. 17)	*Interpretivism* – "concentrates on the meanings people bring to situations and behaviour, [which] are essential to understanding behaviour" (Punch 2014, p. 17) *Constructivism* – "realities are local, specific and constructed ... socially and experientially based" (ibid)
Research methodology preference > Quantitative research (theory verification; deductive) (Bryman 2016)	*Research methodology preference* > Qualitative research (theory generation; inductive) (Bryman 2016)
Nomothetic view > "sees generalised knowledge, universal laws and deductive explanations, based mainly on probabilities derived from large samples, and standing outside the constraints of everyday life." (Punch 2014, p. 33)	*Ideographic view* > "sees nomothetic knowledge as insensitive to local, case-based meanings, and directs attention rather to the specifics or particular cases. It prefers to see knowledge as local and situated." (Punch 2014, p. 33)
Human behaviour is predictable and predicated on knowledge of scientific facts, hard evidence and robust peer reviewed information (Boyes and Stanisstreet 2012)	Human behaviour is unpredictable and psychologically, socially, contextually and situationally constituted (Sapolsky 2017, Kelly and Barker 2016)

Fig. 1 Epistemological Paradigms: Natural Sciences and Social Sciences. *Source* Adapted from Luetz et al. (2020, p. 3); cf. Punch (2014), Bryman (2016), Boyes and Stanisstreet (2012), Kelly and Barker (2016), Sapolsky (2017)

arrive anywhere from nowhere. This principle is well-understood by social scientists who have stressed for years the importance of local context for meaning-making. This is most clearly underscored by way of the epistemological paradigm of "constructivism" (advocated by the social sciences), which stands in marked contrast to the paradigm of "positivism" (advocated by the natural/physical sciences) (Punch 2014; Bryman 2016). The differences between these epistemological approaches to 'truth-seeking' have been explored in the context of behaviour change for sustainability (Luetz et al. 2020) (Fig. 1) and are also deeply ingrained in disciplinary approaches to research and knowledge creation (Chen and Luetz 2020).

Accordingly, social scientists are inherently uncomfortable with the epistemology of positivism (as espoused by the natural/physical sciences), which is premised on objective 'truth' existing independently in the world, waiting to be deductively 'verified' by impartial and detached scientists through hypothesis-testing. In contrast, social scientists prefer the epistemologies of interpretivism and/or constructivism, which are based on the premise that 'truth' about social issues is subjectively 'constructed' by people and may be described qualitatively by locally immersed researchers through a process of inductive theory-generation (Punch 2014; Bryman 2016; Luetz et al. 2020). The disagreements between the two opposing paradigmatic schools of thought have on occasion been fierce, and Punch (2014) has even used the term "paradigm wars" (p. 15) to characterise the struggle for paradigmatic predominance.

The discussion about 'framing' holds important implications for the Pacific Islands region, which the literature highlights as one of the most vulnerable geographic regions in the world, with a high priority for adaptation to climate change (Barnett and Campbell 2010; Nunn 2007, 2009, 2013; Nunn and Kumar 2018; Luetz and Havea 2018). According to Nisbet (2009), the matter of 'messaging' or 'framing' is pivotal:

> If individuals are given an ambiguous or uncertain situation to consider, the different ways in which a message is presented or framed—apart from the content itself—can result in very different responses, depending on the terminology used to describe the problem or the visual context provided in the message. For many members of the public, climate change is likely to be the ultimate ambiguous situation given its complexity and perceived uncertainty. (p. 16)

A closer look at the South Pacific Islands region reveals that climate change has been overwhelmingly presented and 'framed' as a science-informed issue requiring technocratic solutions that tend to ignore local worldviews. The conventional assumption appears to be that "science should set the agenda and that religious worldviews should shift in order to incorporate its narratives" (Veldman 2013, p. 172). This extends to issues of governance; "issues of state policy are foregrounded when analysing climate change and migration … how the various Christian institutions and their representatives influence the policies espoused by the respective governments is seldom made explicit" (Kempf 2012: 236–237). This has proven problematic and woefully inadequate in parts of the world that do not subscribe to technocratic views of the world:

> Given the massive, worldwide problems that are to be expected under climate and environmental change and the fact that a large part of the world's population is not (yet) subject to the technocratic worldview, it is imperative to complement the existing technological and economical oriented problem solutions with alternative perspectives—narratives that integrate the entanglement of humans and their environment. (Bergmann and Gerten 2010, p. 5)

Discounting or ignoring local worldview perspectives in the Pacific Islands region has given rise to a plethora of ineffective and unsustainable adaptation responses (Fazey et al. 2011; Mackay et al. 2018; Piggott-McKellar et al. 2019). The importance of accommodating local worldview perspectives is also highlighted by Fegan (2013) who reports that in Kiribati issues of sea-level rise and climate change adaptation are additionally "compounded by the refusal of many islanders to believe that there is a sea level problem. Many of them, devout Christians, believe that God will raise the islands a little higher, then higher, so that they, and their descendants, will live in peace" (p. 188).

Importantly, adaptation interventions in the Pacific Island region have been mostly externally designed, funded and implemented, thus uncritically privileging a scientific and technocratic worldview that contrasts sharply with that of most Pacific Island people. A pertinent example of technocratic environmental adaptation is offered by Fegan (2013):

> The Tuvalu government is profoundly concerned about global warming and rising sea levels, which threaten to submerge the entire country. Even without sea level rises, the islands are extremely vulnerable to destructive wave action. For instance, the construction of a World War II airfield on the Islands of Funafuti involved the building of piers, infilling beach areas, and excavating deepwater access channels. These humanly made alterations in the 1940s changed local wave patterns, so much so that much less sand now accumulates to form and replenish beaches. Hungry waves now devour and erode the shore even faster than before. (p. 184)

The literature abounds with similar examples of adaptation responses where technocratic worldview-inspired human alterations of the natural environment have failed miserably (Belmar et al. 2016; Nunn and McNamara 2019). Relatedly and importantly, "[m]ost interventions intended to reduce exposure to environmental risk and to enable effective and sustainable adaptation to climate change in the Pacific Islands region have failed to acknowledge influences on decision making of spirituality and connectedness to Nature" (Nunn et al. 2016, p. 1). This has had an overall debilitating effect on activating effective and sustainable adaptation responses in the region.

In the light of the almost total Christianisation of Pacific Islands within the past century, such intervention failures are surprising. The situation cannot continue because every day the need for adaptation to climate change that is effective and sustainable is growing (Hinkel et al. 2018; Nerem et al. 2018; World Meteorological Organization 2019). Given that in the Pacific Islands region "decisionmakers are likely to be influenced more by tradition and local precedent than science" (Nunn et al. 2016, p. 2), makes the purposive exploration of opportunities for faith-engaged approaches to climate change a fertile and long-overdue undertaking.

This paper extends previous research by means of a review of pertinent challenges and arising opportunities. The literature review scoped and analysed all available literature that fell within the purview of its focus on climate change adaptation in the Pacific Island region. The study highlights critical gaps which would benefit from more scholarly attention. Experiences and lessons synthesised in this paper will be useful for both policy and practice serving the cause of climate change adaptation in Pacific Island communities. A better understanding of the science-spirituality nexus in the Pacific will also enable more effective and sustainable adaptation responses.

This paper is organised as follows: Sect. 2 reports the historical context of the South Pacific Island region. Section 3 offers an analysis of selected literature and research. The synthesis offered in Sect. 4 concludes with a call for more consilience in areas of adaptation policy development and research. Section 5 sketches opportunities for future research.

Historical Context and Why It Matters

The first people to settle oceanic-island groups in the western Pacific Ocean did so more than three millennia ago (Carson et al. 2013; Hung et al. 2011) since which time most have been continuously occupied. During this lengthy period, being on comparative small (resource-constrained) and remote landmasses, people evolved ways of living that were well aligned with these island environments, something that explains their unbroken occupation of most (Nunn 2007; McNeill 1994). It is worth noting that terms like 'adaptation' and 'mitigation', recently introduced to the Pacific Islands as novel concepts intended to address the future, are in fact ancient concepts in Pacific island societies.

In the long period before European contact, which became widespread in the early 19th century, much adaptation was material. This ranged from water-conservatory measures like agricultural terracing on high islands and lined-pit agriculture on atolls (Shah et al. 2018; Prasad 2016; McMillen et al. 2014; Thaman 2008) to measures for ensuring sustainable supplies of seafood that involved culturally-grounded declaration (*taboo*) of marine-protected areas (Johannes 2002). But adaptation was also non-material or spiritual, premised on an understanding of the natural world and the ways in which its human inhabitants might influence it. Inevitably, far less is known about this type of adaptation in the Pacific, not least because of the transformation of island societies that followed European arrival and the massive impacts of introduced diseases that set the stage for the rapid uptake of Christianity by indigenous peoples (Campbell 2011).

Spiritual adaptation and understanding of Pacific Island peoples found expression in 'arts' including the stone carvings of Polynesia (including Easter Island *moai*) and the widespread stone arrangements including stone altars (*marae*) and shrines (Lipo et al. 2013; Schifko 2011; Walter et al. 2004). Rituals and traditions characterize most Pacific Island cultures today although their original meanings are often obscure and contested; a good example is Fijian firewalking (Pigliasco 2010).

It is important to understand that, irrespective of how such defining characteristics of Pacific Island societies may have evolved since European arrival in the region, they represent the contemporary legacy of people's (non-western) belief systems that had developed to allow sustainable interactions with island environments. It seems that these belief systems had two major goals—first to ensure a continuous supply of food and water, second to ensure that people survived cataclysmic events like volcanic eruptions and earthquakes.

With respect to the first, there is evidence that Pacific islanders evolved ways of optimising food and water supply, even in the face of prolonged environmental adversity. These ways including methods of food preservation, of water conservation, and of the planting of surpluses (Shah et al. 2018; McNamara and Prasad 2014; Aalbersberg et al. 1988), all contextualized within communal frameworks that pragmatically discouraged individual enterprise that might deprive majorities (McMillen et al. 2017). The cultural practice of offering the 'first fruits' at the start of the harvest to titulary gods (or their earthbound representatives) to thank them for their benevolence is common in the Pacific. Rituals associated with harvesting and planting had a similar purpose; some magic-inducing rituals for breadfruit cultivation have been described recently (Labouisse 2016). Key to much such practice in the Pacific is the concept of *mana*, which "derives from the direct, transpersonal experience of psychic energy … interpreted in cultural traditions as associated with efficacy, healing, transmission of status" and others (Laughlin 2018, p. 410). As elsewhere, the exposition of *mana* and other culturally-grounded wisdom was often aided by the ritual consumption of psychoactive substances like kava (*Piper methysticum*) and betel nut (*Areca catechu*) (Fitzpatrick 2018).

When European/globalized ways of thinking, spearheaded by Christianity, arrived in the Pacific Islands region, they morphed with indigenous spiritual practices in ways that sometimes appear almost invisible today (Kempf 2017). Yet by interrogating Pacific Islanders value systems, it is possible to identify traits that are uncommon in global (Christian) practice yet ubiquitous in the Pacific (Nunn et al. 2016). These include a sense of oneness with Nature, the belief in a God who is actively engaged in maintaining human wellbeing, even a God whose promise to Noah is a reason for denying threats from climate change (Fair 2018; Mortreux and Barnett 2009).

Many external interventions for climate change adaptation by Pacific Island communities over the past few decades have failed to be either effective or sustainable. There are a number of reasons for this but two of the most common are (a) the sidelining of traditional (community-owned) knowledge about climate-driven environmental changes, and (b) the exclusive portrayal of climate change and its solutions as secular and science-informed, both attributes that conflict with most community members' values in the Pacific (see above). An increased acknowledgement of indigenous worldviews in adaptive solutions is likely to improve the uptake and sustainability of adaptation solutions (Klöck and Nunn 2019; Granderson 2017).

Discussion: Epistemologies, Vernacular and Insider-Outsider Perspectives

The critical analysis digested in this section features selected perspectives which these authors deem pertinent for this discourse. These emerged naturally from the review of relevant literature on the topic at hand and include issues of meaning-making, the significance (albeit oft-neglected acknowledgment) of local vernacular and insider-outsider perspectives of 'framing' climate change adaptation issues in the Pacific, and an overview of selected benefits associated with accommodating different complementary epistemologies (Granderson 2017; Rudiak-Gould 2014).

As this literature review has made clear, propagating scientific and technocratic responses to climate change, while ignoring local sociocultural contexts, has significantly hampered the uptake of meaningful and sustainable adaptation measures in the Pacific Island region. According to Fair (2018), the "rejection of religious perspectives appears to emerge from both a misunderstanding (and secular rejection) of religious thought and a desire to enforce the boundaries between the religious and the scientific." (p. 5) As previously alluded to in Sect. 1, the seemingly rigid dichotomy between epistemological paradigms hampers (rather than aides) consilience (Fig. 1). Attempts at creating consilience by integrating the sciences and the humanities (Slingerland and Collard 2012) have been vehemently opposed by epistemological puritanists as an "illicit melange of elements best left separate" (Kempf 2017, p. 23). Regrettably, rather than engaging with *kastom*, spiritually-mindedness and local sociocultural worldviews, scientific narratives have tended to narrowly construe these as barriers (to be overcome), rather than as conduits (to be leveraged). In consequence, there is a kind of perception that extant climate change maladaptation has "to be rectified by an increase in scientific information and a rejection of religious knowledge" (Fair 2018, p. 5; cf. Paton and Fairbairn-Dunlop 2010). Furthermore, invocations to solicit and engage local worldview perspectives in the Pacific Islands region are carelessly dismissed and "delegitimised by reference to climate science, as the latter is treated as unquestionably epistemologically superior" (Fair 2018, pp. 5–6). With reference to Kempf (2017), Fair (2018) synthesises the existence of "epistemological hierarchisation, which places religious knowledge on the very bottom tier" (p. 6), essentially on par with "Christian-inspired deviation, ignorance, passivity and maladaptivity" (Kempf 2017, p. 30).

This overall state of affairs has not served the Pacific Islands region well in terms of activating effective and sustainable climate change adaptation responses. Rather, evidence-based research in the region quite clearly indicates that

> the failure of external interventions for climate-change adaptation in Pacific Island communities is the wholly secular nature of their messages. Among spiritually engaged communities, these secular messages can be met with indifference or even hostility if they clash with the community's spiritual agenda. (Nunn 2017, para 12; cf. Nunn et al. 2016)

In synthesis, propagating the rigid dichotomy between the sciences and humanities is both unfounded, unfortunate and overall unhelpful. In contrast, it seems far more promising and pragmatic to "hold various knowledges (Christian, *kastom* and scientific) in balance, exploring their convergence, connections and tensions" (Fair 2018, p. 7). Using an approach she terms *tufala save* (lit. "double knowledge"), Fair (2018) offers the following example of the description of a rainbow by a Ni-Vanuatu research participant to illustrate how religious and scientific knowledge may *concurrently* be held in balance:

> Oh well that's like kind of climate change, 'cause you know the rainbow came out, that's a promise, but and then scientists say it's 'cause of the water or something … creating a spectrum. It's good to have two beliefs. Both of them are right. (Ruth, youth climate advocate). (Fair 2018, p. 7)

As discussed in Sect. 1, employing context-specific 'framing' of climate change messaging is pivotal. Relatedly, Hulme (2017a) makes the point that religious convictions constitute a potent stimulus or motivational force for engagement on climate change:

> Different regions and diverse groups of stakeholders understand the threat of climate change according to particular and often distinct frames of reference. These religious narratives and rituals shape the nature and credibility of different knowledge claims about climate—what is happening and why—as well as shaping individual and communal ethical and social behaviors. Religious faith communities therefore offer 'thick' accounts of moral reasoning for acting in the world, in response to climate change as much as in response to other social and ecological challenges. Such an approach sits in contrast to secular calls for mitigation and adaptation that rely upon 'thin' global values: widely shared, but culturally non-specific, moral criteria. (p. 245)

Fair (2018) suggests that "[t]his need for 'thick' moral accounts rather than thin evocations of global values seems particularly pressing in the Pacific" (p. 4). Relatedly and importantly, she synthesises that spiritually-shaped local worldviews may be engaged as "a motivational force not mirrored by economics or science" (p. 4). Hence there is an argument that rather than constituting a barrier, religion might alternatively be leveraged as a "cultural resource" (Hulme 2017b, p. 15) because "the most influential messages are those that engage with people's spiritual beliefs"

(Nunn 2017, para 14; cf. Nunn et al. 2016). As noted in Sect. 1, religiosity drives environmental action (or inaction), "rather than the influence working in the opposite direction" (Veldman 2013, p. 178; cf. Roberts 2012; Morrison et al. 2013; Hagevi 2014; Arbuckle and Konisky 2015; Nelson and Luetz 2019). This synthesis holds important implications, given that religious 'worldview' effectively underpins how perspectives on nature, the world and the universe are formed, framed and propagated as being factually sound. Stated differently, climate change adaptation efforts will not eventually succeed 'in spite of' the deep-seated influence of spiritually-shaped worldview in the Pacific Island region, but rather 'because of it' and 'on the back of it'. Exploring pertinent opportunities is therefore long overdue.

An example of expressly engaging spiritual worldview persuasion is offered in Luetz et al. (2018). Intentionally employing the biblically cautious language of "Creation Care" and making it explicit that their treatise "addresses itself to self-professed 'Christians'" (p. 52), the authors stress the epistemological imperative of conserving Nature from the vantage point of a Christian worldview orientation. Stating their premise that according to Scripture the world was created 'by' God,[1] belongs 'to' God[2] and speaks 'of' God,[3] for Believers there would thus arise an important implicational corollary to also protect, conserve and care for the Earth on that basis:

> Expressed in simple language, if God can be known through creation, then creation has inherent worth in and of itself—apart from utilitarian value to humans. And importantly, if God can be 'known' through what He has made, then 'creation care' is an epistemological priority of the highest order. Following this argument, epistemologically speaking, God becomes progressively less 'knowable' from what He has created as the defacing and careless destruction of His creation continues ... Or inversely, the knowledge of God can be preserved through the conservation of His creation. (Luetz et al. 2018, pp. 59, 69)

Leveraging Christian epistemology in this manner therefore holds the promise that adherents to the Faith may be activated to 'care for the Earth' and engage in 'climate change mitigation and adaptation' on the basis of hermeneutics rather than merely on the basis of science or economics. Therefore, tapping into faith convictions holds significant promise as an under-appreciated factor for influencing human behaviour change based on social norms (Boyes and Stanisstreet 2012; Luetz et al. 2018, 2020; Buxton et al. 2020). Hulme (2017b) notes that "climate policies need to tap into intrinsic, deeply-held values and motives if cultural innovation and change are to be lasting and effective" (p. 15). Finally, Nunn et al. (2016) highlight 'local ownership' of climate change adaptation initiatives as an important success factor:

> A major barrier identified to effective sustainable adaptation to climate change in the Pacific Islands region is the 'lack of ownership' that Pacific Island nations have of the climate-change

[1]"In the beginning God created the heavens and the earth." (Genesis 1:1; Holy Bible, NIV).

[2]"The earth is the Lord's, and everything in it, the world, and all who live in it." (Psalm 24:1; Holy Bible, NIV).

[3]"The heavens declare the glory of God; the skies proclaim the work of his hands." (Pslam 19:1; Holy Bible, NIV). "For since the creation of the world God's invisible qualities—his eternal power and divine nature—have been clearly seen, being understood from what has been made, so that people are without excuse." (Romans 1:20; Holy Bible, NIV).

> issue, given that almost all adaptation initiatives have been funded by external (donor) money and have been applied using English (rather than preferred vernaculars) and in unfamiliar cultural contexts. (Nunn et al. 2016, p. 15; cf. Nunn 2009, 2013)

In the literature the need to employ locally meaningful vernacular is noted in relation to discourses on so-called 'climate refugees' where the conceptual language and terminology used almost always bespeaks 'outsider' rather than 'insider' perspectives (Luetz and Merson 2019). Pilot research on climate change related migration in Bougainville has concluded that "local contexts, dialects and expressions (e.g., "Turangu") have much to contribute terminologically with respect to more appropriately informing the definitional and conceptual constructs of policy and research discourses" (Luetz and Havea 2018, p. 23). According to Luetz and Merson (2019, p. 6),

> Inclusivity in coining conceptualizations has already made advances in discourses about disability, and there is the hope that "inclusion" may be similarly normalized in the climate migration domain: "The 'nothing about us, without us' (Charlton 2000) cry within the disability discourse, calling for representation in a bureaucratic system of oppression and disempowerment, is hauntingly relevant". (Luetz et al. 2019, p. 120)

During field research on atolls in Bougainville on the preferred self-descriptions of migrating islander communities there was a tacit sense that locally resident respondents seemingly resisted externally conceived labels and instead referred to themselves using concepts based on their local vernacular. Respondents overwhelmingly rejected the representations "refugees" (79%) and "exiles" (68%) in reference to migration from atolls off the coast of Bougainville and instead preferred local terms from *Tok Pisin* (100%) that they suggested during semi-structured interviewing. Counter-proposed terminological suggestions included *tripman* (male)/*tripmeri* (female), *Turangu*, and *Mekim wokabout*, whose meanings are discussed in Luetz and Havea (2018, pp. 14–15). Given the ubiquitous global media portrayals of the Tulun Atoll islanders as "climate refugees" illustrates the stark difference that exist between insider-outsider perspectives in relation to human migration-as-adaptation to climate change (Luetz 2019a, b). At the same time, "religious environmentalism on the whole is a very interfaith movement. Concern for a shared planet home seems to enable the overcoming of religious differences." (Kearns 2012, p. 135).

In summary, locally meaningful representations can spawn and sustain locally owned and comprehended community responses to climate change. While "belittlement of religious thought seems to mirror wider sentiments in the literature" (Fair 2018, p. 5), there is a compelling sense that "[h]igh levels of spirituality and connectedness to Nature explain the impotence of secular messaging while the high degree of concern about climate change identifies opportunities for intervention" (Nunn et al. 2016, p. 15).

As this discourse has shown, science and technocracy have overwhelmingly conceptualised religion in the Pacific Island region as a barrier (to be overcome), rather than as a conduit (to be leveraged) for meaningful engagement on climate change adaptation. Importantly, such 'epistemological puritanism' has enshrined and entrenched a kind of rigid apartheid (Fig. 1) between the paradigmatic perspectives of the natural/physical and social sciences (or between science and faith). This unfortunate state of affairs has limited the local uptake of climate change messaging and has persisted despite evidence-based research suggesting that

> in the Pacific Islands, the most influential messages are those that engage with people's spiritual beliefs, and the most influential communication channels are often those that involve religious leaders ... Church leaders can heavily influence practical discussions at every level of the community. That makes them an important potential target for agencies aiming to make a real difference in how Pacific Islanders cope with climate change. (Nunn 2017, paras 14, 18; cf. Nunn et al. 2016)

Given the multiplicity and diversity of perspectives in the Pacific Islands region it is high time to ensure that religious knowledge is now also given a seat at the climate change adaptation table, as is slowly becoming the case elsewhere (Havea et al. 2018; Schuman et al. 2018; Schaefer 2017). According to Fair (2018), this holds the promise that resultant perspectives, discourses and initiatives will be more harmonious, holistic, respectful, inclusive and heterogenous but overall, effective and sustainable:

> None of these narratives are the right one: none should be treated as an exclusive vehicle for future climate communication. But the diversity of courses of action they demonstrate suggests the richness and heterogeneity of religious responses to climate change and the potential for fruitful connections between religious and scientific knowledges. They demonstrate the potential for more-than-scientific yet not anti-scientific responses to climate change, which are locally meaningful and morally compelling. (Fair 2018, p. 11)

In synthesis, effective and sustainable climate change adaptation approaches will have the best chance of success whenever they are holistic (both-and) rather than dichotomistic (either-or) (Fig. 2). Towards this end it remains an important priority for climate change adaptation in the Pacific Island region to continue building consilience in local communities through a heightened emphasis on science-faith integration (Slingerland and Collard 2012; Oakes 2019; Austin 2014; Thornton et al. 2012).

Concluding Synthesis: Creating Consilience

Given that in the Pacific Islands region "decisionmakers are likely to be influenced more by tradition and local precedent than science" (Nunn et al. 2016, p. 2), makes the purposive exploration of opportunities for faith-engaged approaches to climate change a fertile and long-overdue undertaking. Towards this end this literature review scoped and analysed literature that fell within the purview of its focus on climate change adaptation in the Pacific Island region. The analysis reveals that climate change is overwhelmingly presented and 'framed' as a science-informed issue requiring technocratic solutions that tend to completely ignore local worldviews. Regrettably, rather than engaging with *kastom*, spiritually-mindedness and local socio-cultural worldviews, scientific narratives narrowly construe these as barriers (to be

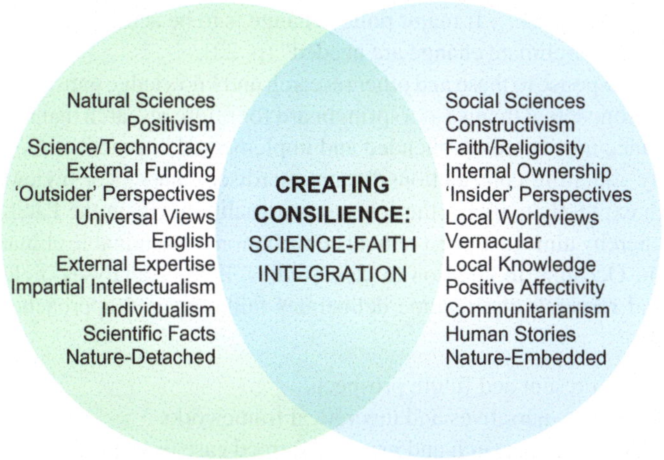

Fig. 2 Creating consilience: science-faith integration. *Source* Adapted from Luetz et al. (2020, p. 12)

overcome), rather than as conduits (to be leveraged) for locally meaningful adaptation to climate-driven environmental change. In consequence, such 'epistemological puritanism' has enshrined and entrenched a kind of rigid apartheid between the paradigmatic perspectives of the natural/physical and social sciences, or between science and faith (Fig. 1). The resultant state of affairs has not served the Pacific Islands region well in terms of activating effective and sustainable climate change adaptation responses. This study concludes that climate change adaptation initiatives in Oceania will have a better chance of being effective and sustainable if they are scientifically sound, spiritually attuned, contextually compelling and locally owned and meaningful. The study recommends fostering consilience through propagating locally sensitive science-faith integration (Fig. 2). Finally, the study identifies and conceptualises opportunities for future research (Sect. 5).

Opportunities for Future Research

This literary review acknowledges opportunities for future research as follows.

First, the surveyed literature highlights a number of knowledge gaps which would benefit from more scholarly attention. Hagevi (2014) notes that "[i]n the future, it is essential to conduct more comparative research regarding the elusive relationship between religiosity and environmental opinion" (p. 105). Furthermore, there is a sense that the literature would benefit from analyses "between insider and outsider accounts" (Veldman 2013, p. 173). Moreover, according to Nisbet (2009), "[a]dditional research using in-depth interviews, focus groups, and sophisticated survey and experimental techniques needs to further explore, identify, and test these frames across audiences ... If major policy change is to be achieved, new meanings and messengers for climate change are needed" (p. 22).

Second, in response to these and other research and knowledge gaps, these authors consider this conceptual treatise as a springboard for future research that may explore how consilience may be comprehended and implemented in practice. As such, they are currently soliciting contributions for an interdisciplinary peer reviewed edited book, which expressly explores the science-spirituality nexus in the Pacific Islands region and thereby aims to make a critical contribution to sustainable climate change adaptation in Oceania. In addition to case studies, literary analyses, field projects and empirical research, the volume deliberates faith-engaged approaches through the prism of:

- *Context*: Past, present and future prospects
- *Theory*: Concepts, narratives and theoretical frameworks
- *Practice*: Empirical research and praxis-informed case examples
- *Doctrine*: Scriptural contributions and perspectives
- *Engagement*: Enlisting religious stakeholders and constituencies.

Comprising peer reviewed works by scholars, professionals and practitioners from across Oceania, the forthcoming book intends to close a critical gap in the literature and thereby envisages making a ground-breaking contribution to holistic climate change adaptation in the Pacific Island region that is scientifically sound, spiritually attuned, locally meaningful and contextually compelling. Prospective contributors who wish to explore participating as chapter authors are encouraged to promptly make contact with these authors. The full call for papers is available at yale.edu.[4] and Mary Rokonadravu/Fiji.[5]

References

Aalbersberg WGL, Lovelace CEA, Madhoji K, Parkinson SV (1988) *Davuke*, the traditional Fijian method of pit preservation of staple carbohydrate foods. Ecol Food Nutr 21(3):173–180

Arbuckle MB, Konisky DM (2015) The role of religion in environmental attitudes. Soc Sci Q 96(5):1244–1263. https://doi.org/10.1111/ssqu.12213

Austin L (2014) Faith-based community radio and development in the South Pacific Islands. Media Int Aust 150:114–121

Barnett J, Campbell J (2010) Climate change and small island states. Earthscan, London, UK

Belmar YN, McNamara KE, Morrison TH (2016) Water security in small island developing states: the limited utility of evolving governance paradigms. Wiley Interdisc Rev Water 3(2):181–193. https://doi.org/10.1002/wat2.1129

Bergmann S, Gerten D (2010) (eds) Religion and dangerous environmental change: transdisciplinary perspectives on the ethics of climate change. Studies in religion and the environment, vol 2. LIT, Münster

[4]http://fore.yale.edu/files/CFP-Climate_Change_Pacific_Islands.pdf.

[5]https://medium.com/@tulani/call-for-papers-climate-change-adaptation-in-the-pacific-islands-opportunities-for-faith-engaged-57917862f965.

Boyes E, Stanisstreet M (2012) Environmental education for behaviour change: which actions should be targeted? Int J Sci Educ 34(10):1591–1614. https://doi.org/10.1080/09500693.2011.584079

Bryman A (2016) Social research methods, 5th edn. Oxford University Press, Oxford

Buxton G, Luetz JM, Shaw S (2020) Towards an embodied pedagogy in educating for creation care. In: Luetz JM , Green B (eds) Innovating Christian education research: multidisciplinary perspectives (Chap. 21). Springer Nature, Singapore

Campbell IC (2011) Worlds apart: a history of the Pacific Islands, 2nd edn. Canterbury University Press, Christchurch

Carson MT, Hung H-C, Summerhayes G, Bellwood P (2013) The pottery trail from Southeast Asia to Remote Oceania. J Island Coast Archaeol 8(1):17–36. https://doi.org/10.1080/15564894.2012.726941

Charlton J (2000) Nothing about us without us. Disability oppression and empowerment. University of California Press, CA

Chen J-M, Luetz JM (2020) Mono-/inter-/multi-/trans-/anti-disciplinarity in research. In: Leal Filho W, Marisa Azul A, Brandli L, Gökcin Özuyar P, Wall T (eds) Quality education, encyclopedia of the UN sustainable development goals. Springer Nature, Cham, Switzerland, pp 1–16. https://doi.org/10.1007/978-3-319-69902-8_33-1

Clarke DM, Wilson C (2011) The Oxford handbook of philosophy in early modern Europe. Oxford University Press, pp. 1–616. https://doi.org/10.1093/oxfordhb/9780199556137.001.0001

Fair H (2018) Three stories of Noah: navigating religious climate change narratives in the Pacific Island region. Geo Geogr Environ 5(2). https://doi.org/10.1002/geo2.68

Fazey I, Pettorelli N, Kenter J, Wagatora D, Schuett D (2011) Maladaptive trajectories of change in Makira, Solomon Islands. Glob Environ Change 21(4):1275–1289. https://doi.org/10.1016/j.gloenvcha.2011.07.006

Fegan B (2013) The attacking ocean: the past, present and future of rising sea levels. Bloomsbury, London, UK

Fitzpatrick SM (ed) (2018) Ancient psychoactive substances. University Press of Florida, Gainesville

Gerten D, Bergmann S (eds) (2012) Religion in environmental and climate change: suffering, values, lifestyles. London, UK, Bloomsbury

Goffman E (1974) Frame analysis: an essay on the organization of experience. Harper & Row, New York

Granderson AA (2017) The role of traditional knowledge in building adaptive capacity for climate change: perspectives from Vanuatu. Weather Clim Soc 9(3):545–561. https://doi.org/10.1175/wcas-d-16-0094.1

Hagevi M (2014) Religion and the environmental opinion in 22 countries: a comparative study. Int Rev Sociol 24(1):1–19. https://doi.org/10.1080/03906701.2014.894333

Hall ET (1976) Beyond culture, 1st edn. Anchor Press, Garden City, NY

Havea PH, Hemstock SL, Des Combes HJ, Luetz JM (2018) "God and Tonga Are My Inheritance!"—Climate Change Impact on Perceived Spirituality, Adaptation and Lessons Learnt from Kanokupolu, 'Ahau, Tukutonga, Popua and Manuka in Tongatapu, Tonga. In: Leal Filho W (ed) Climate change impacts and adaptation strategies for coastal communities. Springer Nature, Cham, pp 167–186. https://doi.org/10.1007/978-3-319-70703-7_9

Hinkel J, Aerts JCJH, Brown S, Jiménez JA, Lincke D, Nicholls RJ, Addo KA (2018) The ability of societies to adapt to twenty-first-century sea-level rise. Nat Clim Change 8:570–578. https://doi.org/10.1038/s41558-018-0176-z

Hulme M (2017a) Climate change. In: Jenkins WJ, Tucker ME, Grim J (eds) Routledge handbook of religion and ecology. Earthscan/Routledge, London, pp 239–248

Hulme M (2017b) Climate change and the significance of religion. Econ Polit Weekly 52:14–17

Hung H-C, Carson MT, Bellwood P, Campos FZ, Piper PJ, Dixon E, Chi Z (2011) The first settlement of remote Oceania: the Philippines to the Marianas. Antiquity 85:909–926

Jenkins W, Berry E, Kreider LB (2018) Religion and climate change. In: Gadgil A, Tomich TP (eds) Annual review of environment and resources, vol 43, pp 85–108
Johannes RE (2002) The renaissance of community-based marine resource management in Oceania. Ann Rev Ecol Syst 33:317–340
Kahneman D (2003) Maps of bounded rationality: psychology for behavioral economics. Am Econ Rev 93(5):1449–1475. https://doi.org/10.1257/000282803322655392
Kearns L (2012) Religious climate activism in the United States. In: Gerten D, Bergmann S (eds) Religion in environmental and climate change: suffering, values, lifestyles. Bloomsbury, London
Kelly MP, Barker M (2016) Why is changing health-related behaviour so difficult? Pub Health 136:109–116. https://doi.org/10.1016/j.puhe.2016.03.030
Kempf W (2012) Climate change, migration, and Christianity in Oceania. In: Hastrup K, Olwig KF (eds) Climate change and human mobility: challenges to the social sciences. Cambridge University Press, Cambridge
Kempf W (2017) Climate change, Christian religion and songs: revisiting the Noah Story in the Central Pacific. In: Dürr E, Pascht A (eds) Environmental transformations and cultural responses. Palgrave Macmillan, New York. https://doi.org/10.1057/978-1-137-53349-4_2
Klöck C, Nunn PD (2019) Adaptation to climate change in Small Island Developing States: a systematic literature review of academic research. J Environ Dev 28(2):196–218. https://doi.org/10.1177/1070496519835895
Labouisse JP (2016) Ethnobotany of Breadfruit in Vanuatu: review and prospects. Ethnobiol Lett 7(1):14–23. https://doi.org/10.14237/ebl.7.1.2016.582
Laughlin CD (2018) Mana: psychic energy, spiritual power, and the experiencing brain. Time Mind J Archaeol Conscious Cult 11(4):409–431. https://doi.org/10.1080/1751696x.2018.1541128
Lipo CP, Hunt TL, Haoa SR (2013) The 'walking' megalithic statues (moai) of Easter Island. J Archaeol Sci 40(6):2859–2866. https://doi.org/10.1016/j.jas.2012.09.029
Luetz JM (2019a) Over-researching migration 'hotspots'? Ethical issues from the Carteret Islands. Forced Migr Rev (61):20–22. The ethics issue (June 2019), Refugee Studies Centre, University of Oxford. https://www.fmreview.org/ethics/luetz
Luetz JM (2019b) Climate refugees: why measuring the immeasurable makes sense beyond measure. In: Leal Filho W, Marisa Azul A, Brandli L, Gökcin Özuyar P, Wall T (eds) Climate action. Encyclopedia of the UN sustainable development goals. Earth and Environmental Science, pp 1–14. Springer Nature, Cham, Switzerland. https://doi.org/10.1007/978-3-319-71063-1_81-1
Luetz JM, Havea PH (2018) "We're not Refugees, We'll Stay Here Until We Die!"—climate change adaptation and migration experiences gathered from the Tulun and Nissan Atolls of Bougainville, Papua New Guinea. In: Leal Filho W (ed) Climate change impacts and adaptation strategies for coastal communities. Springer Nature, Cham, Switzerland, pp 3–29. https://doi.org/10.1007/978-3-319-70703-7_1
Luetz JM, Merson J (2019) Climate change and human migration as adaptation: conceptual and practical challenges and opportunities. In: Leal Filho W, Marisa Azul A, Brandli L, Gökcin Özuyar, Wall T (eds) Climate action. Encyclopedia of the UN sustainable development goals (vol SDG 3—Climate action. Earth and Environmental Science, pp 1–13). Springer Nature, Cham, Switzerland. https://doi.org/10.1007/978-3-319-71063-1_46-1
Luetz JM, Buxton G, Bangert K (2018) Christian theological, hermeneutical and eschatological perspectives on environmental sustainability and creation care—the role of holistic education. In: Luetz JM, Dowden T, Norsworthy B (eds) Reimagining Christian education—cultivating transformative approaches. Springer Nature, Singapore, pp 51–73. https://doi.org/10.1007/978-981-13-0851-2_4
Luetz JM, Bergsma C, Hills K (2019) The poor just might be the educators we need for global sustainability—a manifesto for consulting the unconsulted. In: Leal Filho W, Consorte McCrea A (eds) Sustainability and the humanities. Springer Nature, Cham, Switzerland, pp 115–140. https://doi.org/10.1007/978-3-319-95336-6_7
Luetz JM, Margus P, Prickett B (2020) Human behaviour change for sustainable development: perspectives informed by psychology and neuroscience. In: Leal Filho W, Marisa Azul A, Brandli

L, Gökcin Özuyar P, Wall T (eds) Quality education, encyclopedia of the UN sustainable development goals, pp 1–15. Springer Nature, Cham, Switzerland. https://doi.org/10.1007/978-3-319-69902-8_12-1

McMillen HL, Ticktin T, Springer HK (2017) The future is behind us: traditional ecological knowledge and resilience over time on Hawai'i Island. Reg Environ Change 17(2):579–592. https://doi.org/10.1007/s10113-016-1032-1

McMillen HL, Ticktin T, Friedlander A, Jupiter SD, Thaman R, Campbell J, Veitayaki J, Giambelluca T, Nihmei S, Rupeni E, Orcherton DF (2014). Small islands, valuable insights: systems of customary resource use and resilience to climate change in the Pacific. Ecol Soc 19(4), #44. https://doi.org/10.5751/es-06937-190444

Mackay S, Brown R, Gonelevu M, Pelesikoti N, Kocovanua T, Iaken R, Iautu F, Tuiafitu-Malolo L, Fulivai S, Lepa MA, Mackey B (2018) Overcoming barriers to climate change information management in Small Island Developing States: lessons from Pacific SIDS. Clim Policy 1–14. 10.1080/14693062.2018.1455573

McNamara KE, Prasad SS (2014) Coping with extreme weather: communities in Fiji and Vanuatu share their experiences and knowledge. Clim Change 123(2):121–132. https://doi.org/10.1007/s10584-013-1047-2

McNeill JR (1994) Of rats and men: a synoptic environmental history of the island Pacific. J World Hist 5(2):299–349

Morrison M, Duncan R, Parton K, Sherley C (2013) The relationship between religious persuasion and climate change attitudes in Australia. IDEAS Working Paper Series from RePEc. 2013 Conference (57th), February 5–8, 2013, Sydney, Australia, 152147, Australian Agricultural and Resource Economics Society. http://ideas.repec.org/p/ags/aare13/152147.html

Mortreux C, Barnett J (2009) Climate change, migration and adaptation in Funafuti, Tuvalu. Glob Environ Change 19(1):105–112. https://doi.org/10.1016/j.gloenvcha.2008.09.006

Nelson W, Luetz JM (2019) What can we learn from Pope Francis about change management for environmental sustainability? A case study on success factors for leading change in change-resistant institutional environments. In: Leal Filho W, Consorte McCrea A (eds) Sustainability and the humanities. Springer Nature, Cham, Switzerland, pp 503–524. https://doi.org/10.1007/978-3-319-95336-6_29

Nerem RS, Beckley BD, Fasullo JT, Hamlington BD, Masters D, Mitchum GT (2018) Climate-change-driven accelerated sea-level rise detected in the altimeter era. Proc Natl Acad Sci U S A 115(9):2022–2025. https://doi.org/10.1073/pnas.1717312115

Nisbet MC (2009) Communicating climate change: why frames matter for public engagement. Environ Sci Policy Sustain Dev 51(2):12–23. https://doi.org/10.3200/ENVT.51.2.12-23

Nunn PD (2007) Climate, environment and society in the pacific during the last millennium. Elsevier, Amsterdam

Nunn PD (2009) Responding to the challenges of climate change in the Pacific Islands: management and technological imperatives. Clim Res 40:211–231

Nunn PD (2013) The end of the Pacific? Effects of sea level rise on Pacific Island Livelihoods. Singap J Trop Geogr 34(2):143–171. https://doi.org/10.1111/sjtg.12021

Nunn P (2017) Sidelining God: why secular climate projects in the Pacific Islands are failing. The Conversation, 17 May. https://theconversation.com/sidelining-god-why-secular-climate-projects-in-the-pacific-islands-are-failing-77623

Nunn PD, Mulgrew K, Scott-Parker B, Hine DW, Marks ADG, Mahar D, Maebuta J (2016) Spirituality and attitudes towards nature in the Pacific Islands: insights for enabling climate-change adaptation. Clim Change 136(3–4):477–493. https://doi.org/10.1007/s10584-016-1646-9

Nunn PD, Kumar R (2018) Understanding climate-human interactions in Small Island Developing States (SIDS): implications for future livelihood sustainability. Int J Clim Change Strat Manag 10(2):245–271. https://doi.org/10.1108/ijccsm-01-2017-0012

Nunn PD, McNamara KE (2019) Failing adaptation in island contexts: the growing need for transformational change. In: Klock C, Fink M (eds) Dealing with climate change on Small Islands: towards effective and sustainable adaptation. Gottingen University Press, Gottingen

Oakes R (2019) Culture, climate change and mobility decisions in Pacific Small Island developing states. Popul Environ 40(4):480–503
Paton K, Fairbairn-Dunlop P (2010) Listening to local voices: Tuvaluans respond to climate change. Local Environ 15:687–698. https://doi.org/10.1080/13549839.2010.498809
Piggott-McKellar A, McNamara K, Nunn PD, Watson J (2019) What are the barriers to successful community-based climate change adaptation? A review of grey literature. Local Environ 24(4):374–390. https://doi.org/10.1080/13549839.2019.1580688
Pigliasco GC (2010) We branded ourselves long ago: intangible cultural property and commodification of Fijian Firewalking. Oceania 80(2):161–181
Prasad V (2016) Adopting agroforestry for improved soil health in the Pacific. Xxix International Horticultural Congress on Horticulture: sustaining Lives, Livelihoods and Landscapes (Ihc2014): International symposium on impact of Asia-Pacific horticulture—resources. Technol Soc Welf 1129:35–39. https://doi.org/10.17660/ActaHortic.2016.1129.5
Punch KF (2014) Introduction to social research: quantitative and qualitative approaches, 3rd edn. Sage, Thousand Oaks
Roberts M (2012) Evangelicals and climate change. In: Gerten D, Bergmann S (eds) Religion in environmental and climate change: suffering, values, lifestyles. Bloomsbury, London, UK
Rudiak-Gould P (2014) The influence of science communication on indigenous climate change perception: theoretical and practical implications. Hum Ecol 42(1):75–86
Sapolsky R (2017) Behave: the biology of humans at our best and worst. Penguin Random House, London
Schaefer J (2017) New hope for the oceans: engaging faith-based communities in marine conservation. Front Mar Sci 4
Schifko G (2011) Marae, Moko and Haka. Traditional rituals of the Maori from New Zealand and their meanings in the 21st Century. Anthropos 106(2):707–708
Schuman S et al (2018) Religious beliefs and climate change adaptation: a study of three rural South African communities. Jamba-J Disaster Risk Stud 10
Shah S, Moroca A, Bhat JA (2018) Neo-traditional approaches for ensuring food security in Fiji Islands. Environ Dev 28:83–100. https://doi.org/10.1016/j.envdev.2018.11.001
Slingerland E, Collard M (eds) (2012) Creating consilience: integrating the sciences and the humanities. Oxford University Press, Oxford
Thaman RR (2008) Pacific Island agrobiodiversity and ethnobiodiversity: a foundation for sustainable Pacific Island life. Biodiversity 9(1–2):102–110. https://doi.org/10.1080/14888386.2008.9712895
Thornton A et al (2012) Givers and governance: the potential of faith-based development in the Asia Pacific. Dev Pract 22(5–6):779–791
Veldman RG (2013) Religion in environmental and climate change: suffering, values, lifestyles. J Contemp Relig. https://doi.org/10.1080/13537903.2013.750865
Walter R, Thomas T, Sheppard P (2004) Cult assemblages and ritual practice in Roviana Lagoon, Solomon Islands. World Archaeol 36(1):142–157. https://doi.org/10.1080/0043824042000192614
White L (1967) The historical roots of our ecological crisis. Science 155(3767):1203–1207
World Meteorological Organization (2019) WMO Statement on the State of the Global Climate in 2018. WMO, Geneva

Johannes M. Luetz is Senior Lecturer, Postgraduate Coordinator and Research Chair at CHC, a bespoke private higher education provider in Brisbane, and Adjunct Academic at UNSW Sydney in the School of Social Sciences, where he also earned his Ph.D. in Environmental Policy and Management with a thesis on forced human migration. Dr. Luetz has travelled and consulted extensively for World Vision International on research projects raising awareness of the growing effects of climate change on vulnerable communities in Asia, Africa and Latin America, and the need to meaningfully address vulnerabilities through praxis-informed approaches that work in the real world. Before joining the CHC Faculty he was a Lecturer in Development Studies at UNSW Sydney. He was raised in West Africa and Switzerland, cares deeply about poverty reduction, holistic education, justice and equity, and has researched, written and published conceptually and empirically in these areas. He heads the "Master of Social Science Leadership" program at CHC, which through its research and teaching activities on change leadership seeks to give rise to equitable social transformation.

Patrick D. Nunn is based at the School of Social Sciences, University of the Sunshine Coast, and has expertise in several different fields. After more than 30 years of research, mostly focused in the Asia-Pacific region, his primary field is geography, once largely physical in focus, but now straddling various aspects of sustainability. Professor Nunn has worked for a number of years in climate change, mostly on sea level and on the challenges of effective adaptation in poorer countries. He has also worked on archaeological topics, usually through the lens of palaeoenvironment reconstruction, but also applying his geological training to ceramic mineralogy and radiocarbon chronology. Since 2000, when a coup in Fiji interrupted a planned research programme, Professor Nunn became interested in myths as potential sources of information about geological hazards, particularly earthquakes and tsunamis, volcanic eruptions and abrupt coastal change. Firmly believing in the importance of community awareness, he has ensured that the results of his research have been returned to the people of the land in ways that they can understand its nature and importance, something helped in the case of Fiji by his fluency in the Fijian language and his familiarity with cultural protocols.

Climate Change and the Pacific Region: Some Future Trends

Walter Leal Filho

Abstract This chapter briefly outlines some future trends on climate change in the Pacific region, and suggests some areas where action is needed in order to better prepare the region for the many challenges posed by a changing climate change. In particular, it suggests the creation of a "Marshall Plan for Pacific Small Island Countries", which may offer a long term basis upon which climate change adaptation and resilience efforts may be performed.

Keywords Climate change · Pacific-extreme events · Health-resilience

Introduction

Island nations as a whole, and small island countries in particular, are vulnerable to climate change due to a combination of factors (Nurse et al. 2014). The small islands spread over the Pacific region are no exception, and are known to be especially vulnerable to climate change also as a result of their wide geographical spread, and their remoteness. The impacts of climate change to island nations are manifold and include areas as varied as:

- Human Systems
- Island Settlements and Tourism
- Human Health
- Relocation and Migration
- Economic development.

(Nurse et al. 2014).

The 5th Assessment Report produced by the Intergovernmental Panel on Climate Change (IPCC) in 2014, which looked at climate change on islands under a set of perspectives, outlined the fact that island nations also suffer from trends related to

W. Leal Filho (✉)
Faculty of Life Sciences, Research and Transfer Centre "Sustainable Development and Climate Change Management", Hamburg University of Applied Sciences, Ulmenliet 20, 21033 Hamburg, Germany
e-mail: walter.leal2@haw-hamburg.de

W. Leal Filho (ed.), *Managing Climate Change Adaptation in the Pacific Region*, Climate Change Management, https://doi.org/10.1007/978-3-030-40552-6_16

sea level rise, and various issues whose causes are often not directly related to them (IPCC 2014).

The subsequent parts of this paper handle three main aspects. First, it describes the problems encountered when implementing climate policies in the region. Secondly, it outlines some future trends and areas of action where attention is due, so as to put small island nations in a better position to address the many challenges climate change poses to them. Finally, it provides some conclusions, where the need for integrated approaches is reiterated.

Some Problems in the Implementation of Climate Change Policies

If one considers the fact that Pacific island countries contribute very little to greenhouse gas emissions, but are among the ones to suffer most from the impacts of climate change (Leal Filho 2017), it becomes clear that this disparity needs to be addressed. The consequences of extreme weather events for instance, whose frequency in the region is likely to increase (IPCC 2007, 2014), coupled with the rising sea levels (IPCC 2018), are significant. Apart from damages to properties, they endanger the livelihoods of many communities. In addition, erratic rainfall patterns, frequent flooding, or periods of droughts lead to hardships and human suffering. Moreover, damages to local ecosystems as a result of global warming may also have a "knock on" effect on economic activities such as tourism which, in the Pacific region, similarly to what happens in other parts of the world (Maynard et al. 2019) may be negatively affected by the depletion of coral reefs.

There have been substantial progresses in the Pacific region in respect of addressing climate change. For instance, the process of preparation of the National Adaptation Plans (NAPAs), has given a sense of direction, as to where adaptation initiatives should head to. However, some problems still make it difficult to achieve long-term improvements. Some of these problems are outlined in Fig. 1.

Addressing these problems and their ramifications such as, for instance, migration (Campbell and Warrick 2014) in an effective way, by also learning from the lessons (and mistakes) from the past, needs to become one of the regional priorities. Without due attention being paid to them, there is a risk of wasting value financial and human resources in efforts, whose benefits are limited to the short term.

Some Future Trends

Pacific island nations are among the ones most engaged on efforts to tackle global warming as a whole, and towards addressing the impacts of climate change in particular. As signatories of the UN Framework Convention on Climate Change (UNFCCC)

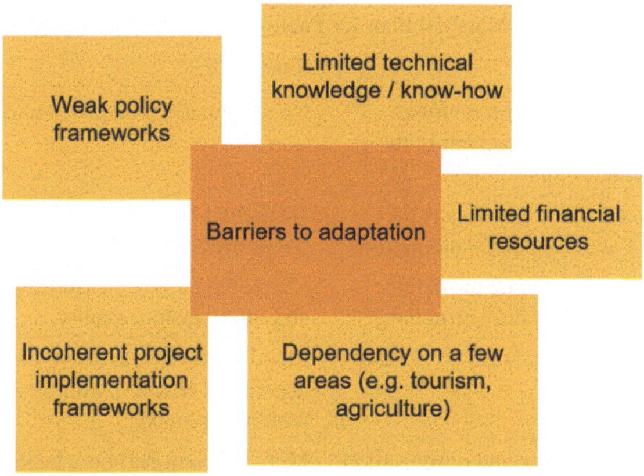

Fig. 1 Some problems hindering the implementation of climate change policies on Pacific islands

on the one hand, and to the Paris Agreement on the other, they have reiterated, time after time, how important addressing climate change is important to them.

The challenges outlined on this paper makes progress difficult. Bearing this in mind, some future trends which should be taken into account are:

1. *Re-focusing international aid*

Whereas global efforts to reduce emissions as part of a global climate mitigation programme are important, it should not be overlooked that for many communities, action is needed immediately. In this context, international aid should also be re-focused, with a view to holistically supporting adaptation strategies and frameworks. This means, in practice, that traditional development aid, usually implemented by means of a specific project funded by one single donor, "integrated funding programmes" are needed, where a programme lasting 5–10 years is supported, instead of a project lasting 1–3 years, as it is largely the case today.

And bearing in mind the nature of the problems faced by small Pacific islands as a result of climate change, it is suggested that a "Marshall Plan for Pacific Small Island Countries" be set-up, with aid targeted at key areas such as:

i. Infra-structure projects to reduce vulnerability of buildings and facilities
ii. Better insurance systems to safeguard incomes due to losses deriving from extreme events
iii. Increased education and capacity building provision in new, innovative areas and the services sector, so as to reduce reliance on subsistence farming or fisheries
iv. Optimisation of sectors such as tourism, so as to reduce its ecological footprint, at the same time preparing the sector for the transformation processes imposed by a changing climate.

Table 1 describes the main features of the proposed plan and the focus of the integrated efforts as part of it. Many countries at the moment suffer from the fact that continuous resource mobilization for concrete initiatives is rather difficult. The

Table 1 Main elements of the Marshall Plan for Pacific Small Island countries

Item	Relevance
Preparation and agreement for a regional construction and reconstruction plan in the aftermath of extreme events	No systematic procedures to date
Use of a common database which lists the vulnerabilities of Pacific Small Island nations	Different vulnerabilities are not well documented, leading to duplications
Systematic procedures for issue of permits (e.g. road building, tourism infra-structure)	Currently, ad hoc measures lack consistency and often, lasting quality
Standardised procedures for building reliable flood defences	Flood defences currently with different degrees of efficacy, with many inefficient ones widely deployed
A strategy for modernisation and optimisation of the use of harbours	More efficient harbours facilitate the transport of people, good and services and are more resilient to extreme events

proposed plan can help to address this problem, since it aims to offer clear targets and indicators, which may facilitate the decision-making process by donors. It is suggested that a sum to be agreed be centrally administered, and be made available by means of integrated multi-year plans. Interested countries would apply for such funds on a competitive basis, under criteria to be agreed (e.g. contribution to vulnerability reduction, support to economic activities, indicators of increased resilience).

It is suggested that the democratically elected national governments of the region discuss and agree on the elements for the “Marshall Plan for Pacific Small Island Countries”, so as to ensure it considers all relevant contexts and needs.

2. *A greater focus on resilience*

The many areas of vulnerability on Pacific island nations makes it difficult to prioritise or concentrate actions in one single area of action. However, an overall focus on resilience-building may help to channel efforts—and funding- in achieving tangible benefits. Figure 2 outlines some areas which should be considered, so as to make resilience-building efforts more robust.

Among the various areas of action, there is a particularly acute need to foster capacity to implement climate change resilience. It is not only about the development and implementation of policies; rather it is about the need to meet information and training needs related to the identification and implementation of strategies and practical initiatives towards the prevention, and management of the impacts of environmental risks and hazards. Here, the Pacific region is unfortunately lagging behind, since there are not many training provisions at the local level on island nations. This need should be immediately addressed, since improved capacity is a key component of sustainable adaptation strategies.

3. *Reducing vulnerabilities*

It is widely known that it is important to identify and pursue the means to protect people, communities and ecosystems from the impacts of climate change. In order

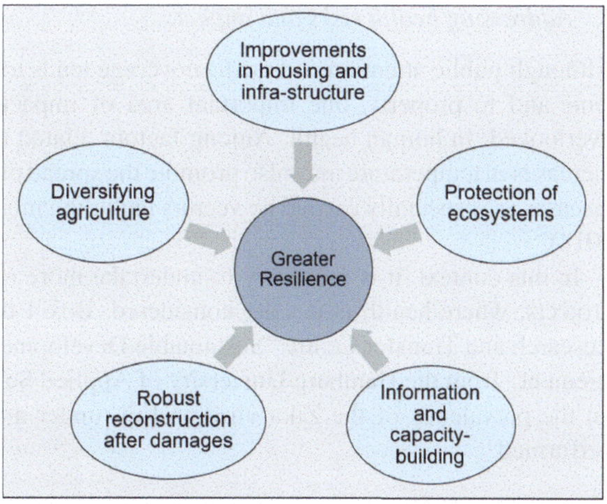

Fig. 2 Some relevant areas to increase the robustness of resilience-building efforts

to yield the expected benefits, this needs to be done in an integrated way. Therefore, initiatives which aim at reducing vulnerabilities need at the same time to consider the nexus people-environment. The "chronic" vulnerability of Pacific Island nations to climate change, especially to inundation and flooding (ADB 2012), also shows very clearly where the key areas for action are. Since a substantial part of their population resides near coastal areas- most of the populations live within 2 km of the shoreline- which are particularly susceptible to the effects of climate change (Mimura 1999, 2007), it is clear that coastal areas ought to be high on the vulnerability reduction list.

4. *Greater access to climate services*

Pacific island nations may substantially benefit from greater access to climate services. This may include the access to meteorological data such as:

(a) temperature
(b) rainfall
(c) wind
(d) soil moisture

and on water conditions (river and ocean), with which graphics, maps and risk assessments may be produced. Climate services may also deploy non-meteorological data. These can be helpful in areas as varied as agricultural production, assessing health trends, evaluating the safety of human settlement in high-risk areas, road and infrastructure maps for the delivery of goods (WMO 2019), and many other uses.

A matter of central relevance when discussing climate services and their usefulness to small island nations, is the fact that the information collected needs to be customized with a view to meeting the specific information needs of different users, settings and communities.

5. *Addressing health risks and impacts*

Although public attention and media coverage tends to focus on damages to ecosystems and to property, one important area of impact of climate change is often overlooked: in human health. Among factors related to heat distress and droughts, increases in temperature may also promote the spread of vector-borne diseases. These diseases are especially carried by vectors including mosquitos, ticks and fleas (WHO 2018)

In this context, it is necessary to undertake more research and execute practical projects, where health issues are considered. Box 1 describes a project led by the Research and Transfer Centre "Sustainable Development and Climate Change Management" from the Hamburg University of Applied Sciences where an investigation on the prevalence of the Zika virus in Fiji, under a climate change context, was performed.

Box 1 The BMBF Project "Climate Change and Prevalence Study of ZIKA Virus Diseases in Fiji"

Climate change represents one of the biggest global health threats of the 21st century. Recent research shows that the global temperature rise, changes in precipitation patterns and an increase of extreme weather events may influence the global distribution of tropical infectious diseases, such as mosquito-borne disease, and may further extend transmission seasons in endemic areas, like the Pacific region. Emerging diseases, such as ZIKV disease, have successfully expanded to geographical areas where only DENV epidemics used to occur, including the Fiji Islands. Although recent findings could only confirm low-level transmission of ZIKV in Fiji during 2013–2017, with highest numbers in 2016, as global temperature rises there is the risk of ZIKV re-emergence of greater extent. The project aims to understand the association between variations in seasonal epidemic size of mosquito-borne diseases (DENV and ZIKV), and environmental conditions in Fiji. The findings of this research may serve as a keystone to evaluate the temporal and spatial vulnerability posed by *Aedes*-borne diseases in the main island of Fiji.

Health efforts related to climate change should be based on three mail pillars:

i. preserving biodiversity and ecosystem services since their very presence and the functions they fulfil and usually to the advantage to the physical and human environment;
ii. equal consideration to the social aspects of climate change such as sanitation, shelter, nutrition and mental health since they may affect people at different levels;
iii. reducing the health risks and likelihood of damages to human well being from disasters and climate-related hazards.

Furthermore, there is a perceived need to undertake research and hence increase the knowledge basis on the links between extreme weather events and climate-sensitive diseases. The list of such diseases is long, and includes malaria, dengue, cholera, filariasis, leptospirosis, schistosomiasis, and ciguatera fish poisoning, whose occurrence is on the rise (WHO 2018).

Conclusions

This paper has tried to summarise some of the main issues raised on this Handbook, and outlined some areas where action is necessary. In the long term and in order to succeed, efforts to increase climate resilience among Pacific island nations should be implemented by means of partnerships, at three main levels:

(a) Macro: this means between international agencies on the one hand, and national governments on the other, orienting funding and technical support to priority areas, and with a long-term dimension in mind, as opposed to focusing on achievements in a couple of years, since these are seldom sustainable.

(b) Meso: this level is characterised by the interactions between national and regional authorities. At the meso-level, national governments consult with the authorities and administration at the regional level (e.g. provincial or district governments) and agree on the measures needed to address climate change and its impacts in a certain region.

(c) Micro: this is one of the levels which matter most, if one considers the fact that the nature of climate change—caused by greenhouse gas emissions—is global and that its impacts are local. It is therefore vital that local communities are engaged and ideally involved in co-production efforts to undertake projects aimed at tackling the local consequences of climate change. In particular, support to re-building and construction of additional infra-structure, needs a strong support basis, at the micro level.

The "Marshall Plan for Pacific Island Countries" proposed here, may act as an enabler to efforts to foster long lasting climate change adaptation, acting under a coherent and consistent framework, which may reduce redundancies and avoid duplications.

It is a fact that the Pacific region needs to cope with more frequent extreme weather events, with erratic rainfall patterns, flooding, droughts and that it regularly needs to cope with natural disasters. It is also a fact that it needs to prepare itself for the long-term impacts of sea level rises. Against this background, the "Marshall Plan for Pacific Island Countries" may offer international agencies, donors and national governments alike with a sound basis, upon which adaptation efforts may be implemented. It may also provide a framework against which progress can be measured, which is not the case at present.

Climate change also needs to be seen as an **opportunity to Pacific island nations**: an opportunity for new investments, with innovative concepts which bear in mind climate change on the one hand, but which also may lead to improvements in the quality of infra-structure and in the quality of lives on the other. It is also an opportunity to seeks ways to diversify economies, to make them less oriented to economic activities which may suffer from climate event, and towards other income and revenue streams which are less vulnerable to extreme events the region suffers.

If duly integrated and if performed under a coherent framework, Pacific island nations may be able to reduce their vulnerability and increase their resilience to climate change, and at the same time reduce the scope and the many problems which it currently faces, as a result of a changing climate.

Acknowledgements The support from the German Federal Ministry of Education and Research (BMBF) to the project "Climate Change and Prevalence Study of ZIKA Virus Diseases in Fiji (Zika under a climate change and epidemiological perspective)" is gratefully acknowledged.

References

ADB (2012) Pacific risks, vulnerabilities, and key impacts of climate change and natural disasters pacific regional strategic program for climate resilience. Manila

Campbell J, Warrick O (2014) Climate change and migration issues in the Pacific. United Nations Economic and Social Commission for Asia and the Pacific. Bangkok, Fiji

IPCC (2007) AR4 climate change 2007: synthesis report. IPCC, Geneva

IPCC (2014) AR5 synthesis report—climate change 2014. IPCC, Geneva

IPCC (2018) Global warming of 1.5 °C. IPCC, Geneva

Leal Filho W (2017) Climate change adaptation in Pacific countries. Fostering resilience and improving the quality of life. Springer, Cham

Maynard J, Gove J, Tracey D, Johnson J, Lecky J, Conklin E, Brainard R (2019) Coral reefs: vulnerability to climate change in West Hawai'i. NOAA, Washington, DC

Mimura N (1999) Vulnerability of island countries in the South Pacific to sea level rise and climate change. Climate Res 12(2–3):137–143

Mimura N (2007) Small islands. In: Parry M, Parry ML, Canziani O, Palutikof J, Van der Linden P, Hanson C (eds) Climate change 2007-impacts, adaptation and vulnerability: working group II contribution to the fourth assessment report of the IPCC, vol 4. Cambridge University Press, Cambridge, United Kingdom

Nurse LA, McLean RF, Agard J, Briguglio LP, Duvat-Magnan V, Pelesikoti N, Tompkins E, Webb A (2014) Small islands. In: Climate change 2014: impacts, adaptation, and vulnerability. Part B: regional aspects. Contribution of working group II to the fifth assessment report of the Intergovernmental Panel on Climate

WHO (2018) Climate change and health. Retrieved 07 Sept 2019 from https://www.who.int/news-room/fact-sheets/detail/climate-change-and-health

World Meteorological Organization (WMO) (2019) What is the Global Framework for Climate Services and what will it accomplish?. WMO, Geneva

The manufacturer's authorised representative in the EU is Springer Nature Customer Service Centre GmbH, Europaplatz 3, 69115 Heidelberg, Germany. If you have any concerns regarding our products, please contact ProductSafety@springernature.com

Printed and bound by CPI Group (UK) Ltd, Croydon, CR0 4YY
30/09/2026
02225458-0014